UNDERSTANDING RADAR SYSTEMS

UNDERSTANDING RADAR SYSTEMS

Simon Kingsley

Department of Electronic and Electrical Engineering
University of Sheffield

Shaun Quegan

Department of Applied and Computational Mathematics
University of Sheffield

SciTech Publishing, Inc
Raleigh, NC
www.scitechpub.com

ISBN 1-891121-05-7
ISBN 13: 978-1-891121-05-0

SciTech books may be purchased at quantity discounts for educational,
business, or sales promotional use. For information, contact the publisher:

SciTech Publishing
911 Paverstone Drive, Suite B
Raleigh, NC 27615
Phone: 919-847-2434
Fax: 919-847-2568
www.scitechpub.com

CONTENTS

PREFACE

Understanding Radar Systems is a book to convey facts and figures and also explain why things are the way they are. It is written for students and young engineers in industry, already competent in electronic engineering or physics, who need to understand modern radar principles, applications and some of the jargon used.

There are three parts to the book. The first three chapters form an easy (and deterministic) guide to the basics of what radar is all about, and what it can do. Chapters 4–6 are possibly more difficult because probability is introduced, but these chapters will be necessary for those who wish to begin calculating the performance of radar systems. The remaining chapters describe different types of modern radar, the problems and the research areas. These applications are used to introduce some more new ideas, so that their relevance is apparent.

This book can be used in several ways: as a 'streetwise' introduction to radar; as a rough guide to calculating the performance of most types of radar; or as a handy reference when starting off on a new radar topic, before going on to the more detailed texts in that area. The most important thing is to understand the subject and enjoy your career in radar.

Simon Kingsley and Shaun Quegan

ACKNOWLEDGEMENTS

We are grateful to those companies who supplied us with details of their systems and especially to Marconi who also supplied the images on the front cover.

The meteor scattering theory in Chapter 9 derives from information supplied by Dr John Milsom (GEC-Marconi Research Centre) who continues to be a fount of knowledge concerning propagation and communications.

Other researchers, past and present, at GEC-MRC who deserve special mentions are Matthew Radford, who provided helpful and informative comments on the draft manuscript and Trevor Blake and John Dawson for their contributions on over-the-horizon radar.

We would like to thank the UK Ministry of Defence for their support of various research programmes that led to some of the original material included in the contents of this book.

Many of the diagrams in Chapter 6 were produced with the expert help of Linda Wilkinson of Sheffield University, to whom we express our thanks.

Finally, we wish to thank all those who offered comments and advice, and also our wives and children for putting up with us while we were writing.

Simon Kingsley and Shaun Quegan

ONE

FUNDAMENTALS

- What is radar?
- What can it do?
- How well can it do it?

A rough-and-ready guide to radar and its capabilities.

1.1 WHAT IS RADAR?

Radar is all about using radio waves to detect the presence of objects and to find their position. The word *radar*, first used by the US Navy in 1940, is derived from *radio detection and ranging*, thus conveying these two purposes of detection and location. Modern radar goes further and is being developed to classify or identify targets, and even to produce images of objects, for example mapping the ground from a satellite.

The principle of radar is that a transmitter sends out a radio signal, which will scatter off anything that it encounters (land, sea, ships, aircraft), and a small amount of the energy is scattered back to a radio receiver, which is usually, but not always, located near the transmitter. After amplification in the receiver, the signals are processed to sort out the required echoes from the 'clutter' of unwanted echoes by a combination of both electronic signal processing and computer software (data processing).

There are many applications for radar, on scale sizes that vary from a few centimetres, such as the measurement of the thickness of furnace walls, to long-range systems probing planets across the solar system. Table 1.1

Table 1.1 Applications of radar

Civil		
Ground-based	Air traffic control	
	Sea traffic control	
	Weather forecasting	
	Speed traps	
	Intruder alarms	
	Radar astronomy	
	Ground probing	
	Industrial measurement	
Sea-borne	Navigation	
	Collision avoidance	
Air-borne	Altimeters	
	Navigation	
	Weather	
Space-borne	Studying Earth resources	
	Sea sensing	
	Manipulating spacecraft	
	Mapping planets and minor bodies	
Military		
Detection	of own forces or enemy forces	
Tracking	of air, sea, land or space targets	
Guidance	of own weapons systems	

gives some idea of the variety of applications, and the list is growing as radar systems find their way into industry, homes and even the motor car.

There are many systems similar to radar, such as sonar (which uses sound instead of radio waves), medical ultrasonics and passive detection systems[†]. Knowledge of radar principles is a good starting point for understanding these other subjects and would help a physicist or electronic engineer, for example, to transfer into medical electronics and body scanning.

Just as in the movie industry, where the apparently immense task of making a film can be broken down into identifying 60 or 70 key scenes, so other problems can be broken down. If you find the prospect of learning all about radar a bit daunting, remember that there are less than 100 key things you really need to know to understand the basics of radar, and these are summarized for you at the end of each chapter, as they occur.

---

† An example of a passive (no transmitter) system is the technique for locating thunderstorms using several receivers to obtain bearings on the radio signals emitted by lightning flashes; the lightning is, in effect, acting as the transmitter for the rest of the system. Biologists tracking animals tagged with tiny radio transmitters have a similar interest in passive detection and location.

1.2 A SIMPLE RADAR EXPLAINED

Figure 1.1 shows a basic radar system. For this example we have chosen a pulse radar, although we shall be looking at other forms of radio transmission later. The principal radio frequency (RF) of the radar, the carrier, is set by the frequency synthesizer. This continuous signal is pulsed on and off (usually spending much more time off than on) by the modulator. The short bursts of radio energy that result are amplified by the transmitter and sent to the antenna via a switch called variously a transmit–receive switch, a T–R switch or a duplexer. There are different designs for these switches, but they all have the same two functions: to connect the antenna to either the transmitter or the receiver at the appropriate times, and to protect the receiver from the full force of the transmitted pulse.

When a pulse is transmitted, the radar clock begins to count time. The radio pulse travels away from the radar at the speed of light, is scattered from a target and returns to the radar. The distance to the target, R (called the *range* of the target), can then be calculated from the time delay. Remembering that velocity = distance/time, we can rearrange this as

$$R = c\tau_d/2 \quad [\text{m}] \tag{1.1}$$

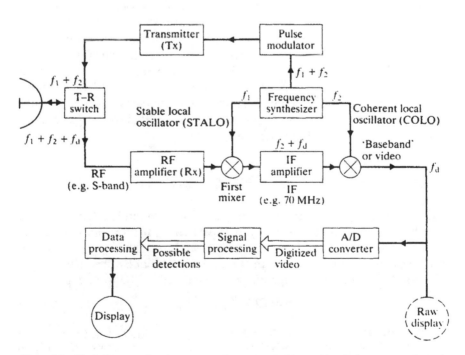

Figure 1.1 Block diagram of an elementary radar system, and some abbreviations commonly used.

where R = range [m], c = velocity of light = 3×10^8 m s^{-1} and τ_d = signal propagation delay [s]. Strictly speaking, an equation such as Eq. (1.1) should not have units attached, because the identity is true whatever units are used. However, it is sometimes clearer if dimensions and units are identified, and so the appropriate SI units are given. Where units do not aid understanding, for example in the probability theory in Chapter 4, we have not included them. When a quantity has no dimensions, we represent this by empty brackets, thus [   ]. In radar practice, units other than SI are often used, especially aircraft height in feet, distance in nautical miles and speed in knots. Conversion factors are given in Appendix III.

   The factor 2 in Eq. (1.1) arises because the distance travelled by the radio pulse is $2R$, i.e. to the target and back. A convenient rule to remember is '150 m per microsecond', meaning that, after transmission, each extra microsecond that elapses before the echo pulse is received implies that the target is 150 m further away. When discussing range, it is common practice to find it being used synonymously with time delay, since they are related by Eq. (1.1). Also, it is quite usual for the duration of radar pulses to be described as lengths, as in 'a pulse length of 1 μs'.

   When the echo pulse returns to the radar, it is absorbed by a receiver, which has been carefully designed to ensure that none of the energy in these very weak signals is wasted. After amplification and conversion to an intermediate frequency (IF) where the electronics is easier to engineer, the signal is detected and then processed ready for display to a human operator. The plan position indicator (PPI) is one type of display that is familiar to many people; a rotating trace produces a circular map in which the radar is represented as being at the centre and range is represented as the distance towards the edge of the display. This display can be 'raw' information containing all the echoes received by the radar, but nowadays it is more common for the display to be synthesized such that only the desired information on targets of interest is displayed.

1.3 OVERVIEW OF RADAR FREQUENCIES

Radio waves are part of the electromagnetic spectrum and so they could be described by either quantum theory or wave theory. It turns out that, at the frequencies used by radar, it is much more useful to think in terms of waves, and we do not refer to quanta in the rest of this book. The frequency f [Hz = s^{-1}]† of an electromagnetic wave is related to the wavelength λ by

$$c = f\lambda \qquad [\text{m s}^{-1}] \tag{1.2}$$

† The hertz [Hz] is named after Heinrich Rudolf Hertz, a professor at Karlsruhe, who is widely credited as having founded radar because of his work in the late 1880s, which included the reflection of radio waves generated by a spark gap generator.

where c is the velocity of light, which is about 3×10^8 m s^{-1} in air or space but may be quite a lot less in other materials such as coaxial cables.

The lowest frequency useful for radar purposes is probably about 1 MHz. Such low frequencies propagate very long distances, but the problems of using them get worse as the frequency gets lower. Some of these problems are as follows:

1. Antennas become very large at low frequencies; for example, at 1 MHz, a quarter-wavelength monopole would be 75 m high, and it becomes impracticable to build high-gain antennas.
2. The ionosphere is a strong scatterer of low frequencies and gives rise to unwanted echoes, which, because the ionosphere is not a static phenomenon, can be confused with moving targets. Given an antenna with a very narrow beam, it might be possible to arrange search patterns that avoided looking in the direction of the ionosphere, but because of the difficulties described in item 1, this is not a practical option.
3. Long radio wavelengths mean that only small changes in frequency occur when signals are scattered from moving targets (the doppler shift), and such small changes take a long time to resolve. Often, targets can only be distinguished from the background of stationary echoes by means of these changes in frequency, and long-wavelength radars can become ineffective at detecting slowly moving targets.
4. At low frequencies there are lots of logistical problems, such as the difficulty of obtaining a transmitting licence and the problem of finding a suitable radio channel free of other users. Also, background noise levels are generally high and only narrow radar bandwidths are available, which, as we shall soon see, restrict the resolution of systems.

Despite all these problems, the rewards for long-range over-the-horizon detection of targets are so attractive that radars do operate at low frequencies, especially in the high-frequency (HF) band, which is from 3 to 30 MHz. Over-the-horizon radar is explained more fully in Chapter 10.

At the higher end of the spectrum used by radar systems, most problems encountered are connected with atmospheric absorption and the limits of available technology. The very highest-frequency radars are laser systems called lidars (derived from *light detection and ranging*), which are usually used for accurate distance measurement, atmospheric studies and in some defence applications. Lidars have a practical upper frequency limit of about 3×10^{14} Hz or 1 μm, set by atmospheric absorption.

The highest operating frequencies used by 'true' radar (using radio frequencies rather than light) are confined to various atmospheric 'windows', which are frequency bands where the atmospheric absorption is not too severe, such as at 35 GHz and near 94 GHz. There are not a great many electronic components available at these frequencies, although this is

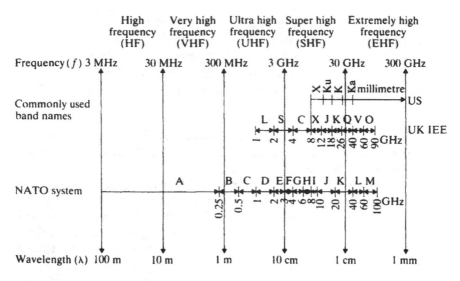

Figure 1.2 Radar frequency band names. The code letters L, S, C, X and K were used for security reasons during World War II and have been widely adopted by radar engineers ever since, despite the introduction of more rational systems.

improving with the development of gallium arsenide technology. The wavelength at 35 GHz is less than a centimetre, and so the term *millimetric radar* is sometimes used to describe these systems. The range of millimetric radars is quite short, because of the absorption problems, but they tend to be compact devices that can have applications such as in the guidance system on a missile. The possibility exists for using optical-type technology at millimetre wavelengths, with the development of devices such as radio lenses and imaging systems, which can be useful when trying to identify targets.

In between the two frequency extremes, the spectrum has been divided up into bands having names that have become part of the jargon of radar. Figure 1.2 explains some of these. The World War II code letters L, S, C, etc., are part of the everyday language of radar engineers, so it is necessary to learn them if you wish to appear knowledgeable! More recently, the alternative NATO band names are increasingly being used.

1.4 ANTENNA GAIN

The antenna is the device for radiating and receiving electromagnetic energy and it has three main functions:

1. To 'beam' power in a given direction in order to increase the radar sensitivity in that direction.

2. To provide for beam steering such that some area of coverage can be provided.

3. To permit the measurement of angular information so that the direction of a target can be determined.

When considering radar antennas it turns out that much of the theory developed in physics courses for optics and optical astronomy is very useful. The concept of reciprocity is especially helpful in allowing us to say that, if transmitter power is fed into an antenna and it spreads out in a certain angular pattern, then, when the same antenna is used for receiving, it has a sensitivity to incoming signals that follows the same pattern. It is common practice to mix these two cases, and you can hear antenna patterns and polar diagrams being discussed one moment as if the antenna were transmitting and the next moment as if it were receiving.

If an antenna were omnidirectional, it would radiate power uniformly over the whole 4π steradians of a sphere. (Do not worry if you have not come across steradians. They are the unit of 'solid' or three-dimensional angle, and are not crucial to what follows.) Such an antenna is called an *isotropic radiator*, but it would not be useful to us because, first, it would not help with the three functions we require of it and, secondly, it can be demonstrated mathematically that it cannot exist in practice. It is more common to build antennas designed to focus or concentrate power in a certain direction. This property of beaming power is known as the directive gain $D(\theta, \phi)$ or the power gain $G(\theta, \phi)$. These terms are often abbreviated to *directivity D* or *gain G*.

While the directive gain describes the radiation pattern of the antenna, it does not give any indication of how lossy the antenna is. The power gain includes the concept of losses, which occur through heating of the antenna itself, the ground plane and any matching devices, as well as power radiated into sidelobes. Power gain may be defined as the ratio of the radiation intensity in the main lobe of the antenna to the radiation intensity from a 100 per cent efficient isotropic antenna having the same power input. The gain of a lossless antenna is given by

$$G \simeq \frac{4\pi}{\Delta\theta \, \Delta\phi} \quad [\quad] \quad (1.3a)$$

where $\Delta\theta$ = width of the beam in the azimuth direction [radians] and $\Delta\phi$ = width of the beam in the elevation direction [radians].

The azimuth angle θ gives directional or bearing information and, as on a magnetic compass, it is measured clockwise from the north. The elevation angle ϕ is measured from the horizon upwards. Any point in the sky can be specified by quoting its azimuth and elevation (see Fig. 1.3). If $\Delta\theta$ and $\Delta\phi$

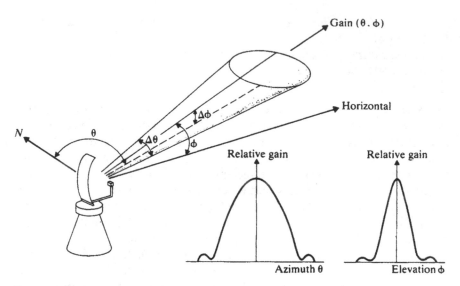

Figure 1.3 Azimuth and elevation angles and beamwidths.

are expressed in degrees, then Eq. (1.3a) may be rewritten as

$$G \simeq \frac{41\,253}{\Delta\theta° \, \Delta\phi°} \quad [\quad] \quad (1.3b)$$

(the constant 41 253 is the number of square degrees on the surface of a sphere). This formula takes no account of the beamshape and yields a result that is about 25 per cent or so too high. For rule-of-thumb calculations, a better constant to choose is 32 000, and when typical losses are included the power gain is usually about

$$G \simeq \frac{25\,000}{\Delta\theta° \, \Delta\phi°} \quad [\quad] \quad (1.3c)$$

Occasionally Eq. (1.3a) is given as $G = \pi^2/\Delta\theta \, \Delta\phi$, which assumes a gaussian beamshape with $\Delta\theta$ and $\Delta\phi$ as the half-power beamwidths (see Skolnik[1]). If you feel comfortable with solid angles, you can also use

$$G \simeq 4\pi/\Omega \quad [\quad] \quad (1.4)$$

where Ω = solid angle of the beam [steradians].

Antenna patterns are directional in three dimensions, and producing one is a bit like squeezing a spherical balloon to produce a protrusion; if you want $G > 1$ in some particular direction, then you have to crush the rest of it and accept $G < 1$ somewhere else. The important point is that the gain integrated all the way round an antenna adds up to unity (or less than unity if efficiency factors are included). The term 'power gain' is most commonly used to denote the maximum gain of an antenna, but beware of the frequent use of the term to describe how the antenna's radiation pattern changes with angle—'the gain falls off rapidly with elevation angle ...', etc.

The maximum gain of an antenna can also be calculated from its size:

$$G = 4\pi A_e/\lambda^2 \quad [\quad] \tag{1.5}$$

where A_e = effective area of the antenna. The effective area is usually less than the real area A (area of a parabolic dish, for example) and is related to it by

$$A_e = \varepsilon A \quad [\text{m}^2] \tag{1.6}$$

Here ε [ ] is an efficiency factor, usually falling in the range 0.4–0.9 for parabolic dishes, which arises because it is difficult to illuminate a dish perfectly from the radio source at the centre. Some antennas, such as those used by domestic televisions, can have effective areas greater than the physical cross-section of the device.

Where does Eq. (1.5) come from? Any source of waves having a linear dimension d, which is large compared to the radio wavelength, tends to have a characteristic beamwidth $\Delta\theta$ in the appropriate plane given by

$$\Delta\theta \sim \lambda/d \quad [\text{radians}] \tag{1.7}$$

We can explain Eq. (1.7) using an optics analogy. Wavelets coming from many small (Huygens) sources along the aperture d interfere at some distant point, as in Fig. 1.4a. Directly in front of the source, we find that the wavefronts add constructively and a beam is formed; but as we begin to move off-axis, they begin to interfere with each other because of the differing path lengths. The actual signal received at a point off-axis depends on the

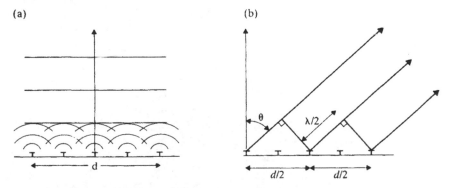

(a)   (b)

Figure 1.4 (a) Wavelets from small Huygens sources add coherently in front of the aperture. (b) At an angle $\sin\theta = \lambda/d$ they interfere destructively and the antenna gain falls to zero, thus giving an approximate characteristic beamwidth to the first nulls of $2\lambda/d$ and λ/d to the -3 dB (half-power) points.

summation of the contributions from all the Huygens sources. We can estimate that, at angles where the contribution from the midpoint of the source has an extra path length of $\lambda/2$ (Fig. 1.4b) over a source near the edge, the process becomes destructive and the beam degrades (the contributions from either side of the centre have smaller and greater differences, which tend to cancel). The off-axis angle at which this occurs is given by

$$\theta \sim \sin\theta = \frac{\lambda/2}{d/2} \equiv \frac{\lambda}{d} \qquad [\text{radians}]$$

As destructive interference occurs at the same angle each side of bore-sight, we can say that the characteristic beamwidth to the first nulls is given by $2\lambda/d$. (Optics books have more details; see for example Chapter 19 on Fraunhofer diffraction in Pedrotti and Pedrotti[2].) It is more common, however, to find radar beamwidths described in terms of the width of the beam at the half-gain points (also called the half-power points or $-3\,\text{dB}$ points). Because of the roughly triangular shape of the beam, the beamwidth at this point approximates to λ/d as in Eq. (1.7). The identity in Eq. (1.7) is well known and is a much used approximation.

Imagine, now, a rectangular aperture having dimensions d_1 and d_2 as shown in Fig. 1.5. The azimuth and elevation beamwidths are given by

$$\Delta\theta \sim \lambda/d_1 \qquad \Delta\phi \sim \lambda/d_2 \qquad [\text{radians}]$$

From the definition of G

$$G \simeq \frac{4\pi}{\Delta\theta\,\Delta\phi} = \frac{4\pi d_1 d_2}{\lambda^2} = \frac{4\pi A}{\lambda^2} \qquad [\quad]$$

Worked example The Siemens–Plessey type 46C weather radar is a C-band system optimized for 5625 MHz and using a 2.44 m diameter

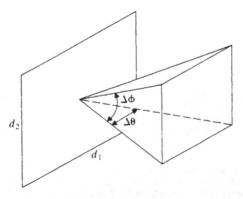

Figure 1.5 The azimuth and elevation beamwidths arising from a rectangular aperture.

parabolic reflector antenna. If the antenna efficiency factor ε is 0.5, what gain and beamwidth would you expect the system to have?

SOLUTION The physical area of the antenna aperture A is given by $\pi d^2/4 = 4.68$ m^2, so the effective area $A_e = 2.34$ m^2. The antenna gain can now be calculated from Eq. (1.5) as

$$G = \frac{4\pi A_e}{\lambda^2} = \frac{4\pi \times 2.34}{(0.0533)^2} \equiv 10\,350 \equiv 40.1 \text{ dB} \qquad [\quad]$$

At first glance the beamwidth might be expected to be given by Eq. (1.7) as

$$\Delta\theta = \Delta\phi \sim \frac{\lambda}{d} = \frac{0.0533}{2.44} \equiv 0.02 \text{ radians} \equiv 1.25°$$

but in practice the effective value of d is less than 2.44 m because of the efficiency factor, and the beamwidth might be expected to be slightly greater than 1.25°.

Siemens–Plessey Radar Limited quote the gain as 40 dB and the beamwidth as 1.5° (to the half-power points) in their technical specifications for this radar[3].

All these arguments involve approximations. More rigorous calculations of antenna gain require a more careful definition of what is meant by beamwidth. There are also many types of radar antenna other than parabolic dishes, and we will be describing some of them later in this book as we look at the different applications of radar.

1.5 THE RADAR EQUATION

The ability of a radar to detect the presence of a target is expressed in terms of the *radar equation*, which is worth deriving, rather than just quoting, because of the insight it gives into the way radars work.

We begin with the transmitter, which has a peak power output P_t [W]. If this power is radiated isotropically by the antenna, then the power flux (the power density per unit area) at a range R is given by

$$\text{Power flux at distance } R = \frac{P_t}{4\pi R^2} \qquad [\text{W m}^{-2}] \qquad (1.8)$$

because $4\pi R^2$ is the area of a sphere of radius R through which all the power must pass.

If the transmitting antenna is not isotropic and concentrates the power towards the target, then we modify Eq. (1.8) by introducing the gain factor

G_t. The power flux in the direction of the beam is now

$$\text{Power flux at the target} = \frac{P_t G_t}{4\pi R^2} \quad [\text{W m}^{-2}] \tag{1.9}$$

The target intercepts a portion of this incident power and re-radiates it. The measure of the incident power intercepted by the target and radiated back towards the radar is called the *radar cross-section*, which is often abbreviated to RCS and is given the symbol σ. The RCS of a target has units of area and indicates how large the target *appears* to be as viewed by the radar. RCS is defined as the power re-radiated towards the radar per unit solid angle divided by the incident power flux/4π steradians.

In reality, the target may be physically much larger in area than the RCS but trying to keep a low radar profile (such as a Stealth aircraft), or it may be very small but trying to make itself appear visible on a radar screen (such as the corner reflecting antennas used on buoys and yachts). There is no fixed RCS for a target, no number that can be painted on the side to say how big it appears to a radar set. The RCS of a target depends on the angle of incidence at which it is viewed, the radar frequency and the polarization used. The RCS also fluctuates with time, as we shall see later.

The power re-radiated by the target is now

$$\text{Power re-radiated} = \frac{P_t G_t \sigma}{4\pi R^2} \quad [\text{W}] \tag{1.10}$$

On the return path this power again spreads out over the sphere of area $4\pi R^2$. Although it does not usually spread out uniformly, the 'gain' of the target is automatically included in the concept of the RCS. The power density at the radar thus becomes

$$\text{Power flux} = \frac{P_t G_t \sigma}{(4\pi R^2)^2} \quad [\text{W m}^{-2}] \tag{1.11}$$

The amount of this returning power that is intercepted by the antenna is determined by its effective area A_e. The mean power received by the radar P_r is thus

$$P_r = \frac{P_t G_t \sigma A_e}{(4\pi R^2)^2} \quad [\text{W}] \tag{1.12}$$

The next move is to substitute for A_e by using Eq. (1.5):

$$G_r = 4\pi A_e / \lambda^2 \quad [\quad]$$

where G_r = gain of the receiving antenna. Finally, the inevitable inefficiencies in a radar system must somehow be introduced and, for now, this is best done by lumping them all together as a system loss factor L_s [ ]. Loss factors may be arranged to appear on either the top or bottom of an equation, but we will adopt the convention that L_s is always less than 1, and therefore appears on the top. Using this definition, the power received by the radar, from the target, is given (with appropriate units also shown) by

$$P_r [\text{W}] = \frac{P_t [\text{W}] G_t [\quad] G_r [\quad] \sigma [\text{m}^2] \lambda^2 [\text{m}^2] L_s [\quad]}{(4\pi)^3 [\quad] R^4 [\text{m}^4]} \qquad (1.13)$$

Although Eq. (1.13) is a complete description of the power received, it is still not useful because it does not indicate whether this power is larger or smaller than the background noise level. Unfortunately, noise is always present, either as *internal noise* from the electronics, or as *external noise* from such sources as the galaxy, the atmosphere, man-made interference or even deliberate jamming signals. All these noise sources are wideband compared to the radar signal, and one of the functions of a radar receiver is to tailor the bandwidth to accept the signal, without permitting any unnecessary further noise to enter. If we were to examine the analogue front end of a radar system using a fast oscilloscope, we would be able to see the noisy signals and perhaps be able to pick out the shape of the echo pulses, if they were large enough. After the analogue-to-digital (A/D) converter, however, there is only one sample for each range gate, which represents the sum of the signal and noise at that point in time.

We will spend more time looking at the properties of noise later, but for now we will simply say that there is an *average* noise power present in the system, to which we will give the symbol N (for typical values see Sec. 2.7). We can now compare the power received from the target with the noise power, in what is variously known as the *signal-to-noise ratio*, SNR or S/N:

$$\text{SNR} = \frac{P_r}{N} = \frac{P_t G_t G_r \sigma \lambda^2 L_s}{(4\pi)^3 R^4 N} \qquad [\quad] \qquad (1.14)$$

This is the all-important radar equation, which is much used in one form or another. It is perhaps a little too complicated to learn, and the best strategy is probably to remember how to derive it quickly from the basics.

Often, the radar equation is used to solve for one unknown. For example, supposing a particular SNR is required for reliable target detection (a typical figure might be 13 dB, which is 20 times noise). The maximum detection range $R_{\max}$ of a given radar can be calculated from

$$R_{\max} = \left(\frac{P_t G_t G_r \sigma \lambda^2 L_s}{(4\pi)^3 N (\text{SNR})} \right)^{1/4} \qquad [\text{m}] \qquad (1.15)$$

Worked example A short-range surveillance radar operates at 3 GHz and uses a 1 m diameter dish for both transmitting and receiving. If the mean transmitter power is 10 kW and the noise level is -140 dB W (see

Table 1.2 A worked example using linear values or decibels

Symbol	Multiply these terms OR Linear value	Add these terms dB	Comments
P_t	10 000 [W]	40 [dB W]	Beware mistaking kW for watts
G_t	1 000 [ ]	30 [dB]	
G_r	1 000 [ ]	30 [dB]	
σ	1 [m²]	0 [dB m²]	
λ^2	0.01 [m²]	−20 [dB m²]	
L_s	0.32 [ ]	−5.0 [dB]	
$1/(4\pi)^3$	1/2000 [ ]	−33.0 [dB]	$(4\pi)^3 \sim 2000$ is a useful approximation to remember
$1/N$	1×10^{14} [W⁻¹]	+140.0 [dB W⁻¹]	
$1/(SNR)$	1/20 [ ]	−13.0 [dB]	
R_{max}^4	8.0×10^{16} [m⁴]	169.0 [dB m⁴]	
R_{max}	~16.8 km	~16.8 km	

Appendix IV), calculate the maximum range at which a small aircraft of radar cross-section 1 m² could reliably be detected†. Assume 5 dB losses and a SNR of 13 dB.

SOLUTION The only difficulty in solving this type of problem is in making the appropriate rearrangement of the radar equation. In this case, it has already been done in Eq. (1.15). A 1 m diameter dish has an area of $\pi r^2 = 0.25\pi$ m². At 3 GHz, $\lambda = 0.1$ m, giving the gain of the antenna as

$$G_t = G_r = \frac{4\pi A}{\lambda^2} = \frac{4\pi \times 0.25 \times \pi}{(0.1)^2} \sim 1000 \equiv 30 \text{ dB} \qquad [\quad]$$

In practice, the gain would be about half this value because of the efficiency factor of the dish. Calculations can now be carried out either linearly or in decibels, whichever you find most convenient. Decibels are explained in Appendix IV, and the advantage of using them is that you have only to add and subtract reasonably sized numbers instead of multiplying and dividing by mind-bogglingly large figures. You can use a table to calculate the answer (see Table 1.2). Those of you familiar with the use of spreadsheets will find this a convenient way to handle such a radar calculation.

† What is meant by reliable detection? We have to consider both the probability of detecting the target P_d and the probability that we might create a false alarm P_{fa} in a sensitive system by declaring a target to be present when in reality only noise is present. In Chapter 4 we discuss how to calculate these probabilities, but as a working rule if P_{fa} is set at 10^{-6} and we require an 85 per cent probability of detection on a single observation, this requires a SNR of around 13 dB.

COMMENTS This type of problem is used quite often as an exercise in familiarization with the radar equation and the size of the numbers involved. The object is not usually to get exact answers but rather to compare one system with another or to find out what happens when one of the parameters is changed. Suppose we wanted to rework this example to find R_{max} for a parabolic dish with twice the diameter (and thus four times the area). We could do this in the dB column by changing both 30 dB values to 36 dB and the total from 169 to 181 dB, giving $R_{max} \equiv 33.5$ km. We could also get the same answer by using the linear column, but it would mean having to do all the multiplying again. Note that doubling the size of the dish doubles the range; why is this?

In the example above, the two really big factors are the R^4 propagation losses and the noise. Very little can be done about either of these and the overall detection range of radars is thus fairly predictable; it is the cleverness and the adaptability of radars that improves with time, rather than the absolute detection range. Also, the radar cross-section used in Eq. (1.14) fluctuates continuously for most real targets and we must use statistical means to describe it. This infers that the radar equation itself is a statistical method of detecting targets and measuring range, rather than being a deterministic calculation; we say more on this in Chapter 4.

1.6 ACCURACY AND RESOLUTION

How well can a radar measure range? To answer this, we must be a bit more careful with our question and define what we mean by the words 'how well'. The *range accuracy* indicates the uncertainty in a measurement of the absolute distance to an object, whereas the *range resolution* tells us how far apart two targets have to be before we can see that there are indeed two targets rather than one large one.

The range resolution of a radar system is fairly straightforward. If the time delay between the echoes from two objects is greater than the pulse duration τ, then two separate echoes are seen (see Fig. 1.6a). If the targets are closer than τ, the echoes will merge (Fig. 1.6b). When the echoes are separated in time by an amount similar to the pulse duration, then they become resolvable (Fig. 1.6c; see also Fig. 6.1). This is like the Rayleigh criterion used in optics, and combining it with Eq. (1.1) gives an expression for the range resolution ΔR:

$$\Delta R = c\tau/2 \quad [\text{m}] \tag{1.16}$$

Usually the radar receiving system samples the receiver output every τ seconds, and each sample represents a distance ΔR called a 'range gate' or a 'range bin'. A radar using a pulse duration $\tau = 1$ μs would thus employ

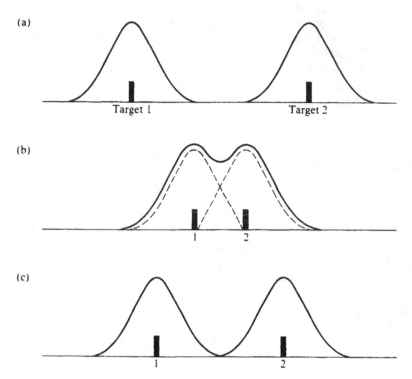

(a)

Target 1   Target 2

(b)

1   2

(c)

1   2

Figure 1.6 Radar range resolution. (a) Two targets are easily resolved when they are more than a pulse length apart, (b) unresolvable when they are much closer than a pulse length, and (c) just resolvable when separated by a pulse length.

a 1 MHz A/D converter to sample the receiver output every 150 m in range, out to the practical maximum range of the system.

Unfortunately, understanding the range accuracy of a system is not as simple as the resolution. Intuition tells us that the accuracy of a range measurement should depend on the 'sharpness' of the pulseshape, and it is therefore rather surprising to discover that the crucial factor determining the range accuracy is the *bandwidth* occupied by the radar. Figure 1.7 gives some insight into why there is this dependence on bandwidth. The system shown in Fig. 1.7a emits a single continuous tone and measures the phase of the echo to find a rough position for the target, but it is ambiguous every wavelength. In Fig. 1.7b a second frequency is added to the transmission to reduce ambiguities and sharpen the position of the target. Adding further frequencies (more bandwidth) eliminates the ambiguities and gives even greater accuracy to the range measurement (Fig. 1.7c).

In practice, pulseshape and bandwidth are related in simple pulse radars. Short pulses take up more bandwidth B of the radio spectrum than long pulses, a result well known to those familiar with Fourier transforms and

(a)

A phase measurement of 90° indicates that the target is in one of these locations

(b)

(c)

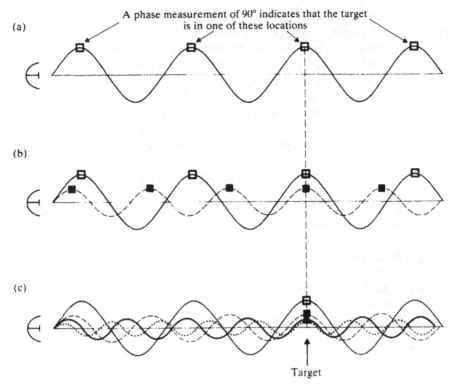

Target

Figure 1.7 Radar range accuracy dependence on bandwidth. (a) With a single frequency, accuracy is poor and the target position is ambiguous. (b) With two frequencies, accuracy and ambiguity are improved. (c) Using more bandwidth, ambiguities are removed and accuracy improves further.

frequency-domain thinking. For those who are not so familiar, it is perhaps sufficient to say that the pulseshape can be built up by adding together a set of sinusoidal waves and that, the shorter the pulse, the higher the frequency of the sinusoids needed to reproduce its shape. For a pulse duration of τ it is a reasonable approximation to say that

$$B \sim 1/\tau \qquad [\text{Hz} \equiv \text{s}^{-1}] \qquad (1.17)$$

To make a more precise statement would require careful definitions of what is meant by 'bandwidth' and 'pulse duration'. We cover these aspects later in the book but, for quick calculations, take τ as the time between the half-power points of the pulse (the 3 dB points) and measure the bandwidth between the half-power points of the spectrum.

Using Eq. (1.17) we can say that a radar with a 1 MHz bandwidth might be expected to be transmitting 1 μs long pulses. In turn, this corresponds to a basic range measurement step size of 150 m using Eq. (1.16). To achieve

15 cm resolution would require a shorter pulse with $\tau = 1$ ns, so that a bandwidth of roughly 1 GHz would be needed.

It is important to be aware that the bandwidth of a radar does not *have* to be limited by Eq. (1.17). As an example, suppose we developed a system that transmitted long pulses during which we swept the frequency of the oscillator deliberately to increase the bandwidth; could we achieve greater range accuracy this way? Radars using such modulation schemes are common, and are known as chirp systems when the frequency sweep during a pulse is linear. By careful processing chirp radars do indeed achieve high range accuracy.

The other factor determining the accuracy of the range measurement is the signal-to-noise ratio, because of the effect that noise has on corrupting the shape of the pulse. Noise is always present and degrades measurements in any system. For example, the length of a line drawn on a piece of paper can be estimated with a centimetre ruler to a lot less than a centimetre under a bright light (SNR > 1). In a dimly lit room (SNR ~ 1) however, it is hard even to estimate it to the nearest line on the ruler and the measurement cannot be made better than the basic step size of one centimetre.

In general, any measurement made with a basic resolution of M will have a root-mean-square (RMS) error δM given by

$$\delta M \simeq \frac{M}{\sqrt{(2 \times \text{SNR})}} \qquad [\text{units of } M] \qquad (1.18)$$

For the range step size ΔR given by Eq. (1.16), substituting for τ using Eq. (1.17) gives the range error as

$$\delta R \simeq \frac{c}{2B\sqrt{(2 \times \text{SNR})}} \qquad [\text{m}] \qquad (1.19)$$

Again, more precise statements could be made if we were to be more careful over definitions, especially the shape and effective bandwidth of the pulse being used. We explore this further in Chapter 6.

As an example, a radar using 1 μs pulses with associated bandwidth of about 1 MHz might be expected to be capable of determining the range of a target observed with a SNR of 20:1 (13 dB) with an error of about 24 m. For a target detected at poor signal-to-noise ratio (SNR ~ 1), not only would the detection be uncertain but we could do no better than to say in which range bin the target occurred, i.e. $\delta R \sim \Delta R$.

1.7 INTEGRATION TIME AND THE DOPPLER SHIFT

Imagine that you are out walking and, glancing briefly over your shoulder, you spot a helicopter flying in the distance. With such a short glimpse it is impossible to tell whether the helicopter is moving or not, and you stop to look more carefully. It soon becomes obvious that the helicopter is not moving quickly, but is it hovering or moving slowly? You stare at it for a long time and decide that it really is hovering. What you have discovered is that measurements of the position of a target can be made quickly, but it takes time to estimate velocities and to distinguish differences in velocities. The smaller the velocity difference, the longer the time needed to estimate it.

The length of time taken to make an observation with a radar set is called the *integration time*, because all the data on a target are integrated or added up until the measurements are sufficiently accurate. It is exactly like photographing moving objects: a night-time photograph of the sky showing a meteor burning up, taken with an exposure of 1/1000th of a second, will show only a faint object, which appears almost frozen in the sky; another photograph taken by leaving the shutter open for half a second will show a much brighter object, whose speed can be estimated from the length of the trail.

How can the speed of an object be measured by radar? There are two methods. The simpler method is similar to watching the helicopter or taking a photograph, in that the speed of a target is estimated from the way its position changes with time (except that optical systems measure transverse velocities and position, whereas radar range information gives the radial component). Observing changes in range is not a very accurate method for a radar to measure speed, although tracking the target for a long time can improve the estimate. This technique is used by *incoherent* radar systems, in which the receiver is tuned to the same frequency as the transmitter but is not phase-locked to it, and so is unaware of any small drifts in frequency between them.

A more accurate way of measuring target speed is to make use of the doppler shift, which is the change in the frequency of the radio signal caused by the motion of the target. The doppler[†] shift is named after Christian Johann Doppler (1803–1853), who pointed out that the colour of a luminous body and the pitch of a sounding body are changed by the relative motions of the body and the observer. In order to detect small changes in frequency, a *coherent* system is needed in which the transmitter and receiver oscillators are phase-locked to reveal any difference in the echo frequency.

Figure 1.8 shows a radar observing a jump-jet. If the jet is hovering, each radio signal sent out by the radar will return with the same phase (as measured with respect to the transmitted signal). If the jump-jet approaches the radar, then each radio signal has to travel a shorter distance and the phase of the echo will change continuously with time. Every $\lambda/2$ through

---

† Although it seems disrespectful to previous generations of scientists, the modern convention is that capital letters are not used to describe the effects named after them (e.g. gaussian noise, doppler shift). The exception is when their name is used as a unit, for example electrical currents are measured in amperes (or amps), but the unit is A.

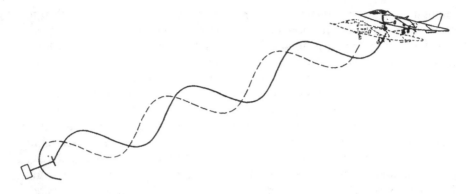

Figure 1.8 The motion of a target causes a change of phase in the radar signal, equivalent to a frequency shift.

which the target moves means that the path length has been shortened by λ and the phase of the echo will have changed by 2π or, in other words, will have rotated by one complete cycle. If the jet were to fly at $\lambda/2\,[\mathrm{m\ s^{-1}}]$ towards the receiver, the radar would detect a 1 Hz change in the radio frequency. If the jet flies at $v_r\,[\mathrm{m\ s^{-1}}]$, then we must work out what this represents in multiples of $\lambda/2$ in order to calculate the corresponding doppler shift f_d; this leads directly to the formula (another one worth learning)

$$f_d = 2v_r/\lambda \qquad [\mathrm{Hz}] \qquad (1.20)$$

where v_r = radial component of the target speed towards the radar.

Transverse components of velocity do not contribute to the doppler shift because the target neither approaches the radar nor recedes from it. Because radars measure only radial components and optical systems, such as television and infrared cameras, measure only transverse information, it is not uncommon to find the two types of system integrated together in 'point' or 'local' defence arrangements.

The expression in Eq. (1.20) is actually an approximation to the full relativistic formula, but it is valid for all cases where $v_r \ll c$, which is almost always the case—very few radar targets approach the speed of light!

Very roughly, the integration time t needed to resolve two doppler frequencies separated by Δf_d is given by

$$t \sim 1/\Delta f_d \qquad [\mathrm{s}] \qquad (1.21)$$

and using Eq. (1.18) again we can say that the error δf_d in the measurement of doppler frequency is given by

$$\delta f_d \sim \frac{1}{t\sqrt{(2 \times \mathrm{SNR})}} \qquad [\mathrm{Hz}] \qquad (1.22)$$

Combining this with Eq. (1.21) means that the error δv_r in the velocity measurement is

$$\delta v_r \sim \frac{\lambda}{2t\sqrt{(2 \times \text{SNR})}} \quad [\text{m s}^{-1}] \tag{1.23}$$

The formulae above are not exact, because it all depends on how the measurements are made. For example, if the doppler measurement is made during a single long pulse, then the exact formula depends on the shape of the pulse; if the measurements are made by integrating many pulses, then the accuracy of the result depends on the 'weighting' or how the information is put together. More details can be found in Chapter 4 and in Skolnik[1].

Worked example A ship sailing at a radial speed of 5 m s^{-1} is observed entering a harbour by a high-frequency (HF) radar operating at 3 MHz and an S-band radar operating at 3 GHz. What observation times would the two systems need to distinguish the ship from the background echoes of the land? If the signal-to-noise ratio is 20 dB in both cases, how accurately could the velocity be measured?

SOLUTION At 3 MHz the radio wavelength is 100 m and the doppler shift of the ship echo can be calculated from

$$f_d = \frac{2v_r}{\lambda} = \frac{2 \times 5}{100} = \frac{1}{10} \quad [\text{Hz}]$$

This frequency must be distinguished from the land echoes, for which $f_d = 0$, and so the frequency difference $\Delta f_d = 0.1$ Hz. From Eq. (1.21) we can say that the observations must be made over 10 s for the ship to be distinguished. A pulsed HF radar usually sends out several hundred pulses per second, and so a few thousand echo pulses would be collected together for doppler processing when the radar was in ship detection mode.

The microwave radar uses $\lambda = 0.1$ m, and when this is inserted in Eq. (1.20) the doppler shift for the ship comes out to be 100 Hz. An observation period of only 0.01 s or 10 ms would therefore be needed to separate the ship and land echoes, and typically this would be achieved using a short burst of about 16 pulses.

A SNR of 20 dB means that the signal power is 100 times the noise power, and putting this into Eq. (1.23) gives velocity errors of $0.5/\sqrt{2} \equiv 0.35$ ms^{-1} for both radar systems. Try it!

1.8 SUMMARY

Radar is used to detect the presence of an object and measure its position and speed. It is possible to make a swift estimate of the performance of a radar system using the formulae given below.

Key equations

- The range of a target:

$$R = c\tau_d/2 \quad [m]$$

- Relation between frequency and wavelength of an electromagnetic wave:

$$c = f\lambda \quad [m\ s^{-1}]$$

- The power gain of a typical antenna:

$$G \simeq \frac{25\,000}{\Delta\theta°\ \Delta\phi°} \quad [\ ]$$

- The maximum gain of an antenna:

$$G = 4\pi A_e/\lambda^2 \quad [\ ]$$

- Characteristic beamwidth:

$$\Delta\theta \sim \lambda/d \quad [radians]$$

- Signal-to-noise ratio:

$$SNR = \frac{P_t G_t G_r \sigma \lambda^2 L_s}{(4\pi)^3 R^4 N} \quad [\ ]$$

- Range resolution:

$$\Delta R = c\tau/2 \quad [m]$$

- Bandwidth:

$$B \sim 1/\tau \quad [Hz]$$

- Range error:

$$\delta R \simeq \frac{c}{2B\sqrt{(2 \times SNR)}} \quad [m]$$

- Doppler shift:

$$f_d = 2v_r/\lambda \quad [Hz]$$

● Error in measuring doppler frequency:

$$\delta f_d \sim \frac{1}{t\sqrt{(2 \times \text{SNR})}} \quad [\text{Hz}]$$

● Error in measuring radial velocity:

$$\delta v_r \sim \frac{\lambda}{2t\sqrt{(2 \times \text{SNR})}} \quad [\text{m s}^{-1}]$$

1.9 REFERENCES

1. *Introduction to Radar Systems*, M. I. Skolnik, McGraw-Hill, New York, 1985. [This is the classic paperback on radar systems, but is more of a reference work than a good read.]
2. *Introduction to Optics*, F. L. and L. S. Pedrotti, Prentice-Hall, Englewood Cliffs, NJ, 1987.
3. *The Plessey Type 46C Weather Radar*, RSL 1479, Issue 8, 1983.
4. *Racal–Decca 2690 BT Marine Radar Series*, Racal Marine Electronics publication reference RME/0008/1187/AD.
5. *Plessey ACR 430 Airfield Control Radar*, RSL, Issue 2, 1987.

1.10 FURTHER READING

There are many specialist technical books on different aspects of radar including a whole 'library' published by Artech House. These are not for the uninitiated, however, and below is a selection of alternative texts that the beginner might consider reading to get different perspectives on the way radars work.

Radar Principles for the Non-specialist, J. C. Toomay, Van Nostrand Reinhold, New York, 1989. [Patchy in coverage, but contains some useful explanations.]
Radar Systems, P. A. Lynn, Macmillan, London, 1987. [Concentrates principally on air traffic control applications.]
Understanding Radar, H. W. Cole, BSP Professional Books, Oxford, 1985. [Readable, concentrating mainly on primary and secondary surveillance radar.]
Radar Principles, N. Levanon, Wiley, New York, 1990.
Technical History of the Beginnings of RADAR, S. S. Swords, Peter Peregrinus for the IEEE, Stevenage, Herts, 1986. [A mine of historical information.]
Radar Design Principles, F. E. Nathanson, McGraw-Hill, New York, 1969. [For the 'advanced' beginner.]

1.11 PROBLEMS

1.1 The *Racal–Decca 2690 BT Marine Radar Series*[4] offers radar options that include (a) a 2.7 m wide antenna operating at a wavelength of 3 cm and (b) a 3.6 m antenna operating at 10 cm. Calculate the horizontal beamwidth of the two antennas, assuming each has a linear efficiency factor of 80 per cent.

1.2 The vertical beamwidth (-3 dB) is quoted as 23° for the 3 cm version of the 2690 BT system. What would you expect the antenna gain to be?

1.3 One of the 2690 BT radar bandwidth settings is 4 MHz. What range resolution would this give?

1.4 The Siemens–Plessey ACR 430[5] is an X-band airfield control radar for the close control of aircraft approach to land in poor weather conditions. The effective aperture size is 3.4 m horizontally by 0.75 m vertically. What would be the beamwidths at an operating frequency of 9.4 GHz?

1.5 What antenna gain would you expect for the ACR 430 antenna?

1.6 During evaluation of the ACR 430, aircraft were positioned for approach by controllers using the radar for the first time. Averaged over 41 approaches to a runway, the worst mean offset from the centreline approach was less than 25 m at a range of 4 n. mile (see Appendix III). How good is this performance, and would it be sufficient to position an aircraft actually on a runway in poor visibility?

1.7 Compare the detection ranges of the two 3 cm radars above for a 10 m^2 target using a single pulse. Assume both radars have a mean noise level of -131 dB W, have losses of -5 dB and require a SNR of 13 dB for reliable detection. The marine radar has a peak transmitter power of 25 kW, but the airfield control radar is more powerful with a peak power of 55 kW.

TWO

DESIGNING A SURVEILLANCE RADAR

- Surveillance
- Choosing parameters
- Radar cross-sections of targets and clutter
- Final design

An attempt to design a surveillance radar reveals how difficult it is to watch the whole sky.

2.1 RADAR AND SURVEILLANCE

Marine radar, air traffic control, air-borne radar defence systems, ground-based radar arrays searching for satellites in space—these are all well known examples of radar surveillance systems that work in all weather conditions, and at all times of day, testifying to the versatility of the radar technique.

Most modern surveillance radars are 'multifunction', jargon for their capacity to carry out other activities at the same time as searching for new targets. One of these functions is to keep track of existing targets until they are within a certain distance of the radar, when an entirely separate system takes over: at a civil airport, for example, a Terminal Area radar guides in the planes; a ship under attack, on the other hand, might engage a close support weapons radar to lock on to the target. In this chapter we are going to concentrate on one of the main applications of surveillance radar, which is to search the sky continuously for new targets.

2.2 ANTENNA BEAMWIDTH CONSIDERATIONS

Surveying the sky presents the radar design engineer with something of a dilemma; the radar needs to scan the whole sky and get back to the starting point as quickly as possible in order not to miss any new developments. However, the radar also needs to spend as much time as possible staring at each part of the sky in order to obtain good results. Perhaps the best way to illustrate the difficulties is for us to follow a typical design process and develop the problem for ourselves.

A common choice of frequency for long-range surveillance is L-band (~ 1.3 GHz) because this avoids the bad-weather problems that can affect higher frequencies. S-band (~ 3 GHz), for example, is often used only for medium-range surveillance up to about 60 nautical miles or 111 km. In practice, many factors would determine the final choice of frequency, including the type of target to be detected and the coverage required. We will assume that a frequency of 1300 MHz ($= 1.3$ GHz) has been chosen,

When considering the antenna design, there are two conflicting requirements. First, there are several good reasons for choosing narrow beams:

1. The angular position of the target can be measured with good precision.
2. The number of unwanted echoes cluttering up the picture at any one time is reduced.
3. The number of interfering signals that can get into the beam at any one time is also reduced.
4. The antenna gain factors G_t and G_r in the radar equation are increased (i.e. the transmitter power is more concentrated in the direction of the target) and, consequently, the signal-to-noise ratio is improved, making the target easier to detect.

Secondly, set against these advantages, is a rather serious drawback. If a given area of sky is to be searched, narrower beams imply that there are more beam positions within that area to be investigated. If we imagine that the whole hemisphere of sky is to be surveyed, this equals 2π steradians of solid angle (4π being the entire sphere). For a beamwidth $\Omega = \Delta\theta \, \Delta\phi$, the number of beam positions required to fill the hemisphere is

$$\text{Number of beam positions} = \frac{2\pi}{\Omega} = \frac{2\pi}{\Delta\theta \, \Delta\phi} \qquad [\quad] \qquad (2.1)$$

But remembering Eq. (1.4), we can write this as

$$\text{Number of beam positions} = G/2 \qquad [\quad] \qquad (2.2)$$

This leads to the general rule that an antenna with a gain of G must probe

G directions to survey an entire sphere; in fact, this is another way of thinking about the meaning of antenna gain.

Worked example If our L-band radar operates with a 6 m diameter dish antenna, how many positions must it search to ensure coverage of the whole sky? Assume an antenna efficiency factor of 0.6.

SOLUTION The physical area of the dish is 28.3 m² but, because of the efficiency factor, this reduces to an effective area of 17 m². The wavelength at 1300 MHz is 23.1 cm and thus the gain of the antenna is given by

$$G = 4\pi A_e/\lambda^2 \simeq 4000 \qquad [\quad]$$

There are therefore 2000 beam positions in the sky to be searched.

The width of the radar beam is given roughly by Eq. (1.7) as

$$\Delta\theta = \Delta\phi \sim \frac{\lambda}{d} = \frac{0.231}{6} \equiv 0.039 \text{ radians} \equiv 2.2°$$

but again efficiency considerations mean that in practice the dish would not be uniformly illuminated by the source antenna at the focus and the effective beam would be broader, perhaps 3° wide.

COMMENT Two thousand beam positions may seem a lot to search, but a 3° beamwidth is wide by today's standards, and radars with narrower beams have even more positions to search.

Each beam position must be inspected at least every 10 s because data at this rate are required to follow the path of targets accurately (a process known as 'tracking', which is carried out by the data processing software). There is a danger that, if the repeat search time is made any longer than 10 s, then not only will the tracking accuracy be poor but the target may manoeuvre during the interval between scans and will be lost by the tracker. If our radar had only a single beam, then the search time per beam position would be (10s)/(2000 positions) giving 5 ms per position. This presents us with a problem—physically it would be difficult to scan a dish at the speed needed to achieve 5 ms per beamwidth and in any case, as we shall see later, more time than this is needed in each beam position to get good results.

Also, while the repeat scan time of 10 s is probably adequate to spot the manoeuvring of civil airliners, there are military aircraft, such as jump-jets, that can make sudden changes in speed and direction capable of confusing radar tracking algorithms. These targets require even more frequent inspection. We will return to this problem of the search time after examining a few more of the surveillance radar system parameters.

2.3 PULSE REPETITION FREQUENCY AND UNAMBIGUOUS RANGE AND VELOCITIES

The greater the rate at which pulses are transmitted by a radar system, the greater the mean power radiated, and it is interesting to ask the question: 'In the design of a radar system, what is the constraint on the maximum rate at which pulses can be transmitted?' The limit occurs when pulses are transmitted so frequently that one pulse is transmitted before the previous pulse has completed the round trip to the target and back. In this situation, it is unclear which transmitted pulse originated which echo pulse, and the target range becomes ambiguous.

The rate at which pulses are transmitted is called the pulse repetition frequency, often abbreviated to PRF. Using Eq. (1.1) and the '150 m per microsecond' rule we can say that, if the time interval T between infinitesimally short transmitter pulses were 1 μs (a PRF of 1 MHz), then ranges up to a maximum of 150 m could be surveyed before the receiver had to be turned off to allow the next transmitter pulse to be sent. More practically, for a PRF of 1 kHz, range ambiguities are 150 km apart (although the range that can be surveyed is a little less than this because of the pulse duration and $T - R$ switching times). The maximum PRF that can be used for an unambiguous range R_{max} is given by

$$PRF \leqslant c/2R_{max} \quad [Hz] \quad (2.3)$$

What would happen if we tried to survey ranges up to 150 km with a 1.5 kHz PRF (which has an unambiguous range of only 100 km)? The answer is shown in Fig. 2.1, in which a continuous sequence of transmitter pulses and the associated echoes from a target are shown. The most obvious assumption is that echo pulses 1 and 2 come from the transmitted pulses A and B respectively, meaning that the target is within the 'first-time-around' range interval of 0-100 km. In the case shown in Fig. 2.1, the first-time-around range is 30 km. However, it is also possible that echo 2 came from

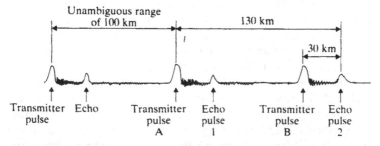

Figure 2.1 A radar system with an unambiguous 'first-time-around' range of 100 km. The target range appears to be 30 km, but it could be 130 km.

transmitter pulse A and echo 1 came from the previous transmission. In this case, the target lies in the 'second-time-around' range interval of 100–200 km and the target range is $30 + 100 = 130$ km. Likewise it is possible to have third-time-around echoes in the range 200 to 300 km, and so on.

Worked example Our surveillance radar is required to have an unambiguous range of 450 km; what is the maximum PRF that may be used? If the pulse length is 3 μs, what is the 'duty cycle' or on/off ratio for the transmitter?

SOLUTION The maximum PRF can be found directly from Eq. (2.3) as 333 pulses/second. Another way to look at the problem is to rearrange Eq. (1.1) as

$$T = 2R/c \quad [s]$$

which tells us that when $R = 450 \times 10^3$ m the round-trip time for a radar signal is $T = 3$ ms. For a pulse emission every 3 ms, the PRF must be $1/T = 333$ pulses/second, often referred to as a PRF of 333 Hz.

The duty cycle is the ratio of the pulse length τ to the inter-pulse period T. In this case

$$\text{Duty cycle} = \frac{\tau}{T} = \frac{3 \times 10^{-6}}{3 \times 10^{-3}} \equiv 10^{-3} \quad [\quad]$$

COMMENT Often, this PRF is too low to adequately sample the doppler frequency. Also, the duty cycle may be too low for certain types of transmitters that need to be worked harder than this if sufficient power is to be developed to give a good performance over 450 km.

There are several ways in which the PRF can be increased without reducing the unambiguous range. One method is to 'colour' or label each pulse in some way so that it can be distinguished from its neighbours. Methods of labelling pulses include transmitting them with different frequencies, phases, polarizations or pulseshapes. There are difficulties with these methods, however, when the target radar cross-section (RCS) fluctuates from pulse to pulse. Labelling pulses also interferes with doppler processing and moving-target indication (MTI)—see Chapter 5—and so has only limited usefulness.

Another method of increasing the PRF is to use bursts of pulses on different PRFs. Each one of these independently may be ambiguous over the range interval, but when used in combination with the others the ambiguities can be eliminated. Two staggered PRFs are demonstrated in Fig. 2.2; when PRF (a) is used the target range is measured as 30, 180, or 330 km, but when PRF (b) is used the range is measured as 105 or 330 km. The true range of the target must therefore be 330 km.

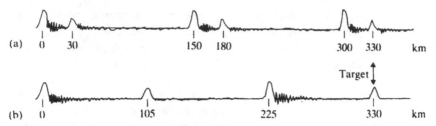

Figure 2.2 The two PRFs (a) and (b) give ambiguous range information when used independently. In combination, only one range is possible.

Historically, *low-PRF radars*, designed to avoid range ambiguity problems and often using MTI, have become known as *MTI systems* to distinguish them from high PRF radars. The main difficulty with MTI systems is that the doppler sampling rate (the PRF) is too low for the speeds of modern aircraft. This is essentially an aliasing problem and leads to ambiguous velocity estimates and, worse, to the occurrence of blind speeds at which a target appears stationary and cannot be resolved against the clutter background. A common example of aliasing occurs in old cowboy movies when the wagon wheels appear stationary because the camera shutter (the sampling device) operates over the same time interval as the spokes take to move from one position to the next. A system designed to detect moving parts of the image would not detect the spokes.

It is possible to overcome the doppler ambiguities with a *high-PRF radar*; in this case the system is often known as a *pulse doppler radar*. High-PRF pulse doppler systems generally suffer from range ambiguities and 'blind' ranges where transmission of other pulses disables the receiver. It is not usually possible to find a compromise PRF that avoids both range and doppler ambiguities. For example, the US AWACS (air-borne warning and control system) uses one mode in which it is ambiguous in range, and a separate mode in which the doppler information is ambiguous. With pulse doppler systems the solution to the ambiguity problem is also to use multiple PRFs, so that the blind ranges occur at different places. Many air-borne radars are *medium-PRF systems*, possessing both range and doppler ambiguities, but using several PRFs to remove these ambiguities.

Note that some older radar systems use *staggered PRFs* in which the pulse intervals vary from pulse to pulse according to some carefully chosen rule. In this case, delays are introduced before processing to permit stationary clutter to be removed by MTI. However, targets beyond the first ambiguity are not properly integrated and second-time-around clutter is not properly cancelled.

In practice, our surveillance radar would probably use multiple PRFs near 333 Hz to resolve doppler ambiguities and blind speeds and also to eliminate any spurious second-time-around echoes. Because it uses a low PRF, our system would be classed as an MTI radar.

2.4 PULSE LENGTH AND SAMPLING

In this section we wish to show that short transmitter pulses imply fast sampling of the receiver output, which can create difficulties for the digital signal processing. Even with the recent rapid growth in computing power we must still be careful not to design a radar that produces digits so quickly that we cannot afford to buy computers large enough to process them on-line.

The pulse length of 3 μs chosen above is fairly typical for a long-range air surveillance radar, and from Eq. (1.16) we know this gives a nominal range resolution of about 450 m. In practice, the resolution would be worse than this because of pulse shaping and other losses in the systems, and the final figure might be nearer 750 m. On the other hand, Eq. (1.19) tells us that the range can be measured more accurately than the nominal 450 m if we have a good signal-to-noise ratio. These improvements in the estimates of range come from two processes. The first is known as 'plot extraction' and involves interpolating between adjacent range samples on each scan. The second process is tracking, which is the smoothing of the apparent path of the target through many observations. The overall range accuracy of the system would be about 50 m after all these processes had been undertaken, and this would normally be adequate for long-range air surveillance.

The output of our radar receiver should be sampled every 3 μs because, as Fig. 2.3a shows, the samples are then separated by a time/distance equal to the transmitted pulse length (as measured at the 3 dB or half-power points). Because the echo pulses must be at least the same length as the transmitted pulses, this sampling rate ensures that no information can be missed. If the sampling were carried out less frequently than every 3 μs, information would be lost and a small target might escape detection by falling between two samples, as shown in Fig. 2.3b. There are some benefits in sampling more frequently than every 3 μs (known as 'oversampling') because it improves the SNR on a target that straddles two range bins (Fig. 2.3c) but it involves a lot more signal processing for only a small reward.

By 'sampling' we usually mean using an analogue-to-digital (A/D) converter to represent the voltages coming out of the receiver by binary numbers that digital and computer hardware can process. If a sample must be taken every 3 μs, this implies an A/D converter speed of 333 kHz.

Radar systems need a high 'dynamic range', meaning that they must be able to process large echoes from nearby objects (often clutter) at the same time as echoes from distant objects that are very faint; this is a natural consequence of the $1/R^4$ term in the radar equation. The bigger this range of voltages in the receiver, the greater the number of bits needed in the A/D converter to represent them and it is roughly true to say that one more bit is needed every time the voltage range is doubled (= 6 dB increase in dynamic range). However, the dynamic range requirement can be eased in practice by using range-dependent gain or *swept gain*. During the transmission of each

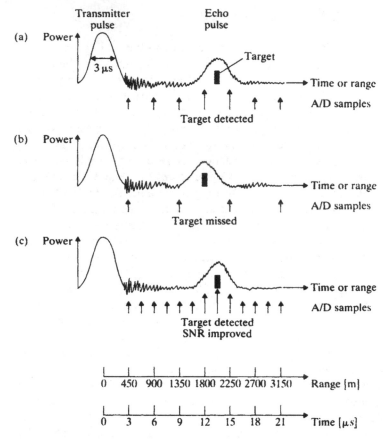

Figure 2.3 (a) The receiver output sampled correctly; (b) undersampled, which may cause a target to be missed; and (c) oversampled, improving the SNR for a target straddling two range bins.

pulse, the receiver is held at zero gain; at short ranges the gain is gradually increased and at long ranges the full receiver gain is used. Beyond 40–50 km, clutter decreases owing to the curvature of the Earth and shadowing effects.

In our surveillance radar we will assume that an 8-bit word is insufficient for each A/D sample and that 2 bytes per sample must be used.

Our receiver output is now being sampled at 666 000 bytes per second, and worse is to come; we need two receiver channels in order to measure the doppler information. A single coherent receiver can reveal the speed of a target but it cannot tell whether it is moving towards, or away from, the radar. A second receiving channel is used to resolve this ambiguity by shifting it 90° in phase from the first channel; these are known as I and Q channels, which stands for In-phase and Quadrature. This second channel must also be sampled at 666 000 bytes per second and the overall data rate produced by our radar is thus about 1.3 Mbytes per second. This sampling rate is not too great for modern computers to store or move around in memory, but it becomes more challenging if complex floating-point calculations have to be carried out. Short-range radars working with much higher bandwidths and shorter pulse lengths than our surveillance system can present the design engineer with some tough and expensive data processing problems to solve.

2.5 RADAR CROSS-SECTION

To make further progress in our design of a surveillance radar we must investigate the radar cross-section (RCS) of typical targets likely to be encountered. We have seen (Sec. 1.4) that the RCS describes the apparent area of the target as perceived by the radar and that it is a measure of how much power flux is intercepted by the target and re-radiated to the radar receiver.

For simple objects, such as a perfectly conducting copper sphere, it is possible to use electromagnetic theory to calculate a single value for the RCS (and this turns out to be quite useful because copper spheres are used to calibrate radar systems, especially on indoor test ranges). However, real targets such as aeroplanes have many reflecting surfaces, which radiate in and out of phase with each other and cause large fluctuations in the RCS. These fluctuations mean that some form of statistical description must be used for the RCS, such as an average value and the extent of the variations about this average. First of all we will look at what influences the average value of the RCS before going on to describe the fluctuations.

The RCS σ of an object is partly dependent on the radar wavelength, and for simple shapes it is possible to give the following guidelines:

1. For target sizes $\gg \lambda$, the RCS is roughly the same size as the real area of the target. This is known as the *optical region* because the RCS approaches the optical value.
2. For target sizes $\sim \lambda$, the RCS varies wildly with changes in wavelength, and it may be greater or smaller than the optical value. This is known as the *resonance* or *Mie region*.
3. For target sizes $\ll \lambda$, the RCS $\propto \lambda^{-4}$. This is known as the *Rayleigh region* after Lord Rayleigh, who discovered that the scattering of light by particles in the atmosphere varies as λ^{-4} (Tyndall, Mie and Debye also contributed to these studies). This wavelength-dependent scattering explains why the sky is blue and the sun appears yellow. When white light from the sun arrives at the Earth, the relatively long-wavelength yellow/red components of the spectrum pass more or less straight through the atmosphere compared with the shorter-wavelength blue

light, which is scattered over the sky. Much the same is true on a larger scale size where low-frequency radar signals are undisturbed by water droplets in rain and clouds, but millimetric radar signals suffer significant scattering. This reasoning lay behind our choice of the relatively low-frequency L-band for surveillance at the beginning of the chapter.

The RCS of a few simple shapes are given in Table 2.1, for the case when the object size is large compared with a wavelength. These simple RCS values turn out to be quite useful for several reasons. First, it is sometimes possible to get a feeling for the RCS of an object by building it up out of a few simple shapes; for example, see the chapter by I.D. Olin in reference 1. Secondly, at long wavelengths, targets often behave as uncomplicated structures because the scattering is not from many tiny surfaces but

Table 2.1 The RCS of some simple shapes when the size of the object is large compared with a wavelength

Target	Aspect[†]	RCS
Sphere		πr^2
More general large curved surface		$\pi r_1 r_2$
Circular flat plate		$\dfrac{16\pi^3 r^4}{\lambda^2}\cos^2\theta\left(\dfrac{J_1(4\pi r\sin\theta/\lambda)}{4\pi r\sin\theta/\lambda}\right)^2$ $= \dfrac{4\pi^3 r^4}{\lambda^2}$ broadside
Square flat plate		$\dfrac{4\pi r^4}{\lambda^2}\cos^2\theta\left(\dfrac{\sin(2\pi r\sin\theta/\lambda)}{2\pi r\sin\theta/\lambda}\right)^2$ $= \dfrac{4\pi r^4}{\lambda^2}$ broadside

Table 2.1 *Continued*

Target	Aspect[†]	RCS
Circular cylinder		$$\frac{2\pi r L^2}{\lambda} \cos^2\theta \left(\frac{\sin(2\pi L \sin\theta/\lambda)}{2\pi L \sin\theta/\lambda} \right)^2$$ $$= \frac{2\pi r L^2}{\lambda} \quad \text{broadside}$$
Circular cone		$$\frac{\lambda^2}{16\pi} \tan^4\psi$$
Dihedral corner reflector		$$\frac{8\pi r^4}{\lambda^2}$$
Trihedral corner reflector		$$\frac{12\pi r^4}{\lambda^2}$$
Short-circuit half-wave dipole		$0.86\lambda^2$
Chaff		$\sim 0.15 N\lambda^2$

† Aspect is considered in the direction of the arrow (where appropriate).

instead involves induced electric currents flowing throughout the target. There are also times during radar systems testing when it is advantageous to have an antenna of known RCS as a calibration target, and so a simple resonant half-wave dipole has been included in Table 2.1.

Although the concept of RCS implicitly includes the contingency that the target of scattering area A has gain G in the direction of the radar, this may be explicitly stated as

$$\sigma = GA \quad [\text{m}^2] \tag{2.4}$$

Combining Eq. (2.4) with $G = 4\pi A/\lambda^2$ gives

$$\sigma = 4\pi A^2/\lambda^2 \quad [\text{m}^2] \tag{2.5}$$

This can be a useful way of thinking about RCS when the target is an antenna, for which the concept of gain is well defined.

When the target is not simple, and cannot be synthesized from simple shapes, there are various well established computer programs that can be applied to the problem of calculating the RCS. If all else fails, a scaled copper model of the target can be built and measured in an anechoic chamber at an appropriate scaled frequency. A good summary of the state of the art was presented in 1989 when an entire issue of the *Proceedings of the IEEE* was devoted to the radar cross-sections of complex objects[2].

In the case of our L-band radar, the radar cross-section of an aircraft will lie in the optical region because the wavelength of 23 cm is much less than the size of the target. We cannot assign a single value to the RCS because it will depend on the aspect angle at which the target is viewed, both in azimuth and in elevation, and also on the polarization angle of the radar. These factors, combined with interference from different scattering surfaces on the target, mean that as we observe the aircraft the RCS will fluctuate. These fluctuations can be treated by finding the mean value of the RCS σ_{av} and a probability density function (PDF) $p(\sigma)$ to describe the variations about the mean. A well known density function for random variables is the chi-squared variable (χ^2) with two, or four, degrees of freedom. The forms are

$$p(\sigma) = \frac{1}{\sigma_{av}} \exp\left(-\frac{\sigma}{\sigma_{av}}\right) \quad \sigma \geqslant 0 \quad [\quad] \tag{2.6}$$

$$p(\sigma) = \frac{4\sigma}{\sigma_{av}^2} \exp\left(-\frac{2\sigma}{\sigma_{av}}\right) \quad \sigma \geqslant 0 \quad [\quad] \tag{2.7}$$

In 1960, P. Swerling[3] used these expressions as the basis for four proposed mathematical models describing different types of RCS fluctuation. These four *Swerling cases* are discussed in Chapter 4 when we examine target detection theory, and they are still in widespread use today, even though more sophisticated models have been developed since. With low-frequency systems, such as over-the-horizon radar, the less complicated scattering mechanisms mean that quite often the RCS of a target remains constant for a considerable period of time; such non-fluctuating targets are sometimes referred to as *Swerling case 5* targets.

The time-averaged (RMS) value of the RCS σ_{av} is sometimes used in the radar equation because this is the only information available. However, the fluctuations mean that the predicted performance will only be achieved for that part of the time when $\sigma \geqslant \sigma_{av}$. The Swerling models predict RCS values lower than average to occur more frequently than above-average values because the amplitudes and phases of two interfering signals vary independently (see Chapter 4). Amplitude and phases must both be nearly equal for above-average values of σ to occur. The PDF for the first case is shown in Fig. 2.4, and it can be seen that values of σ significantly above σ_{av} rarely occur. The use of σ_{av} in the radar equation thus gives an over-optimistic performance estimate, and it becomes important to know which model best describes the RCS fluctuations of the target. Appropriate adjustments can then be made when calculating the probability with which a target can be detected by a particular radar system, or what coherent gain may be assumed when pulses are added together to improve the signal-to-noise ratio.

For our surveillance radar, typical values of RCS that might be expected are given in Table 2.2. Further information may be found in Maffett[7].

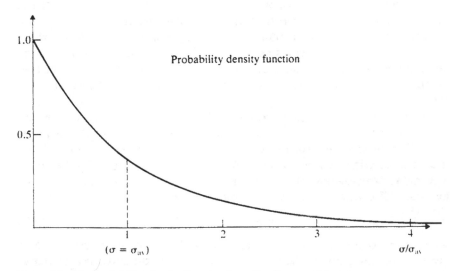

Probability density function

$(\sigma = \sigma_{av})$

σ/σ_{av}

Figure 2.4 The probability density function for a Swerling case 1 target.

Table 2.2 Typical RCS values for some common targets

Target	RCS on linear scale	RCS on log scale
Bird	$0.001 \, m^2$	$-30 \, dB \, m^2$
Cruise missile	$0.01 \, m^2$	$-20 \, dB \, m^2$
Person Small boat Small aircraft	$1 \, m^2$	$0 \, dB \, m^2$
Cabin cruiser Fighter–bomber aircraft	$10 \, m^2$	$10 \, dB \, m^2$
Road traffic Large aircraft	$100 \, m^2$	$20 \, dB \, m^2$
Tankers Large passenger ships	$1000 \, m^2$	$30 \, dB \, m^2$

2.6 CLUTTER

Clutter is a single word used to describe all the unwanted echoes that clutter up the radar picture. Clutter is nearly always present and usually widespread, with echoes arising from hills, buildings, the sea, birds and insects, meteors, the aurora and many other sources. Of course, what is clutter in one application may not be so in another. An example of this is HF radar, which is often used to observe waves on the sea, in which case echoes from ships (ship clutter) confuse the picture; but when HF radar is used for over-the-horizon ship tracking, it is the echoes from waves (sea clutter) that cause the problems.

Clutter may occur as distributed clutter, which increases with the resolution cell size, or as point clutter, which does not. Point clutter arises from discrete scatterers such as electricity pylons. The term *surface clutter* is used to describe land or sea echoes from the area illuminated by the radar. The radar cross-section of the clutter is best described by calculating the average RCS density σ^0, the RCS per unit area, which is given by the ratio

$$\sigma^0 = \sigma_c / A_c \quad [\quad] \tag{2.8}$$

where σ_c = RCS of the area A_c. The symbol σ^0 is sometimes referred to as *sigma zero*, and it can be thought of as representing the radar reflectivity of the terrain.

Volume clutter is similarly used to describe echoes from the atmosphere, where it is the volume illuminated that determines the RCS of the clutter observed. An average RCS per unit volume η is defined as

$$\eta = \sigma_c / V_c \qquad [\mathrm{m}^{-1}] \qquad (2.9)$$

where σ_c = RCS of the volume V_c. We will investigate η and its usefulness for atmospheric studies in Chapter 9.

The value of σ^0 varies with many factors: the type of terrain observed, the direction it is observed from, the weather, the radar wavelength, the polarization used, etc. Detailed studies have been made and databases are available to help the radar designer in any particular location[4]. Typical values for σ^0 are 1/100 (-20 dB) for land, meaning that the RCS observed by the radar is about 1/100th the area actually illuminated, whereas for sea, a reflectivity of 1/1000 (-30 dB) is an average figure.

The echoes from surface clutter are large, often much larger than from the targets of interest. For a given system we can calculate how large they are with the aid of Fig. 2.5. First we must calculate the physical area illuminated, which is a strip of ground one range bin ΔR wide and with a length given roughly by

$$\text{Length of arc} = R \,\Delta\theta \qquad [\mathrm{m}] \qquad (2.10)$$

where R = range [m] and $\Delta\theta$ = beamwidth [radians]. The total area illuminated is

$$\text{Area illuminated} = R \,\Delta\theta \,\Delta R \qquad [\mathrm{m}^2] \qquad (2.11)$$

For a radar looking down to the ground at a depression angle α, the area illuminated is that given by (2.11) multiplied by a factor sec α ($= 1/\cos \alpha$).

We can calculate typical clutter values for our surveillance radar using the beamwidth of 3° (remembering to convert to radians) and the range bin size of 450 m. At a range of 50 km, the area of ground illuminated

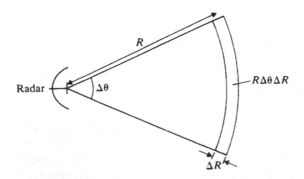

Figure 2.5 The area of ground illuminated by a radar beam at grazing incidence.

(provided the radar was sited high enough to see the ground at this range) would be roughly

$$\text{Area} = 50 \times 10^3 \times 450 \times 0.05 \equiv 1.2 \times 10^6 \quad [\text{m}^2]$$

A reflectivity of 1/100 suggests that the clutter from the ground would be 12 000 m². Assuming the radar is searching for an aircraft of RCS 1 m², the signal-to-clutter ratio would be 1/12 000. In decibels, this figure corresponds to 41 dB, and is one of the factors that determine the dynamic range requirement for the receiver. In practice, at least another 13 dB of dynamic range is needed, so that an adequate SNR is available for target detection, plus some extra allowance for clutter variations around the mean value.

Worked example An HF radar transmits 100 μs pulses at 10 MHz with a 100 m antenna array looking out over the sea. What are the dynamic range requirements for the receiver if a -25 dB m² cruise missile must be detected at a range of 30 km and 60 km?

SOLUTION The wavelength is 30 m and λ/d for the array gives a beamwidth of roughly 0.3 radians. The pulse length implies a range gate size of 15 km. At 30 km range, the physical area of sea illuminated is therefore roughly

$$\text{Area illuminated} = 30 \times 10^3 \times 15 \times 10^3 \times 0.3 \quad [\text{m}^2]$$

$$= 135\,000\,000 \quad [\text{m}^2]$$

Taking the reflectivity of the sea as 1/1000 implies that the clutter has an RCS of 135 000 m² or 51 dB m². The radar receiver must be able to detect noise signals 13 dB lower than the target (for 13 dB SNR detection), and so it must be capable of detecting -38 dB m² signals. The dynamic range required by the receiver, as shown in Fig. 2.6, must therefore be at least 89 dB or 28 000:1 as a voltage ratio. At 60 km range, the clutter will be 3 dB (two times) greater, the target size remains the same and so the dynamic range requirement rises to 92 dB.

COMMENT With increasing range, the signal strength from a target decays as R^4 because of the radar equation but the clutter power falls off as R^3 because the area illuminated has a linear dependence on R. Signal-to-clutter problems therefore get worse at longer ranges.

If target echoes are so much smaller than clutter, how can they be detected at all? The answer is that most targets of interest are moving and can be distinguished from stationary clutter through the use of the doppler effect. Doppler processing to give *sub-clutter visibility* (SCV) is an important part of radar processing and is discussed in Chapter 5. However, the

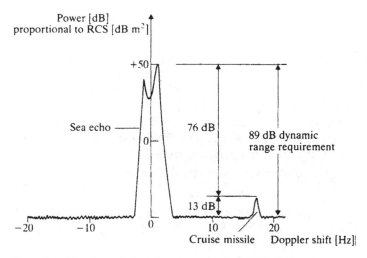

Figure 2.6 The dynamic range requirement of an HF radar receiver.

implication for our surveillance radar design is that we must dwell for long enough in each beam position to gain sufficient doppler resolution to separate target and clutter echoes.

Most surveillance radars also maintain *clutter maps,* which can be static (loaded prior to operation) or dynamically modified during operation. Separate maps are maintained for ground clutter, rain, chaff, etc. The radar antenna beam and the signal processing can then be adaptively modified to suit the environment in a process known as *clutter fixing.*

2.7 NOISE

Before we can proceed with our final design, we need to know what noise level to expect. At L-band the noise is likely to be dominated by internal noise produced by the random motion of thermally excited electrons. Below about 6000 GHz (i.e. for all practical radars), thermal noise can be considered to be *white noise* having a flat power spectral density $S(f)$ given by

$$S(f) = kT_0 \qquad [\text{W Hz}^{-1}] \tag{2.12}$$

where k = Boltzmann's constant = 1.38×10^{-23} J K^{-1} and T_0 = system temperature [K]. T_0 is generally assumed to be 290 K, giving a noise density of -204 dB W Hz^{-1}, a figure engraved on the hearts of most radar engineers. This noise spectrum extends far beyond the bandwidths of the radar system and we are only concerned with *band-limited white noise,* which has been restricted by the passband of the receiver. The mean noise power in the receiver N is given by

$$N = kT_0B \quad [\text{W}] \tag{2.13}$$

where B = bandwidth [Hz]. Note that this figure is independent of the radar operating frequency; an S-band and an L-band radar, both operating with a bandwidth of 1 MHz, would have identical mean noise levels of -144 dB W ($= -204$ dB W Hz^{-1} + 60 dB Hz bandwidth).

In practical receivers, the noise level is found to be worse than in Eq. (2.13) by a factor known as the *noise figure F*. The noise figure is measured over the linear part of the receiver operating range as

$$F = \frac{\text{noise power out of actual receiver}}{\text{noise power out of ideal receiver}} \tag{2.14}$$

In the case of our L-band surveillance radar, the bandwidth can be found from $1/(\text{pulse length})$ = 333.3 kHz = 55.2 dB Hz. Assuming a noise figure for the receiver of 3 dB, the mean noise level is

$$
\begin{aligned}
N &= kT_0FB & [\text{dB W}] \\
&= -204 + 3 + 55.2 & [\text{dB W}] \\
&= -145.8 \text{ dB W}
\end{aligned}
\tag{2.15}
$$

Table 2.3 Losses in radar systems

Loss	Typical value
(a) Losses encountered in many surveillance radar systems	
'Plumbing' loss (in waveguides, T–R switch, rotating joint, phaseshifters, feed)	3.5 dB
Beamshape loss (target not in centre of the beam for duration of the scan time)	2.0 dB
Pulse compression and filter weighting loss	1.0 dB
Sampling loss (target not in centre of the range rate)	0.5 dB
Fast Fourier transform weighting loss	2.0 dB
(b) Components causing losses in some radar systems	
Tx filter	Beamformer
Isolator	Rx protection circuit
Connectors	Rx front-end filter
Radome (protective antenna coating)	
(c) Other losses	
Mismatches to antenna impedance	
Eclipsing loss at short ranges (system not switched back to full receiver sensitivity after transmitting)	
Integration losses (i.e. coherent integration not perfect)	
Constant false-alarm rate processing losses (discussed in Chapter 5)	

The instantaneous noise power in the receiver will fluctuate about this mean value. In Chapter 4 we discuss the statistical properties of noise in more detail, and the way in which this affects the probability of target detection.

2.8 FINAL DESIGN

Having decided most system parameters, we can now work out the transmitter power needed to undertake the task of surveillance. We will assume that the targets will be no smaller than 1 m². Rewriting the radar equation (Eq. (1.14)) gives

$$P_t = \frac{(\text{SNR})N(4\pi)^3 R_{\text{max}}^4}{G_t G_r \lambda^2 \sigma L_s} \quad [\text{W}] \tag{2.16}$$

Next we insert the values chosen during this chapter, i.e.

SNR	20	[ ]	13.0	[dB]
Noise N	2.6×10^{-15}	[W]	-145.8	[dB W]
$(4\pi)^3$	2000	[ ]	33.0	[dB]
R_{max}^4	$4.1 \times 10^{+22}$	[m⁴]	226.1	[dB m⁴] (maximum)
$1/G_t G_r$	6.3×10^{-8}	[ ]	-72.0	[dB]
$1/\lambda^2$	18.7	[m⁻²]	12.7	[dB m⁻²]
$1/\sigma$	1	[m⁻²]	0	[dB m⁻²]
$1/L_s$	3.2	[ ]	$+5$	[dB] (typical minimum)
P_t	16	[MW]	72	[dB W]

A transmitter providing 16 MW of peak power would be very expensive. However, we have so far only applied the radar equation to a single pulse, which would give no sub-clutter visibility. We need to look at the phase difference of the echo from pulse to pulse to see if the target is moving.

The radar equation can be modified to include the integration gain of several pulses added together by including a factor nL_n on the top of the equation; n is the number of pulses and L_n is the loss over perfect integration (i.e. L_n is less than unity). In modern radars, the integration is usually coherent (taking into account phase, as well as amplitude) and L_n can be quite close to unity (near zero loss). Some older systems used incoherent integration by adding up the signals after the detector, where all phase information has been lost. Losses of several dBs were incurred. In fact, there are many potential sources of loss in a radar system, and a summary is shown in Table 2.3 opposite, where suggested typical values for the more regularly incurred losses are also given. Further details of how to calculate losses are to be found in Skolnik[5] and Rohan[6].

One final modification to the radar equation that might be mentioned here is to replace the peak transmitter power P_t by the average power $\bar{P}_t$ using the duty cycle τ/T:

$$\bar{P}_t = P_t \tau / T \qquad [W] \qquad (2.17)$$

i.e.

$$P_t = \frac{\bar{P}_t}{(PRF)\tau} \qquad [W] \qquad (2.18)$$

The radar equation can now be rewritten as

$$SNR = \frac{\bar{P}_t G_t G_r \sigma \lambda^2 n L_n L_s}{(4\pi)^3 k T_0 F (B\tau)(PRF)R^4} \qquad [\quad] \qquad (2.19)$$

The terms B and τ are collected together because their product is approximately unity. This 'pulse radar equation' is easy to convert to an energy equation by replacing the SNR ratio with the energy/noise ratio (E/N) and the transmitted power $\bar{P}_t/(PRF)$ by the transmitted energy E_t. Thinking of the radar equation in terms of energy can be a useful concept when there is a fixed time interval for data collection, as is the case with our surveillance radar.

In our final design, we would probably integrate 32 pulses in each beam position for the following reasons:

1. At 333 pulses per second, 32 pulses gives an integration time of 0.096 s, which, applying Eq. (1.21), corresponds to a useful basic doppler resolution of about 1 m s^{-1}.
2. The 32 pulses give a convenient number on which to carry out fast Fourier transform processing to extract doppler information and achieve sub-clutter visibility.
3. If the antenna were of the revolving dish design with a fixed beam, then with a beamwidth of $3°$ and a rotation time of 10 s, a target would remain in the main beam for a duration similar to the integration time.
4. Using 32 pulses for target detection means that the peak transmitter power can be reduced to 500 kW with a consequent considerable saving in cost.

All these reasons dictate the need for an observation time of nearly 100 ms in each beam position for successful target detection. At the beginning of the chapter we worked out that only 5 ms per beam position is available if the whole sky is to be scanned. This reveals one of the truisms of surveillance—there is never enough time to do everything properly and compromises have to be made. There are several ways out of the dilemma; perhaps the most obvious method is to use several beams simultaneously, for this reduces the rate at which a dish must be scanned. An option available to the most modern radars, where the beam can be steered electronically (Chapter 15), is to make intelligent decisions about how the search pattern should be adapted to the situation, according to a set of priorities. In some cases, the simplest solution is simply to abandon searching certain parts of the sky, usually those at high elevation angles.

In real-life radar design there are many more sophistications and problems to be dealt with. There would be integration losses during the 32 pulses because of RCS fluctuations, a variety of modulation schemes would be used, and so on. But these are all complications that will be easily mastered if you understand the basic design process and the physics of what is happening in radar surveillance.

2.9 SUMMARY

Radar surveillance can be improved through the use of narrow beams, but this may lead to there being more beam positions to be searched than is possible in the time available. Multibeam systems can help with this problem, but they put more pressure on the data processing activities, which are often already stretched, even with today's technology.

The RCS of real targets fluctuates and its statistical nature must be taken into consideration if the radar detection performance is not to be over-estimated. The problem of clutter and the need for sub-clutter visibility is often severe and leads to a need for doppler processing. Careful design of the transmitted waveform is needed to avoid range and/or doppler ambiguities.

Key equations

- Number of beam positions to fill the sky (hemisphere):

$$\text{Number of beam positions} = G/2 \quad [\quad]$$

- Maximum PRF for an unambiguous range R_{max}:

$$\text{PRF} \leqslant c/2R_{max}$$

- Radar cross-section of a target:

$$\sigma = 4\pi A^2/\lambda^2 \quad [\text{m}^2]$$

- Radar cross-section per unit area:

$$\sigma^0 = \sigma_c/A_c \quad [\quad]$$

- Radar cross-section per unit volume:

$$\eta = \sigma_c/V_c \quad [\text{m}^{-1}]$$

- Total area of ground illuminated at grazing incidence:

$$\text{Area illuminated} = R \, \Delta\theta \, \Delta R \qquad [\text{m}^2]$$

- Mean noise level:

$$N = kT_0FB \qquad [\text{W}]$$

- Average transmitter power:

$$\bar{P}_t = P_t\tau/T = P_t\tau(\text{PRF}) \qquad [\text{W}]$$

- Radar equation:

$$P_t = \frac{(\text{SNR})N(4\pi)^3R_{\max}^4}{G_tG_r\lambda^2\sigma L_s} \qquad [\text{W}]$$

- Radar equation in alternative form:

$$\text{SNR} = \frac{\bar{P}_tG_tG_r\sigma\lambda^2nL_nL_s}{(4\pi)^3kT_0F(B\tau)(\text{PRF})R^4} \qquad [\quad]$$

2.10 REFERENCES

1. *Modern Radar Techniques*, M. J. B. Scanlan, Collins, Glasgow, 1987. [Chapter 3 on target characteristics is a useful summary.]
2. IEEE special issue on radar cross sections of complex objects, *Proc. IEEE*, **77**, 5, 1989.
3. Probability of detection for fluctuating targets, P. Swerling, *IRE Trans.*, **IT-6**, 269–308, 1960.
4. *Radar Design Principles*, F. E. Nathanson, McGraw-Hill, New York, 1969. [Contains good chapters on RCS and clutter.]
5. *Radar Handbook*, Ed. M. E. Skolnik, McGraw-Hill, New York, 1990. [Chapter 11 is concerned with the RCS targets.]
6. *Surveillance Radar Performance Prediction*, P. Rohan, Peter Peregrinus for the IEE, Stevenage, Herts, 1983.
7. *Topics for a Statistical Description of Radar Cross Section*, A. L. Maffett, Wiley, New York, 1989. [A detailed study of radar cross section and statistical aspects.]
8. *Technical History of the Beginnings of RADAR*, S. S. Swords, Peter Peregrinus for the IEE, Stevenage, Herts, 1986.

2.11 PROBLEMS

2.1 The Italian company Alenia manufactures the ATCR-44K, an L-band medium-range primary radar designed for use in modern automated air traffic control (ATC) systems. The antenna is a rotating reflector and if a rotation rate of 6 revolutions per minute (RPM) is selected then typically a pulse repetition frequency (PRF) of 480 Hz and a pulse duration of 1.5 μs is used. Alternatively, at 12 RPM the PRF is typically 691 Hz and the pulse duration is 1 μs. Given that the peak transmitter power is 1.2 MW, what is the duty cycle and the mean power for the two modes of operation?

2.2 Derive the expression, given by Alenia, that the number of hits n on a target each time the antenna scans past is given by

$$n = \frac{\Delta\theta(\text{PRF})}{(\text{RPM}) \times 6}$$

where $\Delta\theta$ = azimuth beamwidth [degrees]. If $\Delta\theta = 1.2°$, how many hits per scan are there when the antenna rotation rate is 6 RPM?

2.3 The documentation for the Alenia ATCR-44K radar gives the maximum unambiguous range as

$$R_{max} = \frac{1\,000\,000}{12.4 \times (\text{PRF})} \quad [\text{nautical miles}]$$

What do you think the number 12.4 represents?

2.4 What is the detection range of the ATCR-44K radar for a 2 m^2 target? Assume 5 dB system losses, a further 2 dB for atmospheric absorption, and that a 10 dB SNR is required. Take the elevation beamwidth as 4.7°.

2.5 Does the ATCR-44K radar require staggered PRFs?

2.6 What A/D converter requirement might the ATCR-44K radar be expected to have?

2.7 During World War II the German Navy used a 125 MHz medium-range early-warning radar known as Freya. The rotatable antenna consisted of 12 dipoles mounted in front of a rectangular wire mesh reflector (see Swords[8]). If the peak transmitter power was 20 kW and the pulse duration 2 μs, at what range could the radar detect a 2 m^2 target? Assume an antenna gain of 16 dB and losses of 5 dB. How does this compare with the modern Alenia radar?

THREE

TRACKING RADAR

- Measuring angle
- Monopulse radar
- Tracking accuracy
- Radar guidance systems

Tracking radars are dedicated to a target and are designed to measure angular information accurately.

3.1 INTRODUCTION

A tracking radar continuously measures the coordinates of a moving target in order to determine its path and to predict where it is going. Tracking can be carried out using range, angle or doppler information, but it is the tracking in angle that forms the characteristic feature of tracking radars.

To some extent, all types of surveillance radar, civil or military, could be considering as tracking systems since they form an estimate of the target position each time the scanner returns to look in that particular direction—a process known as *track-while-scan*. Surveillance radars can keep track of many targets simultaneously but the positional accuracy they provide, especially in angle, is not adequate for some purposes.

In contrast, a tracking radar (often a military system) is *dedicated* to a target and observes it continuously and with great precision, usually with a view to engaging a weapons system. Some tracking radars have their own search facilities, but the more usual mode of operation is for a main surveillance radar to warn of any targets posing a particular threat and to download the target coordinates to the tracking system, which then searches the immediate area to *acquire* the target before initiating the tracking process.

The antenna of a tracking radar usually faces the target all the time in order to keep it in the centre of the beam and so maximize the signal-to-noise ratio. If the target moves away from the centre of the beam, this produces an error voltage, which is amplified and fed to servo-motors to drive the antenna back onto the target. The azimuth and elevation of the target are then read from angle transducers, such as synchros, mounted on the antenna axes. The methods of producing error signals from the antenna patterns form one of the main aspects of tracking radar design.

3.2 SEQUENTIAL LOBING

Switching between several beam positions or *sequential lobing* was one of the earliest methods used to derive angular information about a target. An antenna would have two beams displaced slightly left and right of its centreline and the receiver switched rapidly between them. If the target appeared to grow larger in the left-hand beam, this would drive the antenna to the left until the signal was equal in both beams again. Similarly, two beams displaced vertically could be used to give elevation information.

A common method of producing several beams from one antenna was to use a parabolic dish with a block of four microwave horns feeding it at the focus; a fifth horn could be used for transmitting at the centre of the block. At least four pulses had to be transmitted, one for each quadrant, and this is the weakness of the method because rapidly fading targets (Swerling cases 2 or 4) can vary in amplitude from pulse to pulse, thus making comparisons between beams less valid.

3.3 CONICAL SCANNING

An alternative to stepping the antenna beam around the direction of the target is to rotate it continuously. You can simulate the effect by pointing your arm straight out at a distant object and then begin moving your arm in small circles around the object (the angle between the object and the circle you are describing is called the squint angle). If you then deliberately 'miss' the object with the centre of your scan, you will see that once per revolution your arm comes nearer to the object; if this were a radio system, you would get the biggest signal in this position and the information could be used to correct the pointing of the antenna.

One of the first successful applications of conical scanning was Telefunken's 600 MHz 'Würzburg' radar used by the German Luftwaffe in World War II for gun laying and ground-controlled interception of allied aircraft; see details in Swords[1]. The technique is still used today as a reliable and low-cost method of tracking in non-critical situations; for an example see problem 3.1.

Conical scan radars usually process 10 or more pulses per revolution, but even so they suffer from the same handicap as sequential lobing in that the RCS of the target can change between pulses. Also, the repetitive nature of the scanning makes the system vulnerable to electronic countermeasures.

3.4 MONOPULSE RADAR

The problems of pulse-to-pulse variations in echo amplitudes can be overcome by using more than one beam simultaneously to measure the angular position of the target on a single pulse. This technique is known as *monopulse tracking*, a name suggested by Bell Telephone Laboratories in 1946. However, it was originally known as *simultaneous lobing*, which may be a better description, and it was experimented with as early as 1928 (see Rhodes[2]). Monopulse tracking can make use of amplitude information from the antennas, phase information or even both together to give much better precision than the earlier sequential lobing and conical scan systems. Another advantage of monopulse radar is that, in principle, a target can be located from a single pulse measurement, which may be useful when the radar is being jammed and the target is only viewed in glimpses.

The simplest monopulse systems, and in many ways the most reliable, are the *amplitude-comparison monopulse* radars. Two antenna beams are set at an angle to each other and the outputs are connected to a hybrid, which forms sum Σ and difference Δ signals from them. These signals are fed to a pair of matched receivers, mixed to a lower frequency and amplified (see Fig. 3.1).

The sum channel forms a beam that is a combination of the signal power of two individual beams and so has an improved signal-to-noise ratio. This combined beam is used for target detection and to measure the range and doppler information. The gain of the sum channel beam, which directly faces the target, gives monopulse a SNR advantage over the earlier techniques in which the target was viewed off the bore-sight. However, the sum beam is wider than the individual beams and so it is not used to measure angle.

The difference channel produces an error voltage that is roughly proportional to the angular deviation of the target from the bore-sight (see Fig. 3.2), and no output is obtained when the target echo amplitude is the same in both antenna beams. Fading of the echo occurs equally in both beams and so it does not affect this comparison, except for the usual inaccuracies at low signal-to-noise ratio.

The phase-sensitive detector shown in Fig. 3.1 is used to determine the sign of the error voltage so that the servo-motors know which way to drive

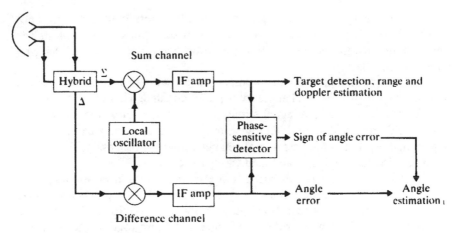

Figure 3.1 Block diagram of an amplitude-comparison monopulse radar system.

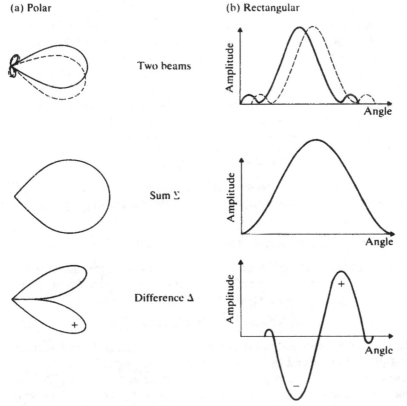

Figure 3.2 Monopulse antenna beam patterns expressed (a) in polar form and (b) in rectangular coordinates.

the antenna to get back on target. It should be stressed that the presence of the phase-sensitive detector does not mean that the system exploits the phase information contained within the radio echo as a means of deriving angular information.

Amplitude monopulse can be improved by taking the ratio Δ/Σ which normalizes the difference channel by the sum channel. This gives a quotient that is independent of the signal strength and linear against the angle error over a wide range of angles. The function of dividing signals can be undertaken digitally, but in the past it has been performed by processing the signals with logarithmic amplifiers and then taking differences.

Perhaps the best way to get to grips with amplitude monopulse is to draw out the antenna patterns in Fig. 3.1 for yourself; try using a pocket calculator to generate a $(\sin \theta/\theta)^2$ pattern, plot two versions of it with one slightly displaced from the other, and then add them together to form the sum channel and subtract them to form the difference channel. Chapter 7 of Hirsch and Grove[3] describes in a particularly helpful way how to use a computer to carry out a more sophisticated simulation of monopulse antenna patterns and differences.

Phase-comparison monopulse is also possible and was tested early on in radar history; in fact, the technique was patented in 1943 (see Rhodes[2]). Two antennas are fixed adjacent and parallel to each other and, by comparing the phase difference of the two outputs, it is possible to derive angular information. An echo arriving along the bore-sight of the antennas will arrive at both of them at the same time (Fig. 3.3a), but a signal arriving at an angle θ to the bore-sight will arrive at one antenna later than at the other because it has had to travel an extra distance x given by

$$x = d \sin \theta \quad [\text{m}] \tag{3.1}$$

where d = separation of the antennas [m]. The distance x can be expressed as a fraction of the radar wavelength λ to give the difference in phase $\Delta\psi$ between the two signals as

$$\Delta\psi = 2\pi d(\sin \theta)/\lambda \quad [\text{radians}] \tag{3.2}$$

The factor 2π in Eq. (3.2) arises because the phase difference increases by 2π radians for every complete wavelength λ travelled by the signal. Note that for small angles $\sin \theta \sim \theta$, leading to the approximation

$$\Delta\psi \sim 2\pi d\theta/\lambda \quad [\text{radians}] \tag{3.3}$$

which is a roughly linear relationship between the angular deviation from the bore-sight direction and the phase-difference error signal.

Phase-comparison monopulse is identical to the Young's slit experiment of physical optics and the interferometry techniques used by radio astronomers. However, the usefulness of phase-comparison methods in tracking radar is somewhat limited because of problems created by multiple signals from

(a)

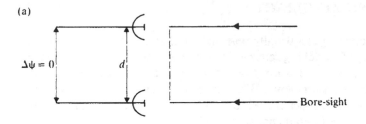

$\Delta\psi = 0$ d Bore-sight

(b)

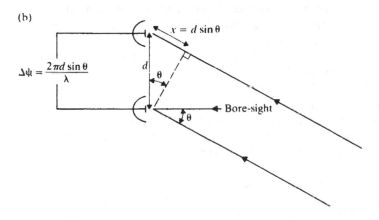

$\Delta\psi = \dfrac{2\pi d \sin\theta}{\lambda}$ $x = d\sin\theta$ d θ Bore-sight θ

Figure 3.3 Phase-comparison monopulse. (a) When the echo arrives along the bore-sight, there is no phase difference between the signals. (b) When the echo arrives at an angle θ, this gives rise to a phase difference $\Delta\psi$.

a single target (a problem known as multipath, described further in Chapter 7). Another problem is concerned with the ambiguities in angular position, called *grating lobes*, that occur when only two widely spaced antennas are used (see Chapter 15).

The best strategy to use when designing tracking radars is to make use of all the information available (amplitude, phase, range and doppler) and to use many antenna elements to reduce grating lobes; this leads on to the design of phased array antennas, which are described in more depth in Chapter 15. Modern phased array antennas can overcome one of the major drawbacks of older tracking radars by making use of electronic beam steering to track more than one target simultaneously. Mechanically driven dishes cannot move fast enough to watch two widely separated areas of sky at the same time, and a weapons system on a ship, for example, could be defeated by a simultaneous attack by two aircraft from different quarters.

3.5 TRACKING ACCURACY

At long range, a target can usually be considered as a point source and the angular accuracy of a tracking radar is determined by both electromechanical factors associated with the control of the antenna turning gear and by the SNR of the radar measurements. Whichever tracking method is used, from Eq. (1.18) we would expect the angular accuracy $\delta\theta$ to be related to the beamwidth and to be fundamentally limited by

$$\delta\theta \sim \Delta\theta/\sqrt{(2 \times \text{SNR})} \qquad \text{[radians]} \qquad (3.4)$$

However, the use of sum and difference channels allows this to be improved upon such that $\delta\theta/k$ may be achievable, where k is the slope of the Δ/Σ curve near $\theta = 0$, see Levanon[4].

The use of sequential lobing, conical scanning or monopulse is merely the means by which the precision of Eq. (3.4) is achieved, but note that the first two techniques will be further degraded by amplitude fluctuations whereas monopulse is not. In practice, RCS fluctuations, multipath and changes in the atmospheric propagation all add together to increase the *tracking noise*. These errors have been elegantly discussed and summarized in Chapter 7 of Berkowitz[5], for those who wish to delve deeper.

> **Worked example** A large ground-based tracking radar has a beamwidth of 1° and detects a target with a SNR of 17 dB on a single pulse. What angular accuracy might be expected if 100 pulses are coherently summed without loss? If the radar bandwidth is 1 MHz and the target were at a range of 10 km, would the transverse error in the target location be greater or smaller than the radial error?
>
> SOLUTION Twice the final signal-to-noise ratio would be 10 000:1 and so the best accuracy obtainable would be 0.01° (this is close to the best achievable in practice).
>
> At a range of 10 km the transverse error $R\,\delta\theta$ corresponds to a distance of 1.7 m. A 1 MHz bandwidth implies a basic range resolution of 150 m, which would improve to 1.5 m for this SNR, so the two accuracies are similar. It would be easier to improve the range accuracy (using more bandwidth) than the angular accuracy, provided the scattering size of the target was small enough to justify this.

At short ranges, the finite size of the target begins to limit the accuracy of the system. Tracking radars need narrow 'pencil' beams and so tend to operate at relatively short wavelengths, which puts the target RCS into the optical region. At these short wavelengths the target behaves as many independent scatterers, each of which contributes in a complex way (i.e. in both amplitude and phase) to the overall RCS and causes an effect similar to optical glinting. We can define a 'centre of gravity' of the target and attempt to track this, but at any given instant the strongest scattering facet may lie elsewhere on the object and the radar may start to wander off the central point. Skolnik[6] shows how certain phase relationships between the scattering elements can even cause the apparent centre of gravity to lie outside the object. The wandering of the apparent target direction, and the attempts to control this effect by increasing the time constant of the antenna drive feedback loop, lead to a form of *angle noise*, which dominates receiver noise tracking errors at short ranges.

3.6 FREQUENCY AGILITY

Frequency agility (similar to frequency diversity in communications) is the process of changing the radar frequency from pulse to pulse. There are considerable advantages to this technique, despite the additional complexity of the radar system (for more details, see Nathanson[7], for example). These benefits include:

1. A reduction in angle noise, because the complex sum of the contributions from the individual target scattering surfaces changes from one frequency to another, and a more accurate mean estimate of the target centre of gravity can be made.
2. A reduction in multipath effects, because the change in wavelength changes the position at which destructive interference occurs.
3. Greater resistance to electronic countermeasures than radars operating on a single frequency.

The frequencies chosen must be sufficiently far apart to give uncorrelated measurements, which means that they must be at least a radar bandwidth apart. The frequency channels must also be a 'target decorrelation bandwidth' apart, defined in a similar way to Eq. (1.16). If the characteristic dimension of the target in range is l, then the separation of frequencies Δf must satisfy

$$\Delta f \gtrsim c/(2l) \quad [\text{Hz}] \tag{3.5}$$

One of the most critical applications of tracking radar is in the defence of ships against incoming missiles, such as Exocets, which have small cross-sections and fly just above the surface of the sea. Compensation for the motion of the ship increases angle noise, the sea creates a strong clutter background and, because salt water is a good conductor, there is a strong multipath image of the missile. In addition to these troubles, the angle noise can be made worse if the missile is programmed to weave (usually to evade gatling gunfire) and so present an ever-changing aspect to the radar. Under these circumstances, frequency agility can be invaluable, as can the contribution from other sensors such as TV systems, lidars and infrared trackers.

3.7 THE TRACKING PROCESS

The first task of a tracking radar is to *acquire* the target allocated for engagement. Tracking radars usually have a search mode to survey a restricted area of sky around the expected target position. Different search patterns are used, but it is common to find options of a horizontal raster scan, similar to a television, a vertical variant of this called a nodding scan, and a low-elevation scan round the entire horizon, as shown in Fig. 3.4.

After the target has been detected, the next step is to estimate its position in a process known as *plot extraction*. We have already seen how the angular information is extracted, but on each pulse the target range and velocity must also be measured.

A rough estimate of the range can be found by locating the range cell in which the echo is largest, but much more precision can be obtained by

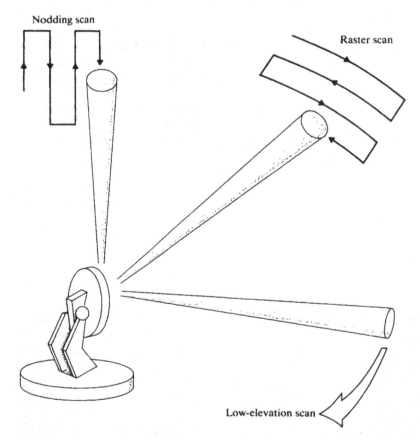

Nodding scan

Raster scan

Low-elevation scan

Figure 3.4 Various methods of scanning for a tracking radar to acquire a target.

interpolating between range cells. If two adjacent cells show equal echo power, then clearly the target lies exactly half-way between them; but if one is larger than the other, then some interpolation formula is required. Usually the interpolation is carried out by fitting the pulse shape or a quadratic curve to the data samples and then differentiating to find the peak. The error in using a simple quadratic rather than the actual pulse shape after matched filtering is quite small. As an example, a quadratic curve can be fitted to the A/D samples shown in Fig. 3.5 using the equation $V = ax^2 + bx + c$. For the central sample, $x = 0$, so that $c = V^0$; at $x = +1$ range cell, $V^+ = a + b + V^0$ and al $- x = -1$ range cell $V^- = a - b + V^0$. Solving these equations gives

$$2a = V^+ + V^- - 2V^0 \text{ and } b = (V^+ - V^-)/2.$$

The maximum of the quadratic occurs at

$$x_{max} = \frac{-b}{2a}$$

The range of the target is then that of the central range cell $+ x_{max}$.

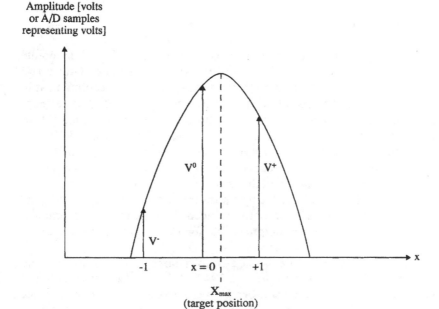

Amplitude [volts or A/D samples representing volts]

V^0

V^+

V^-

-1 $x = 0$ $+1$ x

x_{max}
(target position)

Figure 3.5 Interpolation using a cosine-squared pulse to improve the range accuracy.

Again, the accuracy of the interpolation is limited by the SNR as in Eq. (1.18). Other techniques include hardware range trackers based on the early/late gate synchronizer familiar to most digital communications engineers.

Velocity information can be acquired in a similar manner to range: interpolation between adjacent doppler cells in the fast Fourier transform (or other type of filter) is used to refine the estimate of the target speed. There are, however, complications with velocity measurements; the doppler shift of the target can be spread by weaving and other manoeuvres, and also contaminated by rotating devices such as turbines or helicopter blades. It is sometimes necessary to define a 'centre of gravity' of the doppler measurement and use this in the tracking algorithm. When frequency agility is used, and some of the channels are lost through either jamming or RCS fading, the process of extracting velocity information can become quite interesting!

As the antenna continues to face the target and the radar evaluates its position plot by plot, mathematical algorithms built into the radar software work out the most probable true path of the target through the noisy measurements. One of the most important aspects of the tracking software is to predict where the target is going, the objective being to improve the radar tracking process and to update the aim of any engaging defensive weapons system. Tracking algorithms are common to many types of radar and are covered in Chapter 5.

3.8 RADAR GUIDANCE

A common application of tracking radar is in the defence of local or 'point' targets from incoming air attack. One of the simplest methods (still used) is for the tracking radar to lock on to the target, fire an intercepting missile and then track both objects with a view to narrowing the gap between them by sending control signals up to the intercepting missile on a communication link. This technique is known as *radar command to line of sight* (RCLOS) and is used, for example, by the Rapier missile, which distinguished itself during the Falklands crisis. The drawback of this system is that the missile must stay within line of sight from the radar to the target if it is to be tracked accurately, and this may not be the optimum trajectory. For crossing targets, for example, high lateral accelerations are needed by the intercepting missile just before impact—it would be preferable for the missile to travel outside the beam.

Once a tracking radar has locked onto a target, there is an invisible beam between the two, and in the late 1940s it was realized that it should be possible to fly a missile up this beam; these weapons have since become known as *beamriding missiles*. The missile is equipped with a rearward-facing antenna to provide bearing information, which the autopilot uses to keep the missile in the centre of the radar beam. The attraction of this approach is that it is simple, and a strong SNR is received by the missile because the signal travels only on a one-way path. The disadvantages are that the target must be kept accurately in the centre of the beam and, as with RCLOS, the path up the beam may not be the ideal trajectory for a missile to follow[8].

The disadvantages of beamriding can be overcome to some extent through the use of *semi-active homing* in which the tracking radar acts as a target illuminator. On board the missile a receiver detects the scattered energy from the target, tracks its position and works out the best trajectory for an intercept. The missile does not have to fly within the illuminating beam and the tracking radar is not required to keep the bore-sight exactly on the target, merely to keep the target illuminated.

Although semi-active homing requires more complicated electronics to be carried by missiles, it has proved to be an effective system. During the Gulf War, for example, the Lynx helicopters of the UK Royal Navy illuminated Iraqi ships with their radars and attacked successfully with their semi-active homing Sea Skua missiles. The Patriot missiles, used with such telling effect in the interception of Scud missiles during the same crisis, also use semi-active homing, with the passive radar receivers on board the missiles relaying information back to the main radar processor on the ground, which then feeds commands back up a link to the missile.

The most advanced weapons employ *active homing radar guidance systems* in which the entire tracking radar is carried on board the intercepting missile. Initially the missile is locked on to a target by its host, but once fired it has a high degree of autonomy to pursue its target. Developments in miniaturized analogue and digital electronics, target image processing and the evolution of 35 and 94 GHz radar seeker heads mean that some very sophisticated radar missile guidance technology should be available soon. The first of this new generation of missiles is likely to be the ERINTs (extended-range interceptors), which use active homing and are so small that 16 of them will fit into a launcher that at present holds four Patriots.

Worked example The new generation of anti-armour submunitions are expected to be tiny missiles (0.1 m by 0.6 m long), which are fired in clusters but which contain their own 94 GHz guidance radars for independent targeting[8]. What diameter would an older X-band guided missile have needed in order to obtain the same angular accuracy?

SOLUTION A 10 cm dish at 94 GHz has a nominal angular resolution of $\lambda/d = 1.8°$, although, in practice, this would probably be a little over $2°$.

If, for convenience, we assume an X-band frequency of 9.4 GHz, we can see immediately that the older-type missile would have needed a diameter of the order of 1 m to have a resolution comparable with the proposed new submunitions.

3.9 SUMMARY

Tracking radars are distinguished by their dedication to a target and the precision of their angle measurements. Angular estimation is achieved by comparing the echo in two adjacent beams; the problem of pulse-to-pulse variation in the echo amplitude (or phase) is overcome in monopulse radar by making the comparison simultaneously on each pulse. One of the main applications of tracking radar is the guidance of weapons systems.

Key equation

● The decorrelation bandwidth formula:

$$\Delta f \gtrsim c/(2l) \quad [\text{Hz}]$$

3.10 REFERENCES

1. *Technical History of the Beginnings of RADAR*, S. S. Swords, Peter Peregrinus for the IEE, Stevenage, Herts, 1986.
2. *Introduction to Monopulse*, D. R. Rhodes, McGraw-Hill, New York, 1959.
3. *Practical Simulation of Radar Antennas and Radomes*, H. L. Hirsch and D. C. Grove, Artech House, Norwood, MA, 1988. [A useful book, including software listings.]
4. *Radar Principles*, N. Levanon, Wiley, New York, 1988.
5. *Modern Radar*, Ed. R. S. Berkowitz, Wiley, New York, 1965.
6. *Introduction to Radar Systems*, M. I. Skolnik, McGraw-Hill, New York, 1985.
7. *Radar Design Principles*, F. E. Nathanson, McGraw-Hill, New York, 1969.
8. The evolution of radar guidance, D. A. Ramsay, *GEC J. Res.*, 3(2), 92–103, 1985. [The whole of this special issue is worth reading.]

3.11 FURTHER READING

Artech House produce several books on monopulse radar including:

Monopulse Principles and Techniques, S. M. Sherman, Artech House, Norwood, MA, 1984.
Monopulse Radar, A. I. Leonov and K. J. Fomichev, transl. W. F. Barton and D. K. Barton, Artech House, Norwood, MA, 1986.
Secondary Surveillance Radar, M. C. Stevens, Artech House, Norwood, MA, 1988. [See the second half of Chapter 5.]

3.12 PROBLEMS

3.1 The Siemens–Plessey WF3 Windfinder radar is an X-band tracking radar operating at 9375 MHz that automatically tracks a corner reflector attached to a meteorological balloon. Conical scanning of a 2° beam at 1500 RPM is used and the PRF is 625 Hz; how many pulses per scan are there?

3.2 The radar in problem 3.1 has a peak transmitter power of 55 kW, an antenna gain of 35 dB, a receiver bandwidth of 2 MHz and a noise figure of 10 dB. Assuming 6 dB losses and given the performance indication that a reflector of RCS 120 m^2 can be tracked at a range of 100 km, estimate the angular accuracy of a standard 1 minute observation.

3.3 A weapon control radar system on a modern fighter aircraft operates at I/J-band using a 0.75 m × 0.5 m planar antenna. In dog-fight mode, the radar scans a 20° × 20° field ahead of the aircraft at a medium PRF of 10 kHz. If three pulse bursts of 32 pulses are required to confirm a target detection, what is the repeat time to search the field?

3.4 The Marconi Radar Systems ST802 is a lightweight monopulse naval tracking radar with a 2.4° beamwidth. If a 30 dB SNR echo were received from an attacking missile, what tracking accuracy would you expect? Would this be sufficiently accurate to bring the missile down by radar-controlled gunfire?

3.5 One method of improving azimuth information is to use *doppler beam sharpening* (similar to synthetic aperture radar described in Chapter 11). Here the motion of an air-borne radar can be used to subdivide the antenna beam because the relative velocity of ground clutter varies across the beam. Try these calculations:

(a) For a frequency of 10 GHz (I/J-band), a beamwidth of 4° and an aircraft speed of 300 m s^{-1}, what is the doppler spreading of the clutter across the beam when the azimuth angle is 45° away from the direction of motion and the look-down angle is 10°?

(b) If the doppler resolution of the processing is 10 Hz, to what azimuth resolution would this correspond?

(c) Would doppler beam sharpening be effective looking directly ahead of the aircraft?

FOUR

RADAR DETECTION THEORY

- How likely are we to detect a target?
- How often will we make mistakes?
- The correlation receiver and matched filter
- Detecting fluctuating targets
- Using multiple measurements

Finding targets and making measurements in a noisy, cluttered environment is the essential purpose of radar.

4.1 INTRODUCTION

In constructing a complete radar system, the designer must bear in mind not just the production of signals by the radar receiver, but the interpretation of those signals. Though radar is now put to many diverse uses, its original purpose of detecting objects in some volume of space still constitutes a major part of all its applications. In this case, the user is interested in distinguishing 'targets' in the illuminated volume from the clutter and noise that tend to obscure it. Once a target has been detected, properties such as its range and velocity are likely to be of interest.

For a variety of reasons, the voltage supplied by the receiver is never steady, even if the receiving antenna is fixed. Thermal noise is one source of fluctuation for which there is no cure; the laws of physics are always with us. Other fluctuations may be due to variation within the illuminated volume of space, as can happen if a sidelobe of the main beam is illuminating the ocean or wind-blown vegetation. There may be random emitters contributing to the received signal, such as radio waves from space ('cosmic noise'). For a comparatively small part of the time, under most circumstances, part of the fluctuating signal will be due to the presence of targets.

By examining this fluctuating signal, the operator (human or machine) attempts to find 'events' corresponding to objects of interest. The absence of events may also be important, as can occur when an apparent detection appears in one scan, but fails to appear in the next. Probability theory provides us with criteria for locating events. The approach described here is not the only one possible (see, for example, Shafer's 'theory of evidence'[1]), but it provides one way of answering the most basic question in detection theory: 'Does my signal indicate the presence of a target?'

To answer this question, it is necessary to examine the various stages in the formation of the signal on which we base our decision about the possible presence of a target. In a very simplified form, these are illustrated in Fig. 4.1. After transmitting a known signal $u(t)$, the radar is switched to 'receive' mode. A fluctuating voltage will appear at the front end of the receiver, made up of clutter $c(t)$, noise $n(t)$ and possibly signal, if targets are present. The clutter is the contribution to the return from all extraneous scatterers. The noise is dominated by the contribution from the receiver itself, except at very low frequencies. For the purposes of this chapter, we assume that both types of contribution that tend to cloak the presence of a signal have the same statistical behaviour, and can be treated in the single term $n(t)$, In Chapter 5 we discuss the behaviour of clutter further. If a single scatterer is present in the illuminated volume, embedded in this unwanted voltage will be, in the simplest case, a signal component $Au(t - \tau_d)$ (for the moment we are ignoring any doppler shift caused by target motion). The

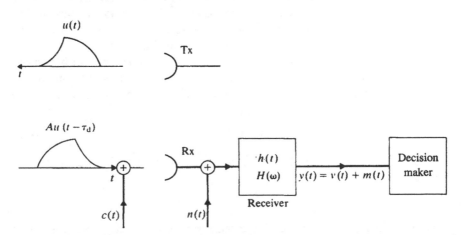

Figure 4.1 The basic system elements affecting target detection.

value A represents the scaling of the signal due to propagation, losses and the RCS of the scatterer (see Eq. (1.13)). For simple hard scatterers, A will be a constant, but in many cases the target RCS will fluctuate (see Sec. 4.12). The value τ_d represents the delay introduced by the propagation of the signal to and from the target, and provides the measure of range.

At this stage, the presence of a signal will normally be difficult to detect, because the energy in the signal is very small compared to the noise power, and to a large extent the voltage trace is dominated by high-frequency noise. It is the task of the receiver to extract all possible information about the presence of a target before any decision is made. In order to make progress with analysing how it should be designed in order to do this, we need to know some basic facts about linear systems. These are covered in any basic text on communications or signal processing, such as Schwartz[2] or Stremler[3], and you may wish to consult one of these texts to remind yourself of their derivation. For the moment, all we need to know is that the behaviour of a linear system is completely described if we know the way it reacts to an input signal that is a short, sharp spike (an impulse or delta function). This is the *impulse response function* $h(t)$, which tells us the response to a delta function $\delta(t)$ at time $t = 0$. Since the response cannot occur before the stimulus, $h(t)$ is causal, i.e. $t < 0$ implies $h(t) = 0$. For a general input signal $f(t)$, the output $g(t)$ is given by a *convolution* operation

$$g(t) = f * h(t) = \int_{-\infty}^{\infty} f(s)h(t - s)\,ds \qquad (4.1)$$

In the integral s is treated as a time variable. If $f(t)$ were the sum of two signals, i.e. $f_1(t) + f_2(t)$, then we can see straight away by substituting into the integral that the response of the system to the sum is the same as the sum of the individual responses. This is what we mean by *linearity*; it is also called the *superposition property*. The other vital property of convolution can be obtained by replacing $f(t)$ by $f(t - \tau_d)$, a delayed version of $f(t)$. If you do this, you find that the response is also delayed by τ_d.

The relations between the input and output signals take on a more attractive form when we take Fourier transforms. This has the remarkable property of converting convolution into multiplication, so that Eq. (4.1) becomes

$$G(\omega) = F(\omega)H(\omega) \qquad (4.2)$$

Here the use of lower- and upper-case letters indicates a Fourier transform pair, i.e.

$$f(t) \leftrightarrow F(\omega)$$

etc. The Fourier transform of the impulse response function $H(\omega)$ provides a complete description of the system, and is called the *system transfer function*.

The receiver will be treated as a linear filter with impulse function $h(t)$ and system transfer function $H(\omega)$, so that the output from the receiver, $y(t)$, can be written as

$$y(t) = v(t) + m(t) = Au * h(t - \tau_d) + n * h(t) \qquad (4.3)$$

where $v(t)$ and $m(t)$ are the responses to the signal and noise inputs respectively, i.e.

$$v(t) = A \int_{-\infty}^{\infty} u(t - \tau_d - s)h(s)\,ds \qquad (4.4)$$

and

$$m(t) = \int_{-\infty}^{\infty} n(t - s)h(s)\,ds \qquad (4.5)$$

Since, in practice, both $u(t)$ and $h(t)$ are of finite duration and $h(t)$ will be causal, we can change the integration limits and write the signal term as

$$v(t) = A \int_{0}^{T} u(t - \tau_d - s)h(s)\,ds \qquad (4.6)$$

where we have assumed that $h(t)$ is non-zero only in the range $0 \leqslant t \leqslant T$. The noise term $m(t)$ can be written similarly.

Both $u(t)$ and $h(t)$ are under the control of the system designer, so that $u * h(t)$ is deterministic and known, but the values of A and τ_d are unknown. However, we expect them to remain approximately constant during the interval over which $u(t)$ is significantly different from 0, i.e. during the illumination of the target by a single pulse. Hence the signal component of $y(t)$ is of known shape, but scaled by A and shifted in time by τ_d. Effectively, there are two degrees of freedom associated with the signal (ignoring frequency shifts for the moment). By contrast, $n * h(t)$ is a random waveform with many more degrees of freedom (which are determined by the bandwidth of the receiver). The detection problem is to tell whether the output $y(t)$ contains noise only or includes a signal term.

In order to assist in this, a major part of the task of the receiver is to bandpass the input. Typically, the input noise has a much wider bandwidth than the signal term, so that bandpass filtering can be performed without losing any information. The bandwidth of the receiver must be matched to that of the signal. If the bandwidth used is too broad, unnecessary noise will corrupt the radar measurement (Fig. 4.2a); if it is too narrow, a significant amount of signal may be lost (Fig. 4.2b). Just having the correct bandwidth is not enough, however. It turns out that the best possible way of extracting information is to shape the system transfer function $H(\omega)$ of the receiver so that the SNR on output is maximized. This optimum spectral shape is called the matched filter, and its time-domain implementation is known as a

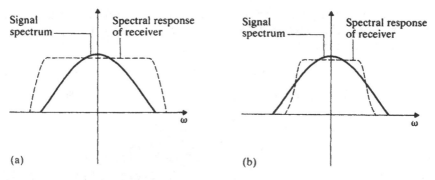

Figure 4.2 The effects of a bandpass filter on target detection. In (a) the passband is too wide, allowing excessive noise to affect detection. In (b) the passband is too narrow, and signal energy is lost.

correlation receiver. These are derived in Secs. 4.7 and 4.8. (The equivalence of optimizing information extraction and maximizing SNR is not as obvious as it may appear; Woodward[4] provides a classic analysis of the problem.) Both before and after matched filtering, the trace is at the intermediate frequency (IF), i.e. it is a narrow-band signal centred on the IF carrier frequency, and it is normal to perform envelope detection to remove the carrier before making any decision. However, since all the essential ideas are unaltered if we ignore the carrier frequency and assume that the input to the filter is at baseband, we have elected to make this simplification. This keeps the mathematical complexity to a minimum. For the reader who prefers a more precise treatment, in Sec. 4.11 we indicate the modifications necessary to deal with the case that occurs in practice.

4.2 THE BASIS FOR DECISION MAKING–PROBABILITY THEORY

The trace $y(t)$ available to the decision maker might appear as shown in Fig. 4.3, and at each instant (corresponding to each range), it is necessary to decide whether the value of $y(t)$ indicates the presence of a target. For example, should the spikes at t_1 and t_2 in Fig. 4.3 be attributed to anything other than noise? This involves a judgement as to whether this value was more likely to have arisen from noise alone or from signal plus noise. The basis for this decision must be a knowledge of the frequency with which different values of $y(t)$ will occur in the two circumstances.

A very useful way to describe how frequently $y(t)$ takes different values is through its *probability density function* (PDF) $p_y(y)$. This function is defined by the property that, for small Δy, values of $y(t)$ in the range $y \leqslant y(t) \leqslant y + \Delta y$ will occur with frequency given approximately by $p_y(y)\,\Delta y$.

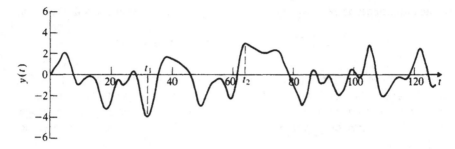

Figure 4.3 A voltage trace on which detection decisions might be made.

Hence, as Δy tends to 0,

$$p\{a \leqslant y(t) \leqslant b\} = \int_a^b p_y(y)\,dy \qquad (4.7)$$

(Read the left-hand-side of this equation as 'the probability that the measured value $y(t)$ lies between a and b'. The right-hand side is the area under the graph of $p_y(y)$ between a and b.) Since $p_y(y)$ is measuring relative frequencies,

$$p_y(y) \geqslant 0 \qquad (4.8)$$

and

$$\int_{-\infty}^\infty p_y(y)\,dy = 1 \qquad (4.9)$$

The second condition simply states that, when a measurement $y(t)$ is made, the value obtained is certain to be a real number.

We have already seen the *exponential PDF* (Eq. (2.6) and Fig. 2.4). It is easy to show that this PDF meets the two conditions. It occurs in a number of important places in radar applications. Of these, we shall meet it as a model for clutter in synthetic aperture radar images, and as a model for targets with fluctuating RCS (Sec. 4.12).

Three particularly useful parameters that can be extracted from the PDF are the *mean* μ, the *variance* σ^2 and the *standard deviation* σ. The mean is one type of representative value for the random function $y(t)$. The other two measure the spread of values taken by $y(t)$. We shall see that they also have very simple physical interpretations for electrical signals.

In order to explain how these parameters are found from the PDF, it is easier initially to do this for a digital signal. Suppose that $y(t)$ can take only a finite set of values $\{y_1, y_2, \ldots, y_M\}$ (which in modern radars is normally the case, since the signal will have passed through an A/D converter). In a long series of measurements, y_i will occur n_i times. Then the total number of measurements is $N = n_1 + n_2 + \cdots + n_M$, and the average value of y is

$$\frac{n_1 y_1 + n_2 y_2 + \cdots + n_M y_M}{N} = \sum y_i p(y_i)$$

Here we have written $p(y_i) = n_i/N$ for the relative frequency of y_i. So the average value of y is found by multiplying each possible value of y by its relative frequency or probability. By the same reasoning, the average value of y^2 would be $\sum y_i^2 p(y_i)$, and similarly for any other average value involving y.

When we deal with a continuous PDF, the only modification needed to these expressions is to replace the sum by an integral. The *mean* or average value is then defined by

$$\mu = \int_{-\infty}^{\infty} y p_y(y) \, dy \tag{4.10}$$

The average square deviation from the mean is known as the *variance*, and is given by

$$\sigma^2 = \int_{-\infty}^{\infty} (y - \mu)^2 p_y(y) \, dy$$

$$= \int_{-\infty}^{\infty} y^2 P_y(y) dy - 2\mu \int_{-\infty}^{\infty} y p_y(y) dy + \mu^2 \int_{-\infty}^{\infty} p_y(y) \, dy$$

$$= \int_{-\infty}^{\infty} y^2 p_y(y) \, dy - \mu^2 \tag{4.11}$$

Here we have used Eqs (4.9) and (4.10) to get from the second to the third line. There are thus two ways to find the variance. The first is to subtract the mean and then average the square differences. The second is to find the average square value, then subtract the square of the mean. Either way gives the same answer. The RMS deviation from the mean is the square root of the variance and is known as the *standard deviation* σ. In these expressions we are implicitly assuming that these parameters do not depend on the time of measurement (the system is stable).

We need to take averages at several points in this chapter, and we do not want to have to write the full integral with the PDF every time. Hence we use the notation $E[\ \ ]$ (for *expectation* or *expected value*) to stand for the average value of whatever appears between the square brackets. As examples, we could write

$$\mu = E[y] \qquad \sigma^2 = E[(y - \mu)^2]$$

It is worth remembering that taking the expectation is a linear operation, so that $E[y + z] = E[y] + E[z]$ and $E[cy] = cE[y]$ if c is a constant.

Worked example In a synthetic aperture radar (SAR) intensity image, an extended target such as an agricultural field may occupy many pixels (picture elements). Each of these pixels is a separate measurement of the energy returned from the target. Because of the coherent nature of SAR processing, each such pixel is affected by a type of noise known as *speckle*. This causes the measurements to have values that are exponentially distributed,

$$p_y(y) = (1/a)\,e^{-y/a} \qquad \text{for } y \geqslant 0$$

Is a single pixel measurement likely to be a good estimate of the mean value of the target?

SOLUTION The mean value is given by

$$\mu = \int_0^\infty (y/a)\,e^{-y/a}\,dy = a$$

It is easy to show that the standard deviation is also given by a (check it!). For this asymmetrical distribution (shown in Fig. 2.4), the probability of finding a value of y below the mean is

$$\int_0^a (1/a)\,e^{-y/a}\,dy = 1 - e^{-1} \approx 0.63$$

So the probability of finding a value of y above the mean is $e^{-1} \approx 0.37$. Hence from a single measurement we are nearly twice as likely to under-estimate the true RCS of the target as to over-estimate it. Because the standard deviation is equal to the mean, single measurements give very unreliable estimates of the true mean value. This creates major difficulties in obtaining precise measurements of the RCS of extended targets in SAR images.

The mean, variance and standard deviation have very direct interpretations in circuit terms. During the time T, $y(t)$ will be expected to have values in the range $y \leqslant y(t) \leqslant y + \Delta y$ for approximately $\Delta t = Tp_y(y)\,\Delta y$ seconds. Hence, if we regard the voltage $y(t)$ as an input to a 1 ohm resistance, then $y(t)$ has a DC value and a mean power given by

$$\text{DC value} \approx \frac{1}{T}\sum y\,\Delta t = \sum y p_y(y)\,\Delta y \qquad [\text{V}] \qquad (4.12)$$

and

$$\text{Power} \approx \frac{1}{T}\sum y^2\,\Delta t = \sum y^2 p_y(y)\,\Delta y \qquad [\text{W}] \qquad (4.13)$$

Allowing Δy to tend to 0, the sums on the right-hand side become integrals, which are $E[y(t)]$ and $E[y^2(t)]$ respectively. Hence we can see that

- DC value $= \mu$ [V]
- Mean signal power $= \sigma^2 + \mu^2$ [W]
- AC signal power $= \sigma^2$ [W]
- RMS AC signal variation $= \sigma$ [V]

We will make extensive use of the gaussian or normal distribution, since the voltage or current due to thermal noise is known on both theoretical and empirical grounds to conform to this distribution. Normally distributed noise $n(t)$ with mean μ and standard deviation σ has a PDF given by

$$p_n(n) = \frac{1}{\sqrt{(2\pi\sigma^2)}} \exp[-(n-\mu)^2/2\sigma^2] \qquad (4.14)$$

so that if we know its mean and standard deviation we know everything about it. (To check that μ and σ really are the mean and standard deviation of the gaussian distribution, we should carry out the integrations of Eqs (4.10) and (4.11); they give the right answers!) Thermal noise has zero mean ($\mu = 0$), so that thermal noise with power σ^2 has PDF

$$p_n(n) = \frac{1}{\sqrt{(2\pi\sigma^2)}} \exp(-n^2/2\sigma^2) \qquad (4.15)$$

For detection problems, we are often concerned with the probability that our measurement exceeds some threshold. If $n(t)$ has a gaussian distribution, the probability that a single measurement $n(t)$ exceeds k standard deviations above its mean is given by the area of the shaded region in Fig. 4.4. This area has the value

$$p\{n(t) > \mu + k\sigma\} = \frac{1}{\sqrt{(2\pi\sigma^2)}} \int_{\mu+k\sigma}^{\infty} \exp[-(n-\mu)^2/2\sigma^2]\,dn$$

$$= \frac{1}{\sqrt{(2\pi)}} \int_{k}^{\infty} \exp(-z^2/2)\,dz$$

$$= \frac{1}{\sqrt{(2\pi)}} \int_{0}^{\infty} \exp(-z^2/2)\,dz$$

$$- \frac{1}{\sqrt{(2\pi)}} \int_{0}^{k} \exp(-z^2/2)\,dz$$

$$= \tfrac{1}{2} - \Phi(k) \qquad (4.16)$$

To get from line 1 to line 2 in this derivation we made the substitution

$$z = (n - \mu)/\sigma$$

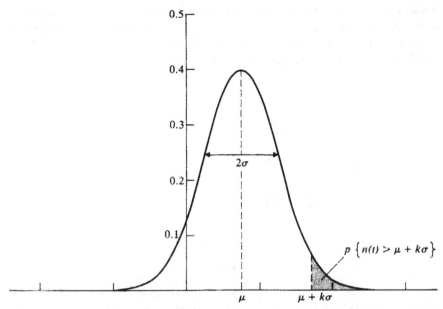

Figure 4.4 The gaussian PDF for mean μ and standard deviation σ; the shaded area corresponds to the probability of observing a noise value k standard deviations above the mean.

In the last line we introduced the function

$$\Phi(t) = \frac{1}{\sqrt{(2\pi)}} \int_0^t \exp(-x^2/2)\,dx$$

which is commonly used in texts on probability and statistics. Values of $\Phi(t)$ are given in Table 4.1; the area corresponding to $\Phi(t)$ is shown as shaded in the Fig. with this Table. The last line of the derivation of Eq. (4.16) uses the fact that $\Phi(\infty) = 1/2$. Another helpful fact is that

$$\Phi(-t) = -\Phi(t)$$

which is clear from the Fig. with Table 4.1 (if you think about it).

Many engineering texts prefer to use a close relation of $\Phi(t)$ known as the *error function*, defined by

$$\text{erf}(t) = \frac{2}{\sqrt{\pi}} \int_0^t \exp(-x^2)\,dx$$

The two functions are related by

$$\text{erf}(t) = 2\Phi(t\sqrt{2})$$

(Be careful if you look up either of these functions that the definition being used coincides with that above, since there are several variations on the definition of both functions.)

Table 4.1 Values of the function $\Phi(k)$, which correspond to the area shown in the diagram for a mean-zero gaussian variable with standard deviation 1. Extreme values of $\Phi(k)$ corresponding to false-alarm probabilities of 10^{-5}, 10^{-6}, 10^{-7} and 10^{-8} are given in Table 4.1(b).

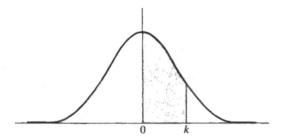

(a)Values of $\Phi(k)$

K	0	1	2	3	4	5	6	7	8	9
0.0	0.0000	0.0040	0.0080	0.0120	0.0160	0.0199	0.0239	0.0279	0.0319	0.0359
0.1	0.0398	0.0438	0.0478	0.0517	0.0557	0.0596	0.0636	0.0675	0.0714	0.0754
0.2	0.0793	0.0832	0.0871	0.0910	0.0948	0.0987	0.1026	0.1064	0.1103	0.1141
0.3	0.1179	0.1217	0.1255	0.1293	0.1331	0.1368	0.1406	0.1443	0.1480	0.1517
0.4	0.1554	0.1591	0.1628	0.1664	0.1700	0.1736	0.1772	0.1808	0.1844	0.1879
0.5	0.1915	0.1950	0.1985	0.2019	0.2054	0.2088	0.2123	0.2157	0.2190	0.2224
0.6	0.2258	0.2291	0.2324	0.2357	0.2389	0.2422	0.2454	0.2486	0.2518	0.2549
0.7	0.2580	0.2612	0.2642	0.2673	0.2704	0.2734	0.2764	0.2794	0.2823	0.2852
0.8	0.2881	0.2910	0.2939	0.2967	0.2996	0.3023	0.3051	0.3078	0.3106	0.3133
0.9	0.3159	0.3186	0.3212	0.3238	0.3264	0.3289	0.3315	0.3340	0.3365	0.3389
1.0	0.3413	0.3438	0.3461	0.3485	0.3508	0.3531	0.3554	0.3577	0.3599	0.3621
1.1	0.3643	0.3665	0.3686	0.3708	0.3729	0.3749	0.3770	0.3790	0.3810	0.3830
1.2	0.3849	0.3869	0.3888	0.3907	0.3925	0.3944	0.3962	0.3980	0.3997	0.4015
1.3	0.4032	0.4049	0.4066	0.4082	0.4099	0.4115	0.4131	0.4147	0.4162	0.4177
1.4	0.4192	0.4207	0.4222	0.4236	0.4251	0.4265	0.4279	0.4292	0.4306	0.4319
1.5	0.4332	0.4345	0.4357	0.4370	0.4382	0.4394	0.4406	0.4418	0.4429	0.4441
1.6	0.4452	0.4463	0.4474	0.4484	0.4495	0.4505	0.4515	0.4525	0.4535	0.4545
1.7	0.4554	0.4564	0.4573	0.4582	0.4591	0.4599	0.4608	0.4616	0.4625	0.4633
1.8	0.4641	0.4649	0.4656	0.4664	0.4671	0.4678	0.4686	0.4693	0.4699	0.4706
1.9	0.4713	0.4719	0.4726	0.4732	0.4738	0.4744	0.4750	0.4756	0.4761	0.4767
2.0	0.4772	0.4778	0.4783	0.4788	0.4793	0.4798	0.4803	0.4808	0.4812	0.4817
2.1	0.4821	0.4826	0.4830	0.4834	0.4838	0.4842	0.4846	0.4850	0.4854	0.4857
2.2	0.4861	0.4864	0.4868	0.4871	0.4875	0.4878	0.4881	0.4884	0.4887	0.4890
2.3	0.4893	0.4896	0.4898	0.4901	0.4904	0.4906	0.4909	0.4911	0.4913	0.4916
2.4	0.4918	0.4920	0.4922	0.4925	0.4927	0.4929	0.4931	0.4932	0.4934	0.4936
2.5	0.4938	0.4940	0.4941	0.4943	0.4945	0.4946	0.4948	0.4949	0.4951	0.4952
2.6	0.4953	0.4955	0.4956	0.4957	0.4959	0.4960	0.4961	0.4962	0.4963	0.4964
2.7	0.4965	0.4966	0.4967	0.4968	0.4969	0.4970	0.4971	0.4972	0.4973	0.4974
2.8	0.4974	0.4975	0.4976	0.4977	0.4977	0.4978	0.4979	0.4979	0.4980	0.4981
2.9	0.4981	0.4982	0.4982	0.4983	0.4984	0.4984	0.4985	0.4985	0.4986	0.4986

Table 4.1 *Continued*

K	0	1	2	3	4	5	6	7	8	9
3.0	0.4987	0.4987	0.4987	0.4988	0.4988	0.4989	0.4989	0.4989	0.4990	0.4990
3.1	0.4990	0.4991	0.4991	0.4991	0.4992	0.4992	0.4992	0.4992	0.4993	0.4993
3.2	0.4994	0.4993	0.4994	0.4994	0.4994	0.4994	0.4994	0.4995	0.4995	0.4995
3.3	0.4995	0.4995	0.4995	0.4996	0.4996	0.4996	0.4996	0.4996	0.4996	0.4997
3.4	0.4997	0.4997	0.4997	0.4997	0.4997	0.4997	0.4997	0.4997	0.4997	0.4998
3.5	0.4998	0.4998	0.4998	0.4998	0.4998	0.4998	0.4998	0.4998	0.4998	0.4998
3.6	0.4998	0.4998	0.4999	0.4999	0.4999	0.4999	0.4999	0.4999	0.4999	0.4999
3.7	0.4999	0.4999	0.4999	0.4999	0.4999	0.4999	0.4999	0.4999	0.4999	0.4999
3.8	0.4999	0.4999	0.4999	0.4999	0.4999	0.4999	0.4999	0.4999	0.4999	0.4999
3.9	0.5000	0.5000	0.5000	0.5000	0.5000	0.5000	0.5000	0.5000	0.5000	0.5000

(b) Extreme values

k	4.26	4.75	5.20	5.61
$\Phi(k)$	0.499 99	0.499 999	0.499 9999	0.499 999 99

4.3 THE EFFECTS OF THE RECEIVER ON THE NOISE DISTRIBUTION

We have already remarked that the thermal noise $n(t)$ entering the receiver has a zero-mean gaussian PDF. This is *not* an adequate description of the noise at the point where the decision is made, because at that point the noise (and any signal component) has passed through the receiver. Hence we need to consider the noise and signal-plus-noise distributions at the output of the receiver. Again, we have to use results from linear system theory (see references 2 and 3). Initially we deal with the case where we have noise $n(t)$ alone on input, giving a noise term $m(t)$ at the output from the receiver.

Though for a general input distribution the calculation of the output PDF can present great difficulties, the situation for gaussian inputs turns out to be remarkably simple. In fact, *gaussian inputs* give rise to *gaussian outputs*. Hence all we need to know to describe the output distribution are the output mean value and standard deviation. The mean is easily taken care of, since it is related to the mean of the input noise by

$$\mu_m = \mu_n \int h(t)\, dt \qquad (4.17)$$

The input noise has mean 0 ($\mu_n = 0$), so $\mu_m = 0$.

We could calculate the standard deviation of $m(t)$ directly, but a proper understanding of the effects of filtering is gained by a more indirect approach. This involves examining the relationship between values of the output separated in time. To do this we use the *autocorrelation function* (ACF), defined as

$$R_m(\tau) = E[m(t + \tau)m(t)] \qquad (4.18)$$

which is assumed to depend only on the separation in time τ (normally called the *lag*) between the measurements. The significance of $R_m(\tau)$ is perhaps best understood by considering the average square difference between the values of $m(t)$ at two times separated by an interval τ. This can be written as

$$E[\{m(t + \tau) - m(t)\}^2] = E[m^2(t + \tau) + m^2(t) - 2m(t + \tau)m(t)]$$
$$= 2[\sigma_m^2 - R_m(\tau)] \qquad (4.19)$$

where we have used the fact that $\mu_m = 0$, and the assumption that the average value of $m^2(t)$ does not depend on t. Since σ_m is independent of the lag (it depends only on the PDF), all the essential information on the rate of change is contained in the ACF. We can obtain the standard deviation from the ACF by

$$R_m(0) = E[m^2(t)] = \sigma_m^2 \qquad (4.20)$$

Rate of change is intimately connected with bandwidth, and the ACF has a corresponding frequency-domain expression in the *power spectrum* $S_m(\omega)$ of the trace, which is obtained by taking the Fourier transform[†] of the ACF

$$R_m(\tau) \leftrightarrow S_m(\omega) \qquad (4.21)$$

From this relation we can write down the ACF if we are given the power spectrum, since, using the definition of the inverse Fourier transform,

$$R_m(\tau) = \frac{1}{2\pi} \int_{-\infty}^{\infty} S_m(\omega) e^{j\omega\tau} \, d\omega \qquad (4.22)$$

We can then also write down the variance, since

$$\sigma_m^2 = R_m(0) = \frac{1}{2\pi} \int_{-\infty}^{\infty} S_m(\omega) \, d\omega \qquad (4.23)$$

The power spectrum can be physically interpreted as giving the distribution of noise power as a function of frequency, in the sense that the contribution

---

† For consistency with most modern books on radar and communications, all the sections of this book dealing with Fourier transform ideas use radian frequency ω rather than cyclic frequency f. By contrast, most engineers in their normal discourse work in Hz, and most of our discussions do the same. Since $f = \omega/2\pi$, conversions are easy enough; translating between the two is just part of the radar game (but still annoying).

to the total noise power from a narrow band $\Delta\omega$ of frequencies centred on ω is well approximated by $S_m(\omega) \Delta\omega/(2\pi)$.

Let us first apply these ideas to the input noise. An important idealization assumes that the input noise can fluctuate infinitely fast, so that knowing the value at any instant of time t gives no information about the value of the noise at any other time, no matter how close it is to t. In correlation terms, this means that the ACF is a *delta function*,

$$R_n(\tau) = (N/2)\,\delta(\tau) \tag{4.24}$$

with corresponding power spectrum

$$S_n(\omega) = N/2 \tag{4.25}$$

The power spectrum is flat, and contains all frequencies with equal power. Noise with this type of power spectrum is known as *white noise*. White noise has infinite bandwidth, which implies infinite power and infinitely fast rate of change. More realistically, *band-limited white noise* has a spectrum that is flat over some finite bandwidth B [Hz] and zero elsewhere, i.e.

$$S_n(\omega) = \begin{cases} N/2 & \text{if } \omega \leqslant 2\pi B \\ 0 & \text{if } \omega > 2\pi B \end{cases}$$

We can relate the constant $N/2$ to the noise power per unit bandwidth and to the system temperature. To do this, note that in either case the total power in the bandwidth B is

$$\frac{1}{2\pi} \int_{-2\pi B}^{2\pi B} \frac{N}{2}\, d\omega = NB \tag{4.26}$$

Hence the noise power per unit bandwidth is given by the constant N. We have already seen in Chapter 2 that the power in bandwidth B is $kT_0 B$, where k is Boltzmann's constant and T_0 is the effective noise temperature of the receiver. Hence

$$N = kT_0 \tag{4.27}$$

For a temperature of 290 K, this gives $N \approx 4 \times 10^{-21}$ W per unit bandwidth, i.e. -204 dB W per unit bandwidth. In essence, the factor 2 on the right-hand side of Eqs (4.24) and (4.25) arises because the bandwidth is defined using the positive frequencies only, while the spectrum is defined for positive and negative frequencies.

To find the power spectrum and ACF (and hence the standard deviation) of the output $m(t)$, we need the following key result from linear system theory. The *input and output power spectra* from a filter with *frequency transfer function* $H(\omega)$ are related by

$$S_m(\omega) = |H(\omega)|^2 S_n(\omega) \tag{4.28}$$

So, if the input is white noise, then

$$S_m(\omega) = (N/2)|H(\omega)|^2 \qquad (4.29)$$

Notice that this relation applies whether the input noise is band-limited or not, as long as the noise bandwidth is equal to or greater than the bandwidth of the receiver. For most practical applications, this will always be the case. The receiver therefore moulds the spectrum of the noise into the shape of its own transfer function, and forces it to occupy the same bandwidth. In Fig. 4.5 we see the effect of this. As input we have white noise (Fig. 4.5a), and as output we have correlated noise (Fig. 4.5b). The rate of change of the output is now controlled by the bandwidth of the receiver, since using Eq. (4.22) we can write the output ACF as

$$R_m(\tau) = \frac{N}{4\pi} \int_{-2\pi B}^{2\pi B} |H(\omega)|^2 \, e^{j\omega\tau} \, d\omega \qquad (4.30)$$

Since $\sigma_m^2 = R_m(0)$,

$$\sigma_m^2 = \frac{N}{4\pi} \int_{-2\pi B}^{2\pi B} |H(\omega)|^2 \, d\omega \qquad (4.31)$$

This has a very simple interpretation, since the energy (or integrated gain)

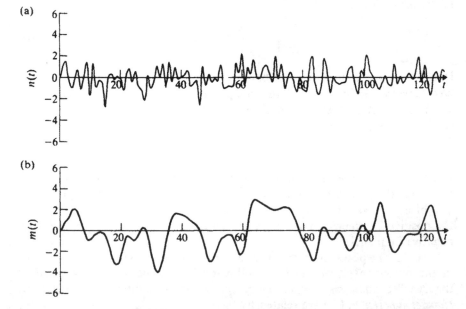

Figure 4.5 The effect of filtering white noise: (a) shows white noise, while (b) shows the output when this is passed through a filter with a cosine-squared impulse response function.

of the receiver is defined by

$$E_h = \int_{-\infty}^{\infty} |h(t)|^2 \, dt = \frac{1}{2\pi} \int_{-2\pi B}^{2\pi B} |H(\omega)|^2 \, d\omega \qquad (4.32)$$

(The equality of the two integrals is a very important result known as *Parseval's theorem*. It is essentially a statement about energy conservation, since it tells us that the total energy can be found by summing the power at each frequency. Calculating the energy in the time or frequency domain will yield the same result.) So

$$\text{Output noise power} = \sigma_m^2 = NE_h/2 \qquad (4.33)$$

This equation illustrates some important properties:

1. The output noise power is independent of the shape of the impulse response of the receiver, but depends simply on its integrated gain
2. For a fixed receiver gain, we can therefore alter the impulse response to meet any conditions on signal detection we choose, without affecting the noise variance.
3. The output noise power is independent of the total input noise power (which is very large; for the white noise model, it is infinite) but instead depends on the input noise power *per unit bandwidth*.

Some of the more general results of this section are needed elsewhere in this book (especially Chapter 5), but if the input is white or band-limited white zero-mean gaussian noise with noise power N W per unit bandwidth, they can be summarized very simply. In fact, the output is also zero-mean gaussian with variance $\sigma_m^2 = NE_h/2$, so has PDF

$$p_m(m) = \frac{1}{\sqrt{(2\pi\sigma_m^2)}} \exp(-m^2/2\sigma_m^2) \qquad (4.34)$$

It is then easy to find the probability that the output noise $m(t)$ exceeds any threshold.

Worked example What is the noise power of a radar operating on 100 MHz, using a 1 MHz bandwidth?

SOLUTION If we assume that the receiver passes all frequencies in this band with a constant gain factor A, then $H(\omega) = A$ if $|\omega - \omega_0| \leqslant \pi \times 10^6$ and is zero otherwise; here ω_0 is the centre frequency (see Fig. 4.6a). Therefore

$$\sigma_m^2 = 2 \times \frac{N}{4\pi} \times 2\pi \times 10^6 A^2 = 10^6 A^2 N \qquad [\text{W}]$$

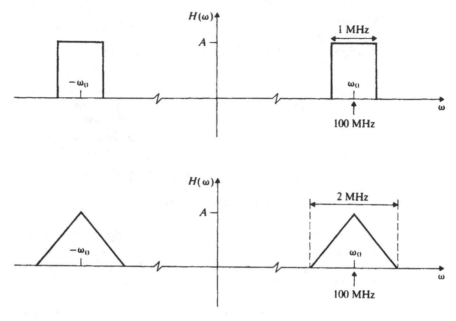

Figure 4.6 Two possible receiver system transfer functions.

For $N = 4 \times 10^{-21}$ and $A = 1$ this gives a noise power of approximately 4×10^{-15} W. Note that the operating frequency of the radar is irrelevant to this calculation; all that matters is the bandwidth.

If the system transfer function of the receiver had a triangular shape (see Fig. 4.6b) with the same bandwidth (defined here as the bandwidth between the two points at which the transfer function drops to half its peak value), then the noise power would be two-thirds of this value (see Problem 4.4).

4.4 THE DISTRIBUTION OF SIGNAL PLUS NOISE

Suppose now that a signal component is present at the input to the receiver. We know that if a delayed signal $Au(t - \tau_d)$ enters the receiver together with additive noise $n(t)$, then the output is given by (see Eq. (4.3))

$$y(t) = v(t) + m(t) \qquad (4.35)$$

where $v(t)$ and $m(t)$ are the signal and noise terms in the output. Since at the point of making a decision the user does not know τ_d or, indeed, whether there is a signal present at all, all that can be said is that, if there is a signal present, the output distribution at time t will be gaussian, but with an unknown mean $v(t)$. Hence the decision maker is trying to separate, on the

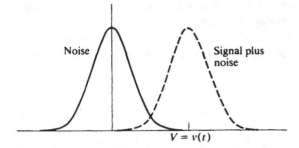

Figure 4.7 The PDFs of noise and signal plus noise at the output of the receiver.

basis of an observation at time t, two possible cases:

1. Noise only: the PDF of a single sample is zero-mean gaussian, with variance $\sigma_m^2 = NE_h/2$.
2. Signal plus noise: the PDF of a single sample is gaussian, with unknown mean $v(t)$ and variance σ_m^2.

In case 2, we can write

$$p_y(y) = \frac{1}{\sqrt{(2\pi\sigma_m^2)}} \exp\left(-\frac{(y-V)^2}{2\sigma_m^2}\right) \qquad (4.36)$$

where V is the value of $v(t)$ at the moment of observation. The expression for case 1 is given by setting $V = 0$. The PDFs in the two cases are as shown in Fig. 4.7. Two things seem clear from this figure:

● A simple threshold is the best way of separating the two cases.
● The greater the value of $v(t)/\sigma_m$, the better the chance of making the right choice[†].

Though we have simplified things by working at baseband, these conclusions also apply at IF (see Sec. 4.11).

4.5 THE SIGNAL-TO-NOISE RATIO

A slight digression is needed at this point, in order to clarify the concept of signal-to-noise ratio. There are two definitions widely used in the literature,

[†] The perceptive reader will realize that for a voltage trace, both the cases of positive and negative V should be considered. This leads to a threshold on $|y(t)|$ and the probability of making the right choice increases as $|v(t)/\sigma_m|$ increases. This refinement complicates the mathematics without clarifying the essential principles, so that both here and in Sec. 4.6 we treat only the case where $v(t)$ is positive.

often without any clear distinction being made between them. For a voltage trace $y(t) = f(t) + n(t)$ consisting of a signal component together with a noise component, the first definition is the ratio of the *energy* in the signal part (integrated over some time T) to the expected energy in the noise over the same time,

$$\text{SNR} = \frac{\text{energy in signal}}{\text{energy in noise}} \qquad [\quad] \qquad (4.37)$$

For mean-zero noise of average power σ^2, the energy received in time T is $T\sigma^2$, so that this definition can be written

$$\text{SNR} = \frac{\text{energy in signal}}{T} \frac{1}{\text{power in noise}} = \frac{\text{power in signal}}{\sigma^2} \qquad [\quad] \quad (4.38)$$

This definition is of most use at IF, so that the integration can be carried out over many cycles of the carrier frequency, while the modulation remains essentially unchanged. In this case, the received signal $v(t)$ effectively has the form $V \cos(\omega_0 t)$, where ω_0 is the carrier frequency and V is a constant. Hence the signal power is $V^2/2$, giving $\text{SNR} = V^2/2\sigma^2$. This is the SNR calculated using the radar equation in the preceding chapters.

The second definition compares the *instantaneous value* of the signal with the RMS noise power,

$$\text{SNR}' = |v(t)|^2/\sigma^2 \qquad [\quad] \qquad (4.39)$$

(some authors use $|v(t)|/\sigma$). This definition is most appropriate when discussing baseband signals. The two definitions are related, since, in practice, the baseband signals will actually be the envelope of an IF signal. Over intervals short compared with the rate of change of the envelope (i.e. much shorter than $1/B$, where B is the bandwidth of the baseband modulation), the IF signal can be written $V \cos(\omega_0 t)$, where the baseband envelope has instantaneous value V. Hence

$$\text{SNR}' = V^2/\sigma^2 \qquad [\quad] \qquad (4.40)$$

and we can see that

$$\text{SNR}' = 2 \times \text{SNR} \qquad [\quad] \qquad (4.41)$$

For consistency, throughout this book we adopt the first definition, which is the one used in the radar equation. Care must therefore be taken in comparing our results with those of other authors, particularly in the sections on the correlation receiver and matched filter (Secs 4.7 and 4.8).

Notice that neither of these definitions is particularly useful to describe the signal-to-noise ratio prior to the receiver, because at this stage we regard the noise as having a very large bandwidth, and hence very large total power (for white noise, this power would formally be considered infinite). It is only after passing through the receiver (or the noise bandwidth is reduced in some way) that the SNR becomes meaningful, since the noise power is then simply $NE_h/2$. Figure 4.7 indicates that, as the SNR increases, the whole of the signal plus noise PDF moves to the right. This implies that the chance of making the right decision when the measured signal is thresholded also increases.

4.6 DETECTION AND FALSE-ALARM PROBABILITIES

A threshold decision may be the right way to separate signal from noise, but how do we choose the threshold? For any threshold Y, four types of events can occur. These are:

- Target present; $y(t) > Y$; correct detection.
- Target present; $y(t) < Y$; missed detection.
- Target not present; $y(t) > Y$; false alarm.
- Target not present; $y(t) < Y$; no action.

The frequency of occurrence of these events can be described using two probabilities. The first is the probability of deciding that a signal was present when in fact there was only noise. This is the probability of a false alarm, given by

$$P_{fa} = \int_Y^\infty p_n(y)\,dy \qquad (4.42)$$

The second is the probability of a correct detection, given by

$$P_d = \int_Y^\infty p_{sn}(y)\,dy \qquad (4.43)$$

The subscripts indicate the PDFs of noise (n) and signal plus noise (sn). Using the PDFs of Fig. 4.7, the areas defining P_{fa} and P_d are shown in Fig. 4.8.

It is clear that, whatever the value of Y, increasing it will cause both P_{fa} and P_d to decrease. This illustrates that there is always a trade-off between improving detections and reducing false alarms. In practice, most radars are operated at a threshold level giving very low false-alarm rates (and hence long times between false alarms), for reasons discussed in Chapter 5. In most places in this book we have adopted a false-alarm probability of 10^{-6} as our working criterion. We now show how this fixes the threshold level, and hence the reliability with which we can detect targets.

Let us calculate the probabilities of a correct detection and a false alarm for a threshold set at k times the RMS noise value. The constant k is known as the (multiplication) *margin*; when expressed in dBs it is additive. If there

(a)

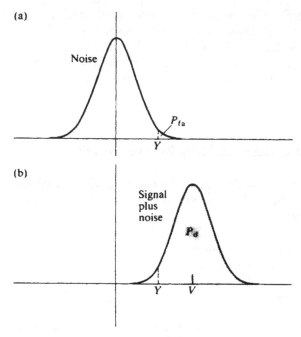

(b)

Figure 4.8 The areas in the PDFs of (a) noise and (b) signal plus noise corresponding to the false-alarm and detection probabilities for a threshold Y.

is a signal value V present, then

$$P_d = p\{y(t) > k\sigma_m\}$$
$$= p\{m(t) + V > k\sigma_m\}$$
$$= p\{m(t) > k\sigma_m - V\}$$
$$= \tfrac{1}{2} - \Phi(k - V/\sigma_m) \qquad (4.44)$$

using the same type of calculation as in Eq. (4.16). Since $\text{SNR} = V^2/2\sigma_m^2$, we can write this as

$$P_d = \tfrac{1}{2} - \Phi(k - \sqrt{(2 \times \text{SNR})}) \qquad (4.45)$$

In Fig. 4.9, we plot P_d as a function of SNR for a range of values of k; this figure shows that, as the SNR rises, the detection probability also increases, as expected.

For fixed k, the probability of a false alarm is fixed, and has the value

$$P_{fa} = p\{m(t) > k\sigma_m\} = \tfrac{1}{2} - \Phi(k) \qquad (4.46)$$

Note that, as V becomes close to 0, $P_d = P_{fa}$, i.e. we are as likely to decide that there is signal present as we are to detect a false alarm.

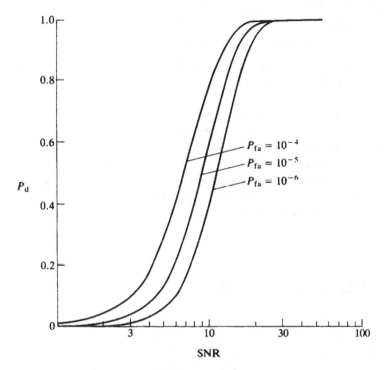

Figure 4.9 A plot of P_d against SNR for a range of values of P_{fa}.

Worked example What would be the false-alarm rate for a threshold set at three times the noise power; what would the single-pulse detection probability be if the SNR for a single pulse was 10 dB?

SOLUTION Since $k = 3$, we can immediately write

$$P_{fa} = \tfrac{1}{2} - \Phi(3) \approx 0.0013$$

Since SNR in power units (as well as in dB) is 10,

$$P_d = \tfrac{1}{2} - \Phi(3 - \sqrt{20}) \approx 0.9292$$

For a false-alarm rate of 10^{-6} we find from Eq. (4.46) that k must have the value 4.75, i.e. the threshold must be set at 4.75 times the noise power. Hence

$$P_d = \tfrac{1}{2} - \Phi(4.75 - \sqrt{(2 \times \text{SNR})})$$

If we work with a detection rate of 80 per cent, this implies that

$$0.8 = \tfrac{1}{2} - \Phi(4.75 - \sqrt{(2 \times \text{SNR})})$$

or

$$\Phi(4.75 - \sqrt{(2 \times \text{SNR})}) = 0.3$$

Hence our working criterion needs a SNR of 15.6 (11.94 dB) for reliable detection. 'Reliable' here means $P_d = 0.8$.

COMMENTS As we have already said, practical receivers do not use the output of the receiver directly, but would normally envelope-detect before making a decision about the presence of a target. This is the subject of Sec. 4.11. In practice, this does not greatly change the value of the detectable SNR. Under the same conditions of 80 per cent detection rate and false-alarm rate of 10^{-6}, we find that a threshold value k of 5.26 times noise power is needed, and the required SNR for detection is 17.9 (12.53 dB).

4.7 THE CORRELATION RECEIVER

Up to now, we have ignored the form of the signal term $v(t)$ coming from the receiver, and have simply examined the implications of thresholding the combined signal plus noise. In practice, the shape of the signal coming from the receiver is largely under the control of the system designer, and its properties are chosen to meet such specified criteria as desirable time or range sidelobe levels, resolution, etc. The most fundamental criterion, against which all other criteria must be balanced, is the ability to detect targets, which, as we have seen, is a function of the SNR. Hence a natural priority is to design the receiver to maximize SNR. By Eq. (4.33), we know that the noise power σ_m is dependent only on the gain of the receiver, not on the shape of its impulse response function. Hence, for fixed gain, the best SNR is obtained by maximizing the response to the signal term. This is achieved by an elegant, conceptually simple and readily implemented processing scheme, known as the correlation receiver, whose frequency-domain implementation is the matched filter.

The scheme relies on a fundamental result known as the *Cauchy–Schwartz inequality*. This states that, given two functions $f(s)$ and $g(s)$ of finite energy, then

$$\left| \int_a^b f(s)g^*(s)\,ds \right|^2 \leq \int_a^b |f(s)|^2\,ds \int_a^b |g(s)|^2\,ds \qquad (4.47)$$

with equality if and only if

$$f(s) = cg(s) \qquad (4.48)$$

for some constant c. Here the asterisk denotes complex conjugate, and we allow complex functions so that we can, if desired, use complex representations of real signals or we can use the result in the frequency domain, where the signals will normally be complex. For time-domain signals, taking $a = -\infty$ and $b = \infty$, the inequality can be interpreted directly as

$$\left| \int_{-\infty}^{\infty} f(s)g^*(s)\,ds \right|^2 \leqslant E_f E_g \qquad (4.49)$$

where

$$E_f = \int_{-\infty}^{\infty} |f(s)|^2\,ds \qquad (4.50)$$

is the energy in $f(s)$, and E_g is defined similarly.

Applying this to the signal component of the output of the receiver (Eq. (4.3)),

$$v(t) = \int_{-\infty}^{\infty} Au(t - \tau_d - s)h(s)\,ds \qquad (4.51)$$

we must replace $f(s)$ by $Au(t - \tau_d - s)$ and $g(s)$ by $h(s)$ (assumed real). Since $Au(t - \tau_d - s)$ is simply a scaled, reversed and shifted version of $u(s)$, its energy is $|A|^2 E_u$, and we can write

$$|v(t)|^2 \leqslant |A|^2 E_u E_h \qquad (4.52)$$

Note that this only depends on the *energy* in the transmitted signal $u(t)$, not its shape. Since the noise variance is $NE_h/2$, the instantaneous SNR satisfies the inequality

$$\text{SNR} = |v(t)|^2/2\sigma_m^2 \leqslant |A|^2 E_u/N \qquad (4.53)$$

The maximum possible SNR at the output of the receiver is equal to the ratio of the total energy of the input signal and the input noise *per unit bandwidth*. This is independent of the integrated gain of the filter, E_h, since both the maximum value of the signal term and the noise power are proportional to the gain.

How do we obtain this maximal value of SNR? The Cauchy–Schwartz inequality tells us that we must choose $h(s)$ to be proportional to $u(t - \tau_d - s)$, so that $h(s)$ is a *reversed, scaled* and *shifted* copy of $u(s)$. The time at which the maximum is attained depends on the delay chosen in the filter. Assuming that $u(t)$ is non-zero only in the range $0 \leqslant t \leqslant T$, let us set the delay equal to T, so that

$$h(t) = u(T - t) \qquad (4.54)$$

(Here we have set the scaling to 1; since the choice of scaling only affects the integrated gain of the receiver, it will not affect the SNR or the detection and false-alarm probabilities. In order to convince yourself of this, carry out the following calculations with $h(t) = Cu(T - t)$, where C is a constant.)

This implies that:

- $h(t)$ is causal
- $E_h = E_u$
- $\sigma_m^2 = NE_u/2$

To find when the peak SNR will occur, substitute this expression for $h(s)$ in Eq. (4.51). Then

$$v(t) = A \int_0^T u(t - \tau_d - s)u(T - s)\,ds$$

$$= A \int_0^T u(t - \tau_d - T + s)u(s)\,ds \qquad (4.55)$$

(using a change of variables). At time $t = T + \tau_d$, $v(t)$ therefore takes the value

$$v(t) = A \int_0^T u^2(s)\,ds = AE_u \qquad (4.56)$$

and

$$\text{SNR} = \frac{|v(t)|^2}{2\sigma_m^2} = \frac{A^2 E_u^2}{NE_h} = A^2 \frac{E_u}{N} \qquad (4.57)$$

As Eq. (4.53) shows, this is the maximum possible value of SNR. This key result means that the maximum response of the correlation receiver gives a signal-to-noise ratio

$$\text{SNR} = \frac{\text{signal energy}}{\text{noise power per unit bandwidth}} = \frac{E}{N} \qquad (4.58)$$

where E is the signal energy. It is important to realize that both E and N are calculated *at the input to the receiver*, so that the details of the receiver impulse response are irrelevant as far as detection is concerned. Useful pulseshapes are affected by other considerations, however, such as resolution; these are discussed in Chapter 6.

The maximum response occurs at the time $T + \tau_d$, owing to the delay of T seconds selected in the receiver. We can therefore find τ_d, as long as the signal is detectable. But we can easily write down the probability of detection at this time, using Eq. (4.45). For a threshold $k\sigma_m$,

$$P_d = \tfrac{1}{2} - \Phi(k - A\sqrt{(2E_u/N)}) = \tfrac{1}{2} - \Phi(k - \sqrt{(2E/N)}) \qquad (4.59)$$

Worked example Assume the radar is operating at a wavelength of 1 cm, and has a gain $G = (50)^2 = 2500$. The individual pulses are of mean power 10 kW and the pulse length is 1 ms. Let us find the single-pulse detection probability of a target with RCS 10 m^2 at a range of 10 km, if the false-alarm rate is 10^{-6}.

SOLUTION Since the energy per pulse is $E_u = 10^4 \times 10^{-3} = 10$ W, and we have already seen that $N \approx 4 \times 10^{-21}$, we simply need to find the value of A. Using the range equation, we need to work in power units, so that

$$\frac{A^2}{2} = \frac{G^2 \lambda^2 \sigma}{(4\pi)^3 R^4}$$

Then

$$A \sqrt{(2E_u/N)} \approx 1.77$$

We already know that we must use a value for k of 4.75, which gives a detection probability

$$P_d \approx \tfrac{1}{2} - \Phi(2.98) \approx 0.001$$

Such very low single-pulse detection rates are not uncommon for radars using multiple-pulse detection techniques. These are discussed in Sec. 4.10. There we show that, for coherent processing, using M pulses gives a SNR improvement by a factor M. In this example, that means that using 10 pulses would bring the detection rate up to 80 per cent.

The operation being carried out by the receiver is perhaps most clearly explained by making the change of variables $s = t' - t + T$ in Eq. (4.55). This gives

$$v(t) = A \int_{t-T}^{t} u(t' - \tau_d) u(t' - t + T) \, dt' \tag{4.60}$$

Here $u(t' - \tau_d)$ is the incoming signal, and $u(t' - t + T)$ is a copy of u, shifted to $t - T$; their product is integrated over the range of values for which the filter takes non-zero values. As t varies, the copy of $u(t)$ moves into alignment with the incoming signal, then out of alignment again (see Fig. 4.10). When it is fully aligned, the maximum possible signal response occurs. At this time, since the receiver energy is fixed at the value E_u, the maximum SNR also occurs. In Fig. 4.11 is shown the response of the correlation receiver to a signal contaminated by white noise (cf. Fig. 4.5). The top two plots show the signal and noise traces; their superposition and the output from the receiver are shown below.

The operation of integrating the product of one function by a shifted version of another is known as *correlation*, and the basis of the operation described above is to correlate the input with a copy of the transmitted signal $u(t)$. This is easily implemented by recycling a copy of the transmitted pulse, and is the principle of the *correlation receiver*.

Input signal

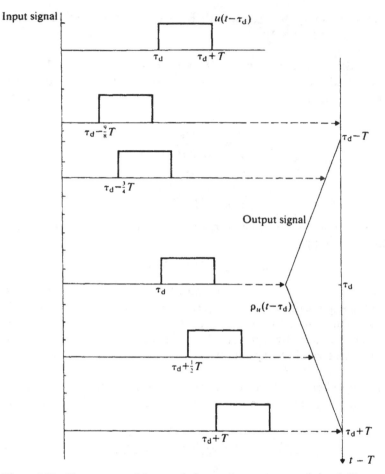

Figure 4.10 The response of the correlation receiver to a returned signal. The top plot shows the return (delayed by τ_d) and the plots below show a succession of positions of the correlating function. On the right is shown the complete time response of the receiver.

The signal term in the output from the correlation receiver is given by (see Eq. (4.55))

$$v(t) = A \int_0^T u(t - \tau_d - T + s)u(s)\,ds \qquad (4.61)$$

which is the correlation of the signal with itself. This is known as *auto-correlation*, and appears so often in radar processing that it deserves its own notation. Since in general we need to consider complex signals $f(t)$, we define the *deterministic autocorrelation function* of a signal $f(t)$ as

$$\rho_f(t) = \int_{-\infty}^{\infty} f(t + s)f^*(s)\,ds = \int_{-\infty}^{\infty} f(s)f^*(s - t)\,ds \qquad (4.62)$$

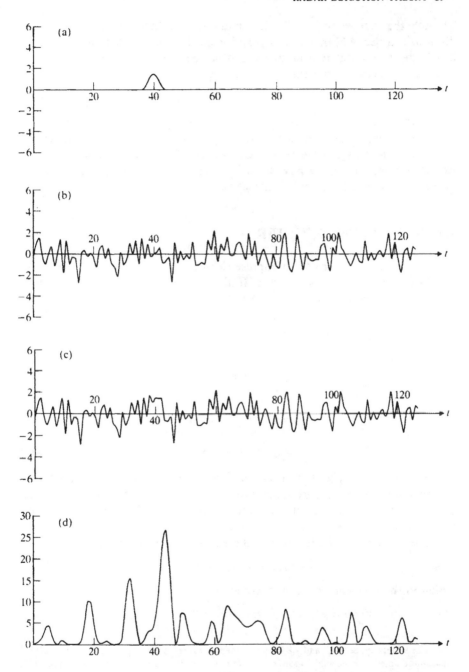

Figure 4.11 The response of the correlation receiver to a signal contaminated with white noise: (a) shows the signal (delayed by τ_d); (b) shows the noise-alone trace; (c) is their summation; and (d) is the squared response of a receiver matched to the signal shape of (a) when the signal-plus-noise voltage trace (c) is input to it.

Though the deterministic ACF is related to the ACF of random signals discussed in Sec. 4.3, they are in practice used in quite different ways, and it is better to use different notation to describe the two concepts.

We can therefore write the signal term as

$$v(t) = \rho_u(t - \tau_d - T) \tag{4.63}$$

The output for a simple pulse $u(t)$ is shown in Fig. 4.10. It is clear that the shape of the input has been drastically changed, but this is irrelevant as far as target (and range) detection is concerned. The form of this change is usually irreversible, as we now show by considering the frequency-domain implementation of the correlation receiver.

4.8 THE MATCHED FILTER

The correlation receiver has an impulse response $h(t) = u(T - t)$, where the constant T simply ensures causality. In the frequency domain, this gives rise to a transfer function

$$
\begin{aligned}
H(\omega) &= \int_{-\infty}^{\infty} u(T - t) e^{-j\omega t} \, dt \\
&= \int_{-\infty}^{\infty} u(s) e^{-j\omega(T - s)} \, ds \\
&= e^{-j\omega T} \int_{-\infty}^{\infty} u(s) e^{j\omega s} \, ds \\
&= e^{-j\omega T} U^*(\omega) \tag{4.64}
\end{aligned}
$$

where we have assumed that $u(t)$ is real[†]. Since $e^{-j\omega T}$ is simply a delay term, we see that the essential part of the transfer function is the complex conjugate of the Fourier transform of the transmitted pulse. This receiver transfer function is known as the *matched filter*.

The response of the receiver to the signal term is given by

$$v(t) = u * h(t - \tau_d) \tag{4.65}$$

which in the frequency domain becomes

$$V(\omega) = e^{-j\omega \tau_d} U(\omega) H(\omega) = e^{-j\omega(\tau_d + T)} |U(\omega)|^2 \tag{4.66}$$

[†] For simplicity, Secs 4.7 and 4.8 have assumed a real transmitted pulse $u(t)$. When $u(t)$ is phase-modulated, it is normally more convenient to treat it as a complex signal. In this case, the impulse response function giving maximum SNR is

$$h(t) = u^*(T - t)$$

The form of the frequency domain filter (Eq. (4.64)) is unchanged.

This normally implies a loss of information about the shape of $u(t)$, since all the phase relations between the various frequencies have been lost. The signal term in the output from the matched filter has a Fourier transform that is the energy spectrum of the transmitted pulse, multiplied by a linear phase term corresponding to range and filter delay.

4.9 KEY ELEMENTS OF SIGNAL DETECTION

Essentially, there are four steps in the treatment of signal detection given above:

1. Formation of the PDFs of noise and signal plus noise.
2. Given a threshold, determination of P_d and P_{fa}.
3. Checking that, for a fixed threshold, P_d depends only on the SNR, and that, as the SNR increases, so does P_d.
4. Hence, verifying that a matched filter is appropriate.

We have carried out this sequence of steps for the simplified case where the incoming signals are at baseband. In practice, they will be at IF. This more realistic case is dealt with in Sec. 4.11.

4.10 DETECTION USING MULTIPLE OBSERVATIONS

Surveillance radars usually combine multiple observations per scan before thresholding to detect the presence of targets. Essentially, this can be done in two ways, either by *coherent* processing of a sequence of M pulses, or by processing each separate pulse and then combining the separate observations in some way (normally simply by averaging). The latter case is called *incoherent* processing. The use of the word 'coherent' here refers to whether the phase relationships (at the carrier frequency) between successive pulses are known and stable.

If they are, then we can regard a sequence of M pulses as one single known waveform. Hence we can construct a matched filter for this waveform. For a non-fluctuating target, the energy returned from a sequence of M pulses will be M times greater than from a single pulse, while the noise power is unchanged. Hence the SNR and peak response of the matched filter will increase by a factor M. The effect that this has on detectability is given by Eq. (4.45).

Worked example A radar system combines eight pulses coherently before making a decision about the presence of a target. We have already seen that this requires a SNR (after coherent processing) of 15.6 if $P_{fa} = 10^{-6}$

and $P_d = 0.8$. Hence the SNR on a single pulse is 1.95. This gives a detection probability of only 0.003 for a single pulse.

For the radar to work on a higher single-pulse detection rate would be a waste of power, given that coherent integration is possible. Multiple pulses are an essential part in designing radar systems to meet specified detection probabilities, while operating at reasonable powers.

It is useful to realize that coherent processing can be achieved by averaging the successive outputs from the receiver. Suppose the radar has available M measurements $y_1, y_2, \ldots, y_M$ per scan, each of which contains a signal term and a noise term, so that

$$y_i = v_i + m_i$$

We will make the assumption that the noise terms m_i are zero-mean gaussian and independent of each other and of the signal terms v_i. As above, we consider the simplest case of a non-fluctuating target so that we can write $v_i = V_0$ for each i. This is where coherence enters, since without it each v_i would be of the form $V_0 \cos \phi_i$, where ϕ_i is an unknown phase term.

Averaging the M measurements gives

$$\bar{y} = \frac{1}{M} \sum y_i = \frac{1}{M} \sum (V_0 + m_i) = V_0 + \frac{1}{M} \sum m_i \qquad (4.67)$$

We can see that $\bar{y}$ is made up of a signal term V_0 and a noise term $\sum m_i / M$. Without coherence, the signal terms would not have added up in phase, and no advantage would be gained by averaging. The average value of M zero-mean gaussian random variables is also zero-mean gaussian, with variance σ_m^2 / M. The effect on the PDFs of noise and signal plus noise for $M = 8$ is shown in Fig. 4.12. The mean values of signal and signal plus noise are unaltered, but the distributions become more sharply peaked about these means.

Averaging leaves the signal value V_0 unchanged, but reduces the noise power by a factor M. Writing SNR_M for the SNR of the average of M observations, we therefore have

$$\mathrm{SNR}_M = \frac{V_0^2}{2(\sigma_m^2/M)} = M \frac{V_0^2}{2\sigma_m^2} = M \times (\mathrm{SNR}_1) \qquad (4.68)$$

and the effective SNR is increased by a factor M. This is the conclusion we reached above on energy considerations.

In the incoherent case, nothing is gained by coherent averaging of the receiver outputs, because the signal terms will be out of phase. We must envelope-detect first. Envelope detection for single pulses is discussed in the next section. We find a more complicated expression for P_d (though P_{fa} is

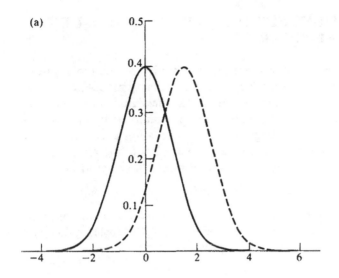

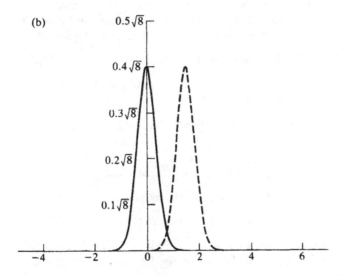

Figure 4.12 The PDFs of noise and signal plus noise for (a) a single pulse and (b) the coherent average of eight pulses.

simpler) for single-pulse detection. Evaluating the effective SNR after averaging an envelope-detected pulse is more complicated than for the gaussian case, but is fully discussed by Marcum[5]. Another useful reference is Levanon[6]. A convenient rule of thumb is that averaging M square-law-detected pulses gives an SNR improvement of the order of $\sqrt{M}$, if M is large.

4.11 MODIFICATIONS FOR INTERMEDIATE-FREQUENCY INPUT TO THE RECEIVER

As we have already indicated, a more realistic model for the radar system and receiver takes account of the fact that radars use a modulated carrier wave of frequency ω_0 on transmission, and that the signal is narrowband-filtered on reception. After reception, the carrier is removed by an envelope (or square-law) detector. The target detection operation is carried out on the ensuing baseband signal. In this case, after reception, the return from a hard (non-distributed, non-fluctuating) target would have the form

$$y(t) = x(t) + m(t) \qquad (4.69)$$

Here $x(t) = v(t)\cos(\omega_0 t)$, where $v(t)$ is a baseband modulation term (corresponding to the baseband signal used in Sec. 4.4 *et seq.*), and $m(t)$ is narrowband noise. Such noise can be represented as

$$m(t) = r(t)\cos[\omega_0 t - \phi(t)] \qquad (4.70)$$

where $r(t)$ and $\phi(t)$ represent the fluctuating amplitude and phase of the noise, both of which have bandwidths comparable to the bandwidth of the receiver and small compared with ω_0. Equation (4.70) can be expanded as

$$m(t) = X(t)\cos(\omega_0 t) + Y(t)\sin(\omega_0 t) \qquad (4.71)$$

where $X(t) = r(t)\cos\phi(t)$ and $Y(t) = r(t)\sin\phi(t)$ are uncorrelated gaussian zero-mean random coefficients, with the properties

$$E[X^2(t)] = E[Y^2(t)] = E[m^2(t)] = \sigma_m^2 \qquad (4.72)$$

(see e.g. Schwartz[2]). Then

$$y(t) = [v(t) + X(t)]\cos(\omega_0 t) + Y(t)\sin(\omega_0 t) \qquad (4.73)$$

The detection operation uses the envelope of this signal,

$$r(t) = \sqrt{\{[v(t) + X(t)]^2 + Y^2(t)\}} \qquad (4.74)$$

The distribution of this envelope was first derived by S.O. Rice of Bell Telephone Laboratories[7]. It has since been discussed by many authors (e.g. Schwartz[2], Papoulis[8]), and has the form

$$p_r(r) = \frac{r}{\sigma_m^2}\exp(-(r^2 + V^2)/2\sigma_m^2)I_0\left(\frac{rV}{\sigma_m^2}\right) \qquad (4.75)$$

where we have written V for the instantaneous value of the slowly varying signal $v(t)$. In Eq. (4.75), $I_0(z)$ is the zero-order modified Bessel function of the first kind, given by

$$I_0(z) = \frac{1}{2\pi}\int_0^{2\pi} e^{z\cos\theta}\,d\theta \qquad (4.76)$$

Tabulated values of this function will be found in most books of mathematical tables (e.g. Abramowitz and Stegun[9]).

Setting $V = 0$ and using $I_0(0) = 1$ gives the distribution of noise alone

$$p_n(n) = \frac{n}{\sigma_m^2} \exp(-n^2/2\sigma_m^2) \tag{4.77}$$

This is the *Rayleigh distribution*. A threshold $k\sigma_m$ gives false-alarm and detection probabilities

$$P_{fa} = \exp(-k^2/2) \tag{4.78}$$

and

$$P_d = \int_{k\sigma_m}^{\infty} p_r(r)\,dr = \int_{k^2/2}^{\infty} \exp\left(\frac{-V^2}{2\sigma_m^2}\right) e^{-u} I_0\left(\frac{V}{\sigma_m}\sqrt{2u}\right) du \tag{4.79}$$

The last expression was obtained from Eq. (4.75) by the substitution $u = r^2/2\sigma_m^2$. Since $SNR = V^2/2\sigma_m^2$, this can be written

$$P_d = e^{-(SNR)} \int_{k^2/2}^{\infty} e^{-u} I_0(2\sqrt{(u \times SNR)})\,du \tag{4.80}$$

so that P_d depends only on the SNR and k. Using Eq. (4.78), it is easy to express k in terms of P_{fa}. This has been used in Fig. 4.13 on page 96 to plot P_d as a function of SNR for a variety of values of P_{fa}. We see that, as the SNR increases, so does P_d, which indicates that matched filtering is required to maximize P_d.

In Levanon[6], a convenient approximation is quoted that effectively eliminates k from Eqs (4.78) and (4.80). This yields the following direct and easy-to-use relation between P_d, P_{fa} and SNR

$$SNR = A + 0.12AB + 1.7B \tag{4.81}$$

where

$$A = \ln\left(\frac{0.62}{P_{fa}}\right) \qquad B = \ln\left(\frac{P_d}{1 - P_d}\right)$$

and SNR in Eq. (4.81) is *not* in dB. This relation is fairly accurate if

$$10^{-7} < P_{fa} < 10^{-3} \qquad \text{and} \qquad 0.1 < P_d < 0.9$$

4.12 TARGET FLUCTUATIONS—THE SWERLING CASES

We have shown that the probability of detection of a target from a single observation depends on the peak response of the matched filter, which is directly proportional to the target's RCS at the moment of observation. In Chapter 2 we noted that this RCS fluctuates for many targets. This comes

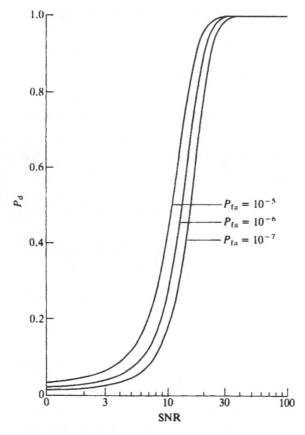

Figure 4.13 Plots of P_d against SNR for a range of values of P_{fa}, after envelope detection.

about if the target is in fact made up of several scattering centres whose orientations (and hence phase relations) relative to the radar change with time. Such changes in phase relationships will give rise to interference effects, so that the target RCS will exhibit fading characteristics. Swerling[10] suggested four types of target model for these complex targets, consisting of two types of variation. The first variation is in the *distribution* of the observed RCS values, and he identified two distributions as of particular interest. For a target of average RCS σ_{av}, these are

$$p_\sigma(\sigma) = \frac{1}{\sigma_{av}} \exp\left(-\frac{\sigma}{\sigma_{av}}\right) \qquad \sigma \geqslant 0 \qquad (4.82)$$

and

$$p_\sigma(\sigma) = \frac{4\sigma}{\sigma_{av}^2} \exp\left(-\frac{2\sigma}{\sigma_{av}}\right) \qquad \sigma \geqslant 0 \qquad (4.83)$$

These are both forms of the chi-squared family of distributions. The first is also known as the *exponential PDF*, and applies to a target comprising many independent scattering centres of approximately equal strength. This distribution has a standard deviation σ_{av} that is equal to the mean. The second PDF is appropriate to targets that have a dominant scatterer together with a number of other smaller scattering centres. In this case, the standard deviation is $\sigma_{av}/\sqrt{2}$.

The other type of variation is in the *rate of change* of the RCS. Again, Swerling suggested two idealizations. The first of these is when the target RCS changes on a timescale comparable to or faster than the PRF of the radar, so that different pulses provide independent samples of the fluctuating RCS. The second is when the target RCS is effectively unchanging between pulses, but fluctuates on the timescale of the radar's scanning pattern. These different timescales affect the probability of detection when several pulses are combined (see Sec. 4.10). Combining the two types of PDF with the two rates of variation yields the four Swerling models. These are normally referred to as Swerling cases 1–4, as set out in Table 4.2. We include the non-fluctuating case as case 5.

Other types of fluctuation have been considered (see, e.g. Berkowitz[11]), but the original Swerling cases still form the basis of most target detection studies.

In order to find out how Swerling fluctuations affect the detection probability P_d, we must follow the steps set out in Sec. 4.9. The basic problem is the derivation of the PDF of signal plus noise, and its subsequent integration. Swerling[10] carried out the necessary calculations for square-law detection. In this case, if $V = v(t)$ is the signal voltage, the variable $V^2/2$ (which measures signal power) will have a PDF of the same form as the Swerling cases, with a mean value that we will write as V_0^2. For single pulses, there is no difference between pulse-to-pulse and scan-to-scan variation. Using a normalized variable

$$z = \frac{y^2(t)}{2\sigma_m^2}$$

Table 4.2 The Swerling cases

PDF	Rate of variation	Case
4.82	Slow	1
4.82	Fast	2
4.83	Slow	3
4.83	Fast	4
Delta function	None	5

which expresses the square-law-detected output in terms of the output noise power σ_m^2, the single-pulse PDF of signal plus noise is

$$p_z(z) = \frac{1}{1 + 2S_0} \exp\left(\frac{-z}{1 + 2S_0}\right) \qquad (4.84)$$

for cases 1 and 2, and

$$p_z(z) = \frac{1}{(1 + S_0)^2}\left(1 + \frac{z}{1 + 1/S_0}\right)\exp\left(\frac{-z}{1 + S_0}\right) \qquad (4.85)$$

in cases 3 and 4. Here $S_0 = V_0^2/(2\sigma_m^2)$ can be regarded as the mean SNR/2. The corresponding detection probabilities for a fixed threshold Y are

$$P_d = \exp[-Y/(1 + 2S_0)] \qquad (4.86)$$

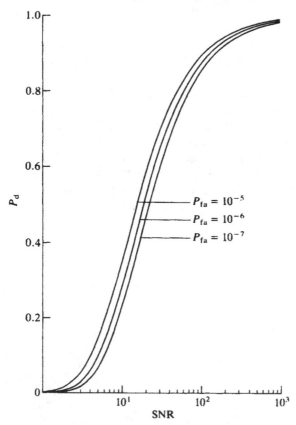

$P_{fa} = 10^{-5}$

$P_{fa} = 10^{-6}$

$P_{fa} = 10^{-7}$

Figure 4.14 Plots of P_d for single pulses after square-law detection, as a function of SNR, for Swerling cases 1 and 2 and a range of values of P_{fa}.

for cases 1 and 2, and

$$P_d = \left(\frac{1}{1 + 1/S_0}\right)\left(1 + \frac{Y}{1 + S_0} + \frac{1}{S_0}\right)\exp\left(\frac{-Y}{1 + S_0}\right) \qquad (4.87)$$

in cases 3 and 4. The PDF of noise alone (normalized to unit standard deviation) can be obtained by setting $S_0 = 0$ in Eqs (4.84) or (4.85). In Fig. 4.14 we plot P_d for Swerling cases 1 and 2 as a function of SNR, for a range of values of P_{fa}.

Swerling[10] also provides expressions for the PDFs of averages of square-law-detected pulses, and of the associated detection probabilities (in some cases the expressions are too cumbersome to be described exactly, and approximations are supplied). These expressions are not reproduced here (Swerling's paper is very clear and readable, and is recommended if you need to make detailed detection performance calculations).

4.13 SUMMARY

Target detection is a probabilistic idea; noise and clutter prevent us from being certain to find the targets we are looking for, and will normally present us with plenty of 'targets' we are not looking for. We can only define the probabilities of detection and of false alarm that we are prepared to live with. These determine the signal-to-noise ratio that is required for detection. Optimal detection performance is achieved by maximizing the SNR at the output of the receiver. The receiver that does this uses the correlation or matched filter principle. As long as a matched filter is being used, the form of the transmitted pulse is irrelevant for detection purposes. All that matters is the ratio of the signal energy to the noise power per unit bandwidth on input to the receiver. By developing expressions for the PDFs of signal plus noise and noise only, relatively straightforward calculations of detection and false-alarm probabilities can be carried out for single pulses. These become more complicated when fluctuating targets and multiple pulses need to be considered, so that graphical or numerical techniques are needed.

Key equations

● For mean-zero gaussian noise:

$$p\{n(t) > k\sigma\} = \tfrac{1}{2} - \Phi(k)$$

● Output noise power from receiver with integrated gain E_h:

$$\sigma_m^2 = NE_h/2$$

● Noise power:

$$N = kT_0 = -204 \text{ dB W/unit bandwidth}$$

- Parseval's relation:

$$E_h = \int_{-\infty}^{\infty} |h(t)|^2 \, dt = \frac{1}{2\pi} \int_{-\infty}^{\infty} |H(\omega)|^2 \, d\omega$$

- Probability density function of signal plus noise (gaussian case):

$$p_y(y) = \frac{1}{\sqrt{(2\pi\sigma_m^2)}} \exp\left(\frac{-(y-V)^2}{2\sigma_m^2}\right)$$

- Probabilities of detection and false alarm (gaussian case):

$$P_d = \tfrac{1}{2} - \Phi(k - \sqrt{(2 \times SNR)})$$
$$P_{fa} = \tfrac{1}{2} - \Phi(k)$$

- SNR at output of matched filter, where E is signal energy on input:

$$SNR = E/N$$

- Probability of detection after matched filtering (gaussian case):

$$P_d = \tfrac{1}{2} - \Phi(k - \sqrt{(2E/N)})$$

- The signal term in the output of the correlation receiver is the auto-correlation function of the transmitted pulse:

$$\rho_u(t) = \int_{-\infty}^{\infty} u(t+s)u^*(s)\,ds = \int_{-\infty}^{\infty} u(s)u^*(s-t)\,ds$$

- The matched filter has a system transfer function that is the complex conjugate of the Fourier transfer of the pulse (ignoring the delay for causality):

$$H(\omega) = U^*(\omega)$$

- Swerling PDFs:

$$p_\sigma(\sigma) = \frac{1}{\sigma_{av}} \exp\left(-\frac{\sigma}{\sigma_{av}}\right) \qquad \sigma \geq 0$$

$$p_\sigma(\sigma) = \frac{4\sigma}{\sigma_{av}^2} \exp\left(-\frac{2\sigma}{\sigma_{av}}\right) \qquad \sigma \geq 0$$

- Coherent averaging of M pulses improves the SNR by a factor M:

$$SNR_M = M \times (SNR_1)$$

- Detection and false-alarm probabilities after envelope detection:

$$P_{fa} = \exp(-k^2/2)$$

$$P_d = e^{-(SNR)} \int_{k^2/2}^{\infty} e^{-u} I_0(2\sqrt{(u \times SNR)})\,du$$

● Relation between P_d, P_{fa} and SNR after envelope detection:

$$SNR = A + 0.12AB + 1.7B$$

where

$$A = \ln\left(\frac{0.62}{P_{fa}}\right) \qquad B = \ln\left(\frac{P_d}{1 - P_d}\right)$$

which can be used if

$$10^{-7} < P_{fa} < 10^{-3} \qquad \text{and} \qquad 0.1 < P_d < 0.9$$

4.14 REFERENCES

1. *A Mathematical Theory of Evidence*, G. Shafer, Princeton University Press, Princeton, NJ, 1976.
2. *Information Transmission, Modulation and Noise*, M. Schwartz, McGraw-Hill, New York, 1980.
3. *Introduction to Communication Systems*, F. G. Stremler, Addison-Wesley, Reading, MA, 1982.
4. *Probability and Information Theory, with Applications to Radar*, P. M. Woodward, Pergamon Press, Oxford, 1953.
5. A statistical theory of target detection by pulsed radar, J. Marcum, *IRE Trans.*, IT-6, 145–267, 1960 [This is the original reference, but the books by Berkowitz[11] and Levanon[6] give useful treatments of this problem.]
6. *Radar Principles*, N. Levanon, Wiley, New York, 1988.
7. Mathematical analysis of random noise, S. O. Rice, *Bell Syst. Tech. J.*, 23 and 24, 1944; reprinted in *Selected Papers on Noise and Stochastic Processes*, Ed. N. Wax, Dover, New York, 1954.
8. *Probability, Random Variables and Stochastic Processes*, A. Papoulis, McGraw-Hill, New York, 1985.
9. *Handbook of Mathematical Functions*, M. Abramowitz and I. A. Stegun, Dover, New York, 1964.
10. Detection of fluctuating pulsed signals in the presence of noise, P. Swerling, *IRE Trans.*, IT-3, 175–178, 1957. [Again, this is the original reference, but the Swerling models are described in many places, such as the *Introduction to Radar Systems*, M. I. Skolnik, McGraw-Hill, New York, 1985.]
11. *Modern Radar: Analysis, Evaluation and System Design*, R. S. Berkowitz, Wiley, New York, 1966.

4.15 PROBLEMS

4.1 A synthetic aperture radar image consisting mainly of grassland is of dimensions 512 × 512 pixels. The image is being scanned to find stationary vehicles thought to have an RCS 10 dB above the mean clutter level. Assuming exponential speckle, find the probability of a false alarm; how many false alarms would be likely to occur in the image? To stress the great difference between exponential and gaussian probabilities, find the signal-to-clutter ratio needed to give the same false-alarm rate if the background could be considered gaussian. (The signal-to-clutter ratio is the ratio of target brightness above the mean and the clutter standard deviation).

4.2 (a) Show that the signal output from a correlation receiver matched to a rectangular pulse is a triangular pulse.

(b) Two possible waveforms to be used for detection purposes are shown as (i) and (ii); (ii) is a binary pulse generated by phase modulation. Each has its own matched filter. Are their detection performances different? Find the outputs from their respective matched filters, and comment on any possible consequences of the differences between them.

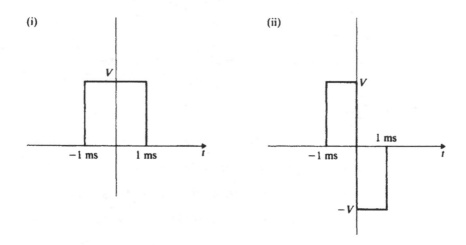

4.3 What would be the expected thermal noise power at the output of a receiver whose impulse response function is

$$h(t) = \begin{cases} \cos^2(\pi t/T) & \text{if } |t| \leqslant T/2 \\ 0 & \text{otherwise} \end{cases}$$

when $T = 10^{-3}$ s? This receiver is matched to pulses of the same shape (because it is symmetrical). Find the SNR at the output of the receiver if the returned signal on input is

$$10^{-8}\cos^2(\pi/T)(t - 10^{-3})$$

Sketch the form of the signal part of the output.

4.4 Confirm the statement in the worked example following Eq. (4.34) that a receiver with a triangular system transfer function generates only two-thirds of the noise power of a receiver with a rectangular passband of the same bandwidth.

How would this result change if we defined the bandwidth for the triangular system transfer function as the bandwidth between the two points at which the system transfer function falls to zero?

4.5 A ground-based surveillance radar with effective aperture 10 m² operates at a wavelength of 25 cm, using 1 μs pulses and a peak power of 0.1 MW. Assuming a matched filter and a noise figure of 10 dB, what would be the single-pulse detection probability from an aircraft with RCS 5 m² at a range of (a) 50 km and (b) 100 km, if the false-alarm rate was set at 10^{-6}?

4.6 The signal-to-noise ratio defined in Chapter 1 involves bandwidth, but this does not appear in the definition of Sec. 4.5. Explain why there is no inconsistency here.

4.7 What SNR would be needed for a 90 per cent single-pulse detection rate and a false-alarm rate of 10^{-6} using (a) the gaussian treatment and (b) the equations for an envelope-detected signal?

4.8 If 16 pulses can be integrated coherently, what single-pulse detection rate is allowable in order to permit 80 per cent detection rate with a false-alarm probability of 10^{-6}?

4.9 We will meet the chirped pulse

$$u(t) = \begin{cases} \exp(-jat^2) & \text{if } |t| \leqslant T \\ 0 & \text{otherwise} \end{cases}$$

(where a is a constant) in Chapter 6. Find its matched filter. (You will need the footnote following Eq. (4.64).) ˙

4.10 For 80 per cent detection rate and 10^{-6} false-alarm rate, what peak power would be acceptable in a radar coherently averaging thirty-two $2\,\mu s$ pulses, if it is operating on a 24 cm wavelength with an effective aperture of $15\,m^2$, has a loss factor of 50 per cent and needs to locate aircraft with RCS down to $20\,m^2$ within a range of 100 km?

FIVE

SIGNAL AND DATA PROCESSING

- Doppler information aids target detection
- Why do we choose low false-alarm rates?
- How do we follow a target once it has been detected?

Separating signals from clutter, and tracking targets have become sophisticated arts.

5.1 INTRODUCTION

The reliability and low cost of modern digital electronics have revolutionized radar engineering. Over the years, the A/D converter has moved progressively up the receiver chain towards the antenna, increasingly replacing analogue electronics by digital hardware and computer software. This engineering activity has two objectives: to remove unwanted clutter as far as possible, and to detect targets against the residual clutter and noise.

Radar engineers make a distinction between *signal processing*, the fast hardware processing developed mostly by electronic engineers, and *data processing*, which is concerned mainly with target-detection software and is very much the preserve of mathematicians. The dividing line between the two is neither clear-cut nor stationary, as it gradually moves up the receiver chain with the increasing ability of modern computers to replace hard-wired circuits. The current position for a typical radar system is shown in Fig. 5.1.

The objective of this chapter is to take a guided tour through the signal processing and data processing shown in the block diagram of Fig. 5.1, but

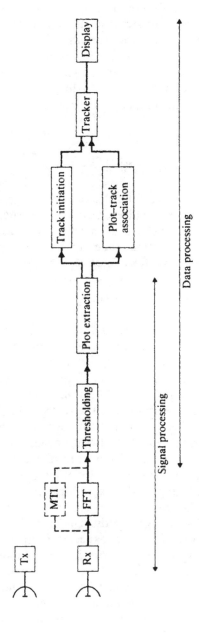

Figure 5.1 Signal and data processing in a modern radar system.

105

we cannot begin until we know more about the properties of the clutter we are trying to filter out. The simplistic case might be the problem of detecting an aircraft flying against a bare mountainside in the background. In this case the clutter is stationary. But suppose the mountainside is covered in trees blowing about in the wind, and the aircraft is a helicopter nearly hovering; our problem becomes more difficult. Perhaps the ultimate challenge is presented by an air-borne radar in look-down mode trying to distinguish a tank from wind-blown vegetation and moving clouds of rain. Clearly it becomes important to study the properties of ground clutter before we begin signal processing.

5.2 PROPERTIES OF CLUTTER

Typically, clutter arises from area- or volume-extensive regions illuminated by the main beam or one of the sidelobes of the antenna pattern. In Chapter 4 we made the assumption that such clutter has the same (gaussian) statistics as thermal noise at the pre-detection stage. This means that after envelope detection the clutter values have a *Rayleigh distribution*

$$p_c(c) = \frac{c}{\sigma_c^2} \exp\left(\frac{-c^2}{2\sigma_c^2}\right) \tag{5.1}$$

The expressions for detection and false-alarm probabilities are exactly the same as those derived in Sec. 4.11, with clutter taking the place of noise.

This clutter model is valid as long as, within the illuminated region, the clutter can be regarded as consisting of many independent random scatterers of comparable strength. Such conditions normally hold in weather clutter or chaff, where the scattering is from very many droplets or particles. They are also encountered in other common situations, such as low-resolution marine radars operating in calm sea conditions, or radars illuminating desert or agricultural areas. When the radar is operating under conditions where the return may be dominated by a few brighter scatterers within each resolution cell (such as can happen when a high-resolution radar illuminates a patch of ocean whose dimensions are comparable with the water wavelength), then this model no longer holds. Under such conditions, larger values of clutter occur more often than expected from the Rayleigh distribution, so that false alarms will occur more frequently unless the detection threshold is raised.

Analysis of the detection/false-alarm rates for non-gaussian clutter requires a model for the clutter and signal-plus-clutter PDFs at the detection stage. In practice, this normally relies on empirical distributions, supplemented by physical reasoning. Several types of model are in use for different circumstances. For sea clutter, log-normal, Weibull and K distributions have been used *inter alia*. With different parameters, the same models also have

Table 5.1 Values of v_{rms} and doppler spread for 1 GHz transmitter

Target type	v_{rms} [m s^{-1}]	Doppler spread [Hz] $= \frac{2\Omega}{3}v_{rms}$
Sea clutter, windy day	0.9	6
Forest, moderate wind	0.2	1.3
Chaff	1	6.7
Rain clouds	1.9	12.7

their place in describing land clutter from a variety of terrain types. (See reference 1 for a description of these distributions and for further references.)

When the targets being sought by the radar are moving, doppler information can also be used to separate them from the clutter. This is complicated by the fact that many forms of clutter, such as the sea surface, wind-blown vegetation, chaff and rain clouds, are themselves in motion, and hence also give rise to doppler shifts. As a result, the clutter spectrum is normally spread about zero doppler frequency. For most purposes, the spectrum $C(\omega)$ can be treated as having a gaussian shape, i.e.

$$C(\omega) = C(0) \exp\left[-\frac{a^2}{2}\left(\frac{\omega}{\omega_0}\right)^2 \right] \qquad (5.2)$$

where ω_0 [rad s^{-1}] is the transmitter frequency and a is a dimensionless constant controlling the width of the spectrum. For a gaussian function $\exp(-\omega^2/2\sigma^2)$, the RMS spread has the value σ, so that the RMS spread of doppler frequencies corresponding to Eq. (5.2) is ω_0/a. Since doppler frequency is related to radial velocity by $f_d = 2v_r/\lambda$ [Hz] or $\omega_d = 2v_r\omega_0/c$ [rad s^{-1}], the spread in doppler frequencies will be related to the apparent RMS velocity spread (v_{rms}) of the clutter by the relation

$$\omega_0/a = 2v_{rms}\omega_0/c \qquad (5.3)$$

so that $a = c/(2v_{rms})$. It can be seen that this clutter spectrum model predicts that the spectrum width should be proportional to the operating frequency, which is approximately true in practice. Typical values of v_{rms} and the corresponding doppler spread for a 1 GHz transmitter are given in Table 5.1.

5.3 MOVING-TARGET INDICATOR PROCESSING

Most radar systems need some form of doppler processing to filter out clutter and thereby reveal faster-moving targets. These days, such filters are implemented digitally, either as some form of fast Fourier transform (FFT) algorithm or as a set of transversal filters.

In digital signal processing, the data $f(t)$ are sampled at discrete instants separated by a fixed timestep Δt. This gives rise to a digital signal $\{f_l\}$, where $f_l = f(l\,\Delta t)$. Transversal filters simply take a weighted sum of blocks of length M of the sampled data to give output

$$g_p = \sum_{l=0}^{M-1} a_l f_{p-l}$$

Here the a_l are the weighting constants. Digital approximations to, for example, the matched filter described in Chapter 4 can be constructed in this way. (There are many good books dealing with this topic; Stremler[2] is an example.)

A digital approximation to the Fourier transform can also be constructed in a similar way, using complex weights. This is the discrete Fourier transform (DFT). It is related to the sequence $\{f_l\}_{l=0}^{M-1}$ by the equation

$$F_p = \sum_{l=0}^{M-1} f_l\, e^{-2\pi j l p/M} \qquad \text{for } 0 \leqslant p \leqslant M-1 \tag{5.4}$$

where we have followed the convention of using upper-case letters to represent the Fourier transform. The importance of the DFT as an approximation to the continuous Fourier transform stems largely from the fact that, if M is a power of 2, there is a very efficient way to perform the M weighted summations implied by Eq. (5.4). This implementation of the DFT is known as the fast Fourier transform (FFT). (See, for example, Brigham[3] for a very good treatment of the FFT and its use in signal processing.) It makes possible efficient calculations of, for example, convolution or correlation, which can be prohibitively expensive on computing time if carried out in the time domain. The FFT was known to Gauss, but came to prominence when it was rediscovered by J.W. Cooley and J.W. Tukey in 1965, and has since revolutionized signal processing. There are now many ways of implementing it, and many other algorithms that may be used to compute the doppler spectrum (such as the fast Hartley transform), but the FFT remains the backbone of modern doppler processing.

With these techniques, the modern radar design engineer can arrange the signal processing to produce filters of the desired characteristics and speed of operation. In the past, doppler processing was not easy to achieve. Some systems made use of analogue bandpass filters to separate those signals arriving at the carrier frequency (no doppler shift) from those displaced in the spectrum by the target motion. By far the most common technique was the moving-target indicator (MTI), still in use in many radars round the world.

The principle of MTI is shown in Fig. 5.2. Each echo from a given range gate is subtracted coherently from a delayed version of the previous echo from that range gate. If nothing has changed, cancellation occurs, which

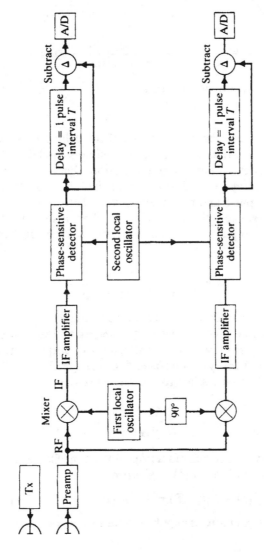

Figure 5.2 Block diagram of MTI processing.

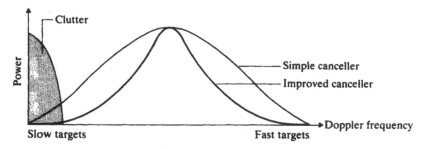

Figure 5.3 Clutter cancellation by MTI processing.

would be complete in the absence of noise. If the echo has changed phase slightly (and amplitude strictly, but limiting amplifiers can be used to remove the amplitude dependence) because of its motion, then cancellation will be less complete. For a target in uniform motion there is a constant change in phase from pulse to pulse and cancellation does not occur. The two-pulse cancelling MTI is therefore acting as a high-pass filter, as shown in Fig. 5.3.

The impact of two-pulse cancellation on the clutter can be quantified by the *clutter attenuation factor*

$$CA = \frac{\text{input clutter power}}{\text{output clutter power}} \tag{5.5}$$

We saw in Chapter 4 that noise power and noise variance are the same thing, and that we can calculate noise power by integrating the power spectrum (see Eq. (4.23)). We also saw how to calculate the output spectrum given the input spectrum and the system transfer function (see Eq. (4.28)). Exactly the same principles can be applied to clutter. This gives

$$CA = \frac{\int C(\omega)\,d\omega}{\int C(\omega)|H(\omega)|^2\,d\omega} \tag{5.6}$$

where $H(\omega)$ is the system transfer function of the canceller. Since the output of the canceller from an input $f(t)$ is given by

$$g(t) = f(t) - f(t - T) = f(t)*[\delta(t) - \delta(t - T)] \tag{5.7}$$

where T is the pulse separation, its impulse response function is $\delta(t) - \delta(t - T)$, and hence

$$H(\omega) = 1 - e^{-j\omega T} = 2je^{-j\omega T/2}\sin(\omega T/2) \tag{5.8}$$

Using Eqs (5.2) and (5.8) in Eq. (5.6) gives, after simplification,

$$CA = \frac{0.5}{1 - \exp(-2\omega_0^2 T^2 v_{rms}^2/c^2)} \tag{5.9}$$

(a)

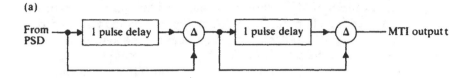

(b)

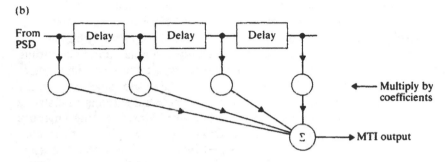

Figure 5.4 (a) A three-pulse cancelling system. (b) A more general multipulse cancelling scheme.

Normally $\omega_0 \ll c$, so that the exponential term can be approximated by the first two terms in its power series. This gives

$$CA = \left(\frac{c}{2\omega_0 T v_{rms}} \right)^2 \tag{5.10}$$

For a radar operating at 1 GHz, with a PRF of 1 kHz, the power in stable clutter with $v_{rms} = 1$ m s^{-1} would be attenuated by a factor in excess of 570 (27.6 dB).

The performance of the two-pulse canceller can be improved by extending the principle to a three-pulse canceller as shown in Fig. 5.4a, or to the more general scheme of Fig. 5.4b. The attenuation of low-velocity clutter is improved, as shown in Fig. 5.3. As Fig. 5.4b shows, multipulse cancelling schemes are well on the way to becoming a modern digital transversal filter.

Historically, the time delay was implemented using an ultrasonic delay line. A delay line is really only an analogue first-in, first-out memory device, and, when digital electronics arrived, digital delays were introduced. Digital delays permit increased flexibility in the PRF, which can be used to remove blind speeds.

Blind speeds are caused by the same aliasing effect discussed in Chapter 2. If the target speed is such that it moves a distance of $\lambda/2$ (or $n\lambda/2$) between pulses, then it will appear to have stationary phase and will be cancelled in the same way as zero-velocity clutter. Small changes in the PRF alter the blind speed to prevent a target remaining undetected, but require the MTI delay to be changed to keep in step.

The other problem of MTI, known as tangential fading, has no remedy because it is something of an own-goal. When the path of the target carries it tangentially through the radar beam, the radial component of its speed goes through zero and the echo is filtered out by the signal processor.

On the whole MTI, and the modern transversal filter equivalents, are very successful because the calculations are simple and sequential, and the answer is available as soon as the last digit arrives. In contrast, FFT-based processing requires blocks of data to be collected before processing can begin.

Despite its simplicity, MTI processing needs to be carefully engineered. Local oscillators must have a frequency stability and phase noise better than the frequency separations that the MTI is seeking to impose. The system must have sufficient mechanical stability to cause no detectable phase changes; this may sound easy, but with a dish antenna rotating in a strong wind it can be difficult to avoid detrimental vibrations. Rotating antennas have the additional problem that, at angles off-bore-sight, there is contamination of the doppler spectrum caused by the motion of the dish itself. In many cases, the dish motion is the factor limiting MTI performance, and is a further reason why modern radars are more and more tending to use flat antenna arrays with electronic beam steering (see Chapter 15).

5.4 FAST FOURIER TRANSFORM PROCESSING

Transversal filters and MTIs are often modified to null out moving clutter, such as rain clouds, with a clutter notch that can be controlled adaptively. The whole of MTI thinking is geared towards removing clutter in this way, so that a target is left competing only with system noise.

A different approach is to concentrate on target detection by developing a special filter that allows only the target through. The target would then be detected in the narrowest possible bandwidth, so minimizing the noise. Unfortunately, until the target has been detected, we do not know its speed and we cannot construct the filter. The solution to this dilemma is to use a bank of filters, just overlapping, such that the target must appear in one of them. The slow-speed filters will be contaminated with clutter, but we can search through the remainder to find the target. The output of this filter bank is known as the doppler spectrum (see Fig. 5.5).

There are many ways of constructing a filter bank, including analogue methods, but these days the quickest way is to use a FFT algorithm. To do this, a set of samples $\{u_l\}$ from a fixed range gate are gathered, using the multiple pulses available within a single scan. The effect of the FFT is most easily explained if we assume a non-fluctuating target with doppler frequency ω_d. We can therefore write the signal sequence as

$$u_l = V \exp(j\omega_d l T) \tag{5.11}$$

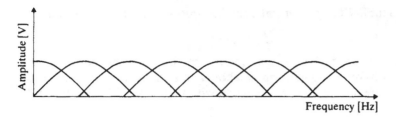

Figure 5.5   The output of the FFT can be thought of as a bank of bandpass filters.

where T is the time spacing of the samples. The amplitude of each sample is the constant V, and for simplicity we have ignored the carrier frequency. The DFT of this sequence (using Eq. (5.4)) is

$$U_p = V \sum_{l=0}^{M-1} \exp(j\omega_d lT) \exp(-2\pi j lp/M) = V \sum_{l=0}^{M-1} \exp[j(\omega_d T - 2\pi p/M)l]$$

(5.12)

Writing

$$z = \exp[j(\omega_d T - 2\pi p/M)]$$ (5.13)

we can see that this is a geometric progression. Hence we can write down the sum as

$$U_p = V \frac{1 - z^M}{1 - z}$$ (5.14)

Since $z^M = \exp(j\omega_d MT)$, the quantity on the top line of Eq. (5.14) is constant; it does not depend on p. Hence as p varies, U_p is maximized when we make the bottom line of Eq. (5.14) as small as possible, i.e. set $z = 1$. Using Eq. (5.13), this means that the *maximum term* in the FFT occurs when

$$p = \omega_d MT/2\pi$$ (5.15)

From this, we can find the doppler frequency ω_d.

Equation (5.15) implies that the possible solutions for the doppler frequency are also digital (they are all multiples of $2\pi/MT$). Hence there may be some error introduced in the estimate of the doppler frequency, with the correct value lying between the maximum and one of its neighbours. At this point, you may object that this error is needed to stop the expression in Eq. (5.14) becoming infinite when z is exactly 1. This does not happen, because if z is exactly 1, the summation in Eq. (5.12) gives the value MV. (You should check this.) So the *maximum value* in the FFT is MV. (It is also worth noting, if the argument above makes you uneasy, that the limiting value of Eq. (5.14) as p gets close to $\omega_d MT/2\pi$ is MV.)

To find the SNR improvement from this process, we can write the input as

$$f_l = u_l + n_l$$

where u_l and n_l are the signal and noise terms in the input sequence. The signal terms have the same form as above, and we assume that the noise terms are all mean-zero, uncorrelated and with variance σ_n^2. Taking the DFT gives

$$F_p = \sum_{l=0}^{M-1} u_l \exp(-2\pi j l p / M) + \sum_{l=0}^{M-1} n_l \exp(-2\pi j l p / M) \qquad (5.16)$$

The first summation is deterministic, and is given by Eq. (5.14), but the noise summation is random. We need the average noise power, where the noise term is given by the sum in Eq. (5.16), i.e. by

$$N_p = \sum_{l=0}^{M-1} n_l \exp(-2\pi j l p / M) \qquad (5.17)$$

This requires a slight modification of the methods used in Chapter 4, since the noise term here is *complex*. In this case, the noise power is found by averaging the *squared modulus* of N_p, so that

$$\text{Noise power} = E[|N_p|^2] = E[N_p N_p^*] \qquad (5.18)$$

Therefore

$$\text{Noise power} = E\left[\sum_{l=0}^{M-1} n_l \exp(2\pi j l p / M) \sum_{k=0}^{M-1} n_k \exp(-2\pi j k p / M) \right]$$

$$= E\left[\sum_{k,l=0}^{M-1} n_l n_k \exp[-2\pi j p (l - k)/M] \right]$$

$$= \sum_{k,l=0}^{M-1} E[n_l n_k] \exp[-2\pi j p (l - k)/M] \qquad (5.19)$$

Since the noise is uncorrelated,

$$E[n_l n_k] = \begin{cases} \sigma_n^2 & \text{if } l = k \\ 0 & \text{otherwise} \end{cases}$$

Hence

$$\text{Noise power} = \sum_{l=0}^{M-1} \sigma_n^2 = M\sigma_n^2 \qquad (5.20)$$

The noise power does not depend on p, so is the same for each term in the DFT sequence.

If we now look at the SNR before and after the DFT, we find that the signal energy for each sample is V^2, and the SNR is therefore given by

$$\text{SNR (before)} = V^2/2\sigma_n^2$$

At the maximum response of the DFT, however, the signal gives a value of MV. Hence the SNR is

$$\text{SNR (after)} = M^2 V^2/2M\sigma_n^2 = MV^2/2\sigma_n^2$$

Thus there is a gain in SNR by a factor M at the peak response of the DFT.

The analysis has only considered the case of uncorrelated noise. For clutter, we would need to take account of the correlation of the clutter samples (including the phase correlation, which gives rise to the doppler spectrum of the clutter). This complicates the analysis, and is not carried out here. However, in the range of doppler frequencies where the clutter contribution is negligible and noise is the dominant factor, the analysis carried out above is valid.

5.5 THRESHOLDING

Radar data processing has been described as 'the ruthless abandonment of data'. Somehow the overwhelming radar data rate, which we calculated in Chapter 2, must be reduced to a few numbers representing observations of known and new targets. The key operation to achieve this data reduction is the thresholding process, where the data are compared with a reference level, and only those few signals exceeding this level are investigated further.

Early radars made use of voltage comparator devices in which the reference voltage was under the control of the operator. When the voltage exceeded the threshold, the data were displayed or recorded. A system using an externally imposed reference is only as good as the speed and experience of the operator, and it has the limitation that this threshold cannot be adjusted continuously, as the antenna rotates, to compensate for variations in the clutter and noise levels in different directions.

The introduction of digital processing permitted the reference level to be generated internally from the observations themselves, thereby permitting more sensitive and faster reacting thresholds to be used. There are two basic steps to setting the threshold. First, the mean value of the noise or clutter is found, when no targets are thought to be present. As described in Chapter 4, the threshold is then normally set at some multiple of this noise power. This multiple is the factor k introduced in Eq. (4.44), and is known as the (multiplicative) margin. Usually the margin is added to the mean in dB, i.e.

$$\text{Threshold [dB W]} = \text{mean [dB W]} + \text{margin [dB]} \qquad (5.21)$$

(see Fig. 5.6). For envelope detection, a false-alarm probability of 10^{-6} requires a value of k equal to 5.26 (see Sec. 4.6), so that the margin would

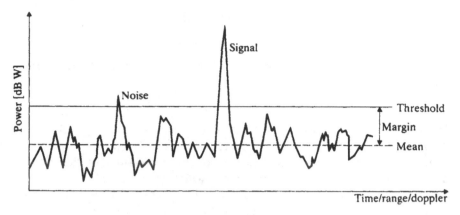

Figure 5.6 The mean and margin used in a thresholding system.

be 7.2 dB. This addition of a margin in dB implies that, as the mean noise level varies, the threshold also varies in such a way as to keep a constant false-alarm rate, as long as the noise model described in Sec. 4.11 is valid. In the case of some rarer noise models, a more sensitive detection system can be designed (for the same false-alarm rate) by using other ways of creating a threshold, such as the linear addition of a margin to the mean noise power.

There are only two main decisions to be made in thresholding: the first is the choice of the basis on which to calculate the mean power, and the second is the setting of the margin to achieve the correct sensitivity.

The mean power is calculated by finding the average power in *reference cells*, which are near the test cell, but which are not so close that they are contaminated by the presence of the target. The usual procedure is to leave guard cells either side of the test cell, which are not included in the mean, as shown in Fig. 5.7. The number of cells contributing to the mean, and the width of the guard band, depend on how quickly the background is changing and how much the target is spread across the cells. The method must therefore be optimized for each radar system, but typical schemes are shown in Fig. 5.8.

So far we have not explained what we mean by a cell adjacent to the test cell. It could be a radar resolution cell close to the test cell in range, doppler, azimuth, elevation or even time (i.e. what happened in that particular cell in previous scans). All of these parameters have been used at one time or another, but one-dimensional thresholding in the range dimension, or two-dimensional thresholding in the doppler/range plane are probably the most commonly used schemes.

Having found a method of determining the mean, the next task is to add a suitable margin. This is the problem we analysed in Chapter 4, where we derived the relationship between SNR, the margin k and the false-alarm and detection probabilities. We showed that, if the margin is set very low,

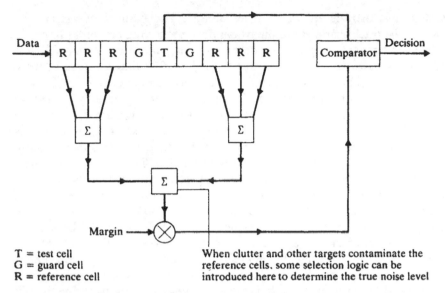

T = test cell
G = guard cell
R = reference cell

When clutter and other targets contaminate the reference cells, some selection logic can be introduced here to determine the true noise level

Figure 5.7 Guard cells G are used to prevent large signals in the test cell T from contaminating the mean of the reference cells R.

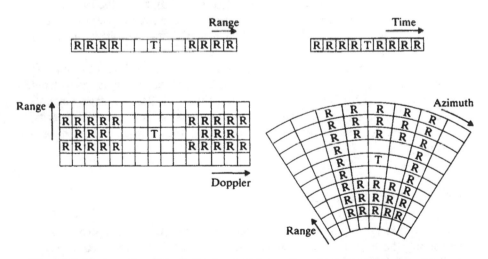

Figure 5.8 Various schemes for finding local reference cells to determine the mean level around the test cell.

the probability of detection will be good (even faint target echoes are likely to be detected) but the probability of false alarm will also be high because a significant number of noise spikes will trigger the system (see Fig. 4.8). Frequent false alarms can be very inconvenient if they trigger a defence system, but there is an even greater danger. Too many false alarms will swamp the tracker and cause the whole data processing computer to crash; this could temporarily paralyse a defence system completely. As a result, low false-alarm rates are used; the value for P_{f_a} of 10^{-6} that we have adopted throughout this book is comparatively large for many systems.

The solution to the problem of setting the margin correctly is to leave the job to the computer. If there are too few threshold crossings, the margin can be lowered to make the system more sensitive, so that the system is triggered more frequently. If the number of threshold crossings increases too far, exceeding a set limit, the margin is automatically increased to control it. To the beginner this seems a rather empirical approach to the problem, which leads to the probability of detection varying with the level of clutter/noise/jamming. In practice, there really is no other way to proceed, except possibly in the most advanced computer systems where more processing power can be allocated to those areas where higher clutter levels are experienced. Keeping the number of threshold crossings constant is effectively the same as keeping the false-alarm rate constant (assuming that very few of the crossings are due to real targets). Hence this technique has become known as *constant false-alarm rate* (CFAR) processing and is widely used in radar practice. If the false-alarm rate is kept constant and the clutter statistics change, it is inevitable that this leads to changes in the probability of detection. This is because changes in the clutter power are equivalent to changes in SNR; Eqs (4.45) and (4.80) describe quantitatively how this affects P_d.

The CFAR process can be improved to cope with contamination of reference cells by clutter or other targets. Sometimes it is considered worthwhile to make a second pass through the data after the largest signals have been identified and removed by the first pass. This *double thresholding* improves the sensitivity of the system, but at the cost of increased processing effort.

5.6 PLOT EXTRACTION

Once a potential target, or plot, has been identified by a signal crossing the threshold, the next step is to extract all the available information about it. Azimuth and elevation information are usually extracted by monopulse techniques, and the range by interpolation between range gates, as described in Chapter 3. Similarly, the target speed is found by interpolation between doppler cells.

The purpose of collecting this information at this stage is because a true target will behave in a fairly predictable manner in each of the range/angle/speed dimensions and will be said to 'track' correctly. Plots arising from noise or clutter spikes (false alarms) usually behave randomly and do not track; for example, the rate of change in range will not agree with the velocity estimate obtained from the doppler information. However, this is not always the case; on occasion, sea clutter spikes due to persistent waves may recur sufficiently often for false tracks to be initiated.

5.7 PLOT–TRACK ASSOCIATION

When a radar returns to search a particular area, a number of plots will be detected. These plots may form part of a known track in the track list, part of a new emerging track, or be random false alarms. The first part of the tracking operation is to sort the plots into these three categories, in a process known as plot–track association, or association–correlation.

Plot–track association can be thought of in terms of the matrix shown in Fig. 5.9. Each of the new plots is considered as a possible candidate for association with each of the existing tracks. The matrix may be thinned somewhat because, as Fig. 5.9 shows, some plots cannot be associated with existing tracks because they are too far away. The remaining possibilities are then considered by minimizing the function:

$$\sum \frac{(\text{forecast position})_m - (\text{actual position})_n}{\text{tracking error variance}} \tag{5.22}$$

where the sum is over all tracks m and all plots n. Some plots will associate with known tracks and the remaining free plots are tested as potential new tracks (see next section). Any remaining free plots, failing to form tracks, are regarded as false alarms and are discarded at this stage.

There are more complications in plot–track association than might be imagined. Figure 5.10a shows two tracks that are about to cross; does

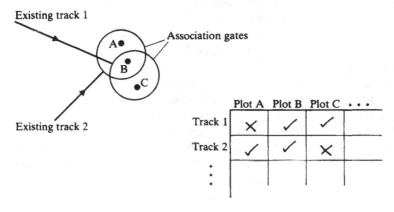

Figure 5.9 Associating plots to tracks.

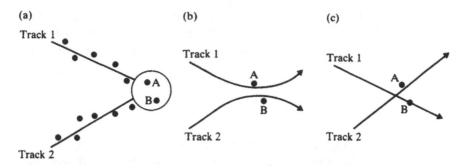

Figure 5.10 There are potential complications when tracks appear to cross.

plot A go with track 1 and B with track 2? The answer depends on our track model. If the two targets were ships, we would assume this association to be correct, as they might be expected to diverge, as shown in Fig. 5.10b; but if the targets were aircraft at different heights, our belief might be that they continued as straight tracks, as in Fig. 5.10c.

Other information is also contained within the track model. For example, the plots should be distributed randomly either side of a straight track, assuming the observations are corrupted by zero-mean gaussian noise. The run sequence of '1's (above the track) and '0's (below) can be tested for randomness and used to help decide which plot best fits with which track.

5.8 TRACK INITIATION

When a new sequence of plots appears on a radar display and begins to form a straight line, or other recognizable trajectory, an operator is likely to become increasingly convinced that a true target is present. How could this process be automated? The answer is that there are almost as many schemes as there are radar systems, and that every engineer has a favourite, but most methods are based on simple up/down counting logic.

Imagine a surveillance radar with a rotating scanner. On the first revolution a free plot is identified that could form part of a track and so an up/down counter is incremented. On the next revolution the plot is there again and the counter is incremented further. On the next scan the plot is missing and the counter is decremented. This process continues until either the count reaches zero and the track is deleted from the track list, or the count reaches some predetermined value and the track is declared to be genuine and the target existence is reported to the radar operators.

Simple track initiation schemes might use $+3$ for a hit, -2 for a miss and declare the target to be present when the count exceeds 7. A weak, fading target might therefore take some time to be detected as it approaches the radar, as the following sequence shows:

Scan	Hit/Miss	Count	Total	
1	H	$+3$	3	
2	M	-2	1	
3	H	$+3$	4	
4	M	-2	2	
5	H	$+3$	5	
6	M	-2	3	
7	M	-2	1	
8	H	$+3$	4	
9	H	$+3$	7	
10	H	$+3$	10	Track declared present

Some of the limitations of such simple schemes can be avoided by the Boolean combination of rules such as 'seven out of ten scans must be hits AND two of the first three must be hits'.

Track initiation schemes tend to be empirical and they turn out to be quite hard to compare analytically, because of the Bayesian nature of the problem. In the long term they may be replaced by artifical intelligence methods of searching for patterns in the data (see Chapter 15).

How should you decide whether a plot is near enough to a track to be included as a possible candidate for that track? The answer is shown in Fig. 5.11. The first plot defines the position of the target, a civil airliner for example, at a moment in time. Ten seconds later the radar returns to scan the same sector, and the aircraft could be anywhere within a circle of radius equal to the maximum possible speed of the aircraft times the 10 s interval, giving a typical radius of 3 km. After the second plot has been identified

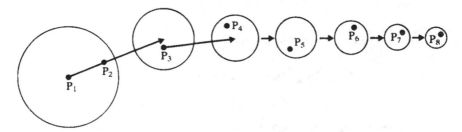

Figure 5.11 The plot–track association gate decreases with increased observations.

within this circle, there is some indication of the aircraft heading, so the next association gate need not be so large. As Fig. 5.11 shows, the association gates can eventually be reduced to a size determined by the accuracy of the tracker. Usually, this gate size is about one standard deviation for a straight track, but nearer 3 standard deviations for a manoeuvring target.

5.9 TRACKING

The actual process of tracking a target is composed of a track-smoothing algorithm, identical to a conventional digital filter, together with track-maintenance algorithms to cope with missing data or apparent branching caused by multiple choices of plots.

The simplest smoothing algorithm is the $\alpha-\beta$ filter shown in Fig. 5.12. The object of the algorithm is to find a compromise between the two extremes of a track composed of measurements joined together (no smoothing, in which case the target appears to move in a random zig-zag manner) and a track drawn as a straight line that ignores the data points completely.

Radar plots are assumed to be related to the true target position by

$$\text{Measured position} = \text{true position} + \text{plot noise} \qquad (5.23)$$

where the plot noise is assumed to have a zero mean and a constant noise power of σ^2.

When tracking in cartesian coordinates, suppose the predicted next x value of the target is x_p and the measured position is x_n; then we need to take some account of the difference by using $\alpha(x_n - x_p)$, where α is a

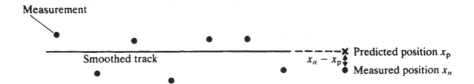

Measurement

Smoothed track

Predicted position x_p

$x_n - x_p$

Measured position x_n

New smoothed position $x_s = x_p + \alpha(x_n - x_p)$

$\alpha = 1$ Joins measurements

$\alpha = 0$ Ignores data

Figure 5.12 The $\alpha-\beta$ tracker.

number between 0 and 1. When $\alpha = 1$ we are joining the dots, and when $\alpha = 0$ we are ignoring the data. In this way we can use α to define the new smoothed position of the target x_s to be

$$x_s = x_p + \alpha(x_n - x_p) \tag{5.24}$$

Likewise, the new velocity estimate can be found from

$$v_n = v_{n-1} + \frac{\beta}{T}(x_n - x_p) \tag{5.25}$$

where $v = $ speed $[\text{m s}^{-1}]$ and $T = $ time interval between observations $[\text{s}]$.

Once the track has been updated by the addition of the new smoothed positions, the new predictions are made for the next scan using

$$x_p = x_s + Tv_n \tag{5.26}$$

For a straight-line track, a tedious but uncomplicated proof (based on minimizing the sum of squares of the differences between the track model and the measurements) can be used to show that the optimal values of α and β are given by

$$\alpha = \frac{2(2n-1)}{n(n+1)} \tag{5.27}$$

$$\beta = \frac{6}{n(n+1)} \tag{5.28}$$

where $n = $ number of observations.

Putting a few values of n into Eqs (5.27) and (5.28) shows that there is almost no smoothing of the initial data, but the damping factor is progressively increased as the observations continue. This increased confidence in the true track of the target means that the tracking error is reduced with increasing time.

The error variance of the smoothed positions, which tells us the tracking error, is given by

$$\sigma_n^2 = \frac{2(2n-1)}{n(n+1)}\sigma^2 \tag{5.29}$$

The error variance of the predicted position, used to set the size of the association gate, is given by

$$\sigma_{n|(n-1)}^2 = \frac{2(2n-1)}{(n-1)(n-2)}\sigma^2 \tag{5.30}$$

Two independent trackers are used in the x and y coordinates and the algorithm can also be extended to track in acceleration in an $\alpha - \beta - \gamma$

tracker. The algorithm may also be developed to take advantage of doppler measurements of velocity.

The advantage of the $\alpha-\beta$ tracker is that it is very easy to implement. The disadvantages are that it provides optimal smoothing only for linear tracks, evenly spaced observations and gaussian noise.

A more general-purpose tracker is the Kalman–Bucy filter. An analogy can be formed between tracking algorithms and the suspension on a motor car, which attempts to smooth out the irregularities of a nominally straight road. The $\alpha-\beta$ tracker learns how bumpy the road is and adjusts the damping factor of the suspension accordingly. The Kalman–Bucy filter is more sophisticated; it attempts to learn the 'pattern' in the sequence of bumps in order to predict what will happen next. It then adjusts the suspension for the predicted event.

Like the $\alpha-\beta$ tracker, the Kalman–Bucy filter relies on minimizing the square difference between the track model and the measurements. At each step, the best estimate of the whole track up to that time is made, using all previous observations and the evolving track model. The prediction of the parameters (position, velocity and perhaps acceleration) of the next position along the track uses all the available information, with more recent measurements weighted more heavily. This leads to a superficially complicated algorithm (detailed descriptions of Kalman filtering can be depressing). However, the beauty of this algorithm is that it can be applied *recursively*, i.e. the solution to the minimization problem at each step makes as much use as possible of the solution at the previous step. This continual updating provides a numerically stable and computationally efficient means of predicting the evolving track, in a manner consistent with all the available information.

Even after tracking has been successfully initiated, there are still a number of things that can go wrong, and track-maintenance algorithms are required to ensure that the target is followed successfully. One of the commonest problems is when an aircraft, which has previously been flying in a straight line, suddenly makes a manoeuvre. This must be both detected and allowed for in the tracker.

One method of detecting manoeuvres is the *integrated bias detector*, shown in Fig. 5.13. If the straight-track model is correct, the measurements will fall equally on either side, and so a running summation of the errors will be roughly zero (when the sign of the error is included). If the target begins to turn, then the measurement errors will quickly build up on one side of the track, and when the total exceeds some predefined level, a manoeuvre is declared to have been detected. In an $\alpha-\beta$ filter the values of α and β can be increased to improve the ability of the tracker to follow the target (but with less accuracy). With a Kalman–Bucy filter the procedure is usually to add plant noise to the filter (increasing elements in the predicted error covariance matrix) to achieve the same effect. This reflects the increased

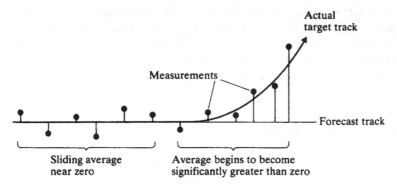

Figure 5.13   The integrated bias detector.

uncertainty (and hence error variance) in the track model when the target is manoeuvring.

Another problem is to decide what to do if the track appears to branch. Track branching occurs when there are so many false plots that the true one cannot be identified with certainty. Some systems attempt to cope with this problem by tracking both branches for a while and then trying to select the most promising looking branch; this is not unlike the *Viterbi algorithm* used in digital communications to find the decoding route through the trellises used in convolution encoding. Another approach is to use *probabilistic data association* (PDA), formulated originally by Bar-Shalom and Tse[4] and extended by others to take account of multiple targets, manoeuvres, track initiation, etc. (see e.g. Colgrave *et al.*[5]). PDA builds on the Kalman–Bucy filter to use an average of all the validated plots, weighted by the probability of their having originated from the target. Although PDA introduces inaccuracies from the (weighted) inclusion of incorrect plots, it avoids the risks of an increasing number of track branches eventually jamming the computer and causing a system crash.

If the target track enters a particularly dense area of clutter or jamming, it may not be possible for the software to maintain the track and operator intervention may be necessary. Part of the effectiveness of radar systems is determined by the skill, intelligence and experience of operators. In future, it may be possible to use expert systems to replicate the local knowledge of experienced operators, but it is unlikely that high-level decisions in radar (e.g. whether to fire on a target) will be taken without human control, or at least a human right of intervention.

5.10 SUMMARY

The target detections that initiate a track must take advantage of whatever gains are possible in SNR, because viable radar systems must operate at very low false-alarm rates. Making use of doppler information is a crucial way of increasing the SNR. Once a track has been initiated, a variety of methods of increasing sophistication are possible to retain the track. All of them run into trouble as tracks become more complicated and the radar environment becomes more congested.

Key equations

● The Rayleigh clutter model:

$$p_c(c) = \frac{c}{\sigma_c^2} \exp\left(\frac{-c^2}{2\sigma_c^2}\right)$$

● The gaussian clutter spectrum:

$$C(\omega) = C(0) \exp\left[-\frac{a^2}{2}\left(\frac{\omega}{\omega_0}\right)^2\right]$$

where $a = 2c/v_{rms}$.

● The clutter attenuation factor:

$$CA = \left(\frac{2c}{\omega_0 Tv_{rms}}\right)^2$$

● The equations for α–β tracking:

$$x_s = x_p + \alpha(x_n - x_p)$$

$$v_n = v_{n-1} + \frac{\beta}{T}(x_n - x_p)$$

$$x_p = x_s + Tv_n$$

$$\alpha = \frac{2(2n - 1)}{n(n + 1)}$$

$$\beta = \frac{6}{n(n + 1)}$$

where n = number of observations.

5.11 REFERENCES

1. *Introduction to Radar Systems*, M. I. Skolnik, McGraw-Hill, New York, 1981.
2. *Introduction to Communication Systems*, F. G. Stremler, Addison-Wesley, Reading, MA, 1982.
3. *The Fast Fourier Transform*, E. O. Brigham, Prentice-Hall, Englewood Cliffs, NJ, 1974.
4. Tracking in a cluttered environment with probabilistic data association, Y. Bar-Shalom and E. Tse, *Automatica*, 11, 451–461, 1975.

5. Track initiation and nearest neighbour incorporated into probabilistic data association, S. B. Colgrave, A. W. Davis and J. K. Ayliffe, *IE Aust. IREE Aust.*, 6(3), 191–198, 1986.

5.12 PROBLEMS

5.1 If the signal and noise were roughly equal at the output of a radar receiver, what SNR would you expect after averaging four 32-point FFT blocks of observations? Assume the target RCS and speed to be constant. What would happen to the SNR if the target began to manoeuvre?

5.2 A surface-wave HF radar uses a PRF of 275 Hz and observes an aircraft for 3.72 s. What is the potential coherent gain and how would this improve the range estimate? These radars use vertical polarization (see Chapter 10). Would the target remain visible through a manoeuvre?

5.3 A radar system is required to have a high probability of detection of 95 per cent on envelope detection of a single pulse. This P_d is to be achieved at the expense of a high false-alarm rate of 10^{-5}. What margin should be chosen for the thresholding and what SNR is required?

5.4 The following series of x-coordinate position measurements of a target were made at the rate of one per second:

t	1	2	3	4	5
x_n	0	35	88	118	158

Tabulate the values of α, β, x_s and v_n for the $\alpha-\beta$ tracker and predict the x value for the next position.

5.5 In problem 5.4, what improvement in the target position results from the track smoothing? Suggest a size for the association gate in the x dimension for the sixth observation.

SIX

DESIGNING RADAR WAVEFORMS

- How accurately can we measure range and velocity?
- Can we distinguish closely spaced targets?
- How much freedom do we have to improve radar performance?

How do we choose waveforms that tell us as much as possible about the properties of the target?

6.1 INTRODUCTION

In Chapter 4, we established that the correlation receiver (or matched filter) provides the optimum method of detecting a stationary target. Detectability is unaffected by the form of the transmitted pulse. All that matters is the ratio of the signal energy E to noise power per unit bandwidth N, both calculated at the input to the receiver. The signal-to-noise ratio at the output of the correlation receiver is then given by E/N. However, there are other important aspects of system performance for which the signal shape does become important. These include the resolution, ambiguity and accuracy[†] of the measurements made by the radar.

The resolution of a radar system is a measure of its ability to separate closely spaced targets in range or velocity. These can be treated separately, but simultaneous measurement of both range and velocity is often required. In this case there is an unavoidable uncertainty, so that two targets at different ranges and velocities may in principle and practice be indistinguishable.

[†] Throughout we assume no bias in the measurement, so that accuracy and precision will be equivalent.

Ambiguity occurs if the output of the receiver from a single target contains multiple peaks that can be mistaken for other targets. Such peaks may be caused by noise, but may also be produced by the shape of the transmitted waveform. An obvious example is if there are significant sidelobes in the radar antenna pattern. Objects illuminated by the sidelobes will be interpreted as though they were in the main lobe. This will lead to angular positioning errors or multiple detections generated by a single target. It can also cause clutter to be interpreted as target.

Accuracy refers to the expected spread of measurements about the true value. We have seen that detection and ranging are essentially the same problem. The range of a target depends on the time delay at a peak in the output from the matched filter, as long as this peak exceeds some threshold. Noise in the output will cause the peak to be displaced randomly from its true position. The standard deviation of this variation will be our adopted measure of range accuracy. We adopt a similar definition for the accuracy of measurements of doppler frequency.

6.2 BANDWIDTH AND PULSE DURATION

All these aspects of system behaviour are affected by the radar bandwidth, whose definition we need to make more precise. Several different definitions of this important concept are in use. The simplest definition is applicable if the signal $u(t)$ is *band-limited*, i.e. it has a Fourier transform $U(\omega)$ for which $U(\omega) = 0$ if $\omega > \Omega$ rad s^{-1}. Then it is natural to take the bandwidth as Ω rad s^{-1}, or $\Omega/2\pi$ Hz. None of the commonly used pulse modulations satisfy this relation exactly, but it is of considerable theoretical value because of its relation to the Shannon–Whittaker sampling theorem[†] (band-limited signals can be reconstructed exactly from their samples as long as the sampling frequency exceeds the Nyquist rate of Ω/π samples per second.) In practice, many real signals are deliberately band-limited by the use of anti-aliasing filters.

The most commonly used engineering definition of bandwidth is the 3 dB width, which is the separation in frequency of the half-power points in the energy spectrum of $u(t)$, i.e. it is obtained by solving

$$|U(\omega)|^2 = \tfrac{1}{2}|U(0)|^2 \tag{6.1}$$

Clearly this definition is only applicable if $|U(\omega)|$ has a maximum at $\omega = 0$, which is the case for most commonly used pulseshapes.

From a theoretical point of view, perhaps the most useful definition is

[†] This chapter uses a number of results from Fourier transform theory. A good reference is Bracewell[1].

the *effective bandwidth* σ_ω used by Woodward[2]. It is given by the formula

$$\sigma_\omega^2 = \frac{\displaystyle\int_{-\infty}^{\infty} \omega^2 |U(\omega)|^2 \, d\omega}{\displaystyle\int_{-\infty}^{\infty} |U(\omega)|^2 \, d\omega} = \frac{\displaystyle\int_{-\infty}^{\infty} \omega^2 |U(\omega)|^2 \, d\omega}{2\pi E} \tag{6.2}$$

This definition assumes that $\int \omega |U(\omega)|^2 d\omega = 0$, which can always be arranged. (For real signals, it will always be the case, since then $|U(\omega)|$ is an even function of ω; more generally, since $u(t)$ is in fact the modulation of a carrier frequency, it will arise by defining a suitable centre frequency for the modulated signal.) This means that σ_ω^2 is the second moment of the Fourier transform of the unit energy signal $u(t)/\sqrt{E}$ about its mean, 0. It is therefore a measure of the spread of energy in the spectrum of the signal, and is analogous to the variance in probability theory or moment of inertia in mechanics. For calculation purposes, it can be very convenient to use the Fourier transform pair

$$u'(t) = du/dt \leftrightarrow j\omega U(\omega)$$

Then using Parseval's theorem, we can write

$$\sigma_\omega^2 = \frac{\displaystyle\int_{-\infty}^{\infty} |u'(t)|^2 \, dt}{E} = \frac{\text{energy in } u'(t)}{\text{energy in } u(t)} \tag{6.3}$$

There are analogous definitions of pulse duration. For time-limited pulses, such as a rectangular or raised cosine pulse (see Table 6.1b and d), we can take the actual pulse length. The 3 dB duration is frequently used, and for theoretical purposes we can use the *effective pulse duration* σ_t, defined by

$$\sigma_t^2 = \frac{\displaystyle\int_{-\infty}^{\infty} t^2 |u(t)|^2 \, dt}{\displaystyle\int_{-\infty}^{\infty} |u(t)|^2 \, dt} = \frac{\displaystyle\int_{-\infty}^{\infty} t^2 |u(t)|^2 \, dt}{E} = \frac{\displaystyle\int_{-\infty}^{\infty} |U'(\omega)|^2 \, d\omega}{2\pi E} \tag{6.4}$$

In Eq. (6.4) we have assumed that the centroid of the pulse is at 0, i.e. $\int t |u(t)|^2 \, dt = 0$, and we have used the Fourier transform pair

$$tu(t) \leftrightarrow jU'(\omega)$$

Though the definition of effective bandwidth and pulse duration may seem unnecessarily complicated, they are of considerable value because they can be handled analytically. From them, we can derive a number of fundamental relations between accuracy, resolution and ambiguity that clarify our understanding of the inherent limitations of any radar system, even though we may in practice use simpler definitions. Hence we adopt these definitions in the following sections.

The value of E, σ_ω and σ_t for some basic pulseshapes are given in Table 6.1. In this table the values for the rectangular and triangular pulses (b and c) are found from the values of the trapezoidal pulse (a) by setting $a = T$ and $a = 0$ respectively. Notice that, as the trapezoidal pulse becomes more like the rectangular pulse, the required effective bandwidth increases without limit. This reflects the fact that instantaneous jumps in signal value are not physically possible. However, because of the great value of idealized rectangular pulses in discussing system performance, it is common to assign a less stringent definition of bandwidth to this pulse. Often this is taken as the distance to the first null in its Fourier transform. For the rectangular pulse shown in Table 6.1, this gives a bandwidth of π/T rad s^{-1}.

A fundamental property of the bandwidth and duration of a pulse arises from the scaling relation of Fourier transforms. This states that, for any Fourier transform pair and non-zero constant A, $f(t) \leftrightarrow F(\omega)$ implies

$$[f(At) \leftrightarrow (1/|A|)F(\omega/A)]$$

Since $f(At)$ is a pulse of the same shape as $f(t)$, but reduced in width by a factor $1/A$, the scaling relation says that, for any sensible definition of bandwidth and pulse duration, decreasing the duration of a pulse by a factor A increases its bandwidth by the same factor, and vice versa. So, for pulses of the same shape, the time × bandwidth product is constant. For many simple pulses, this product is of the order 1 (see items c, d and e in Table 6.1). This is one of the many rules of thumb employed by radar engineers, but must be used cautiously, as it depends on the definition of bandwidth. Often a definition is used in which it becomes true, as we have just seen for the rectangular pulse. However, for more complex pulses, this relation is far from the case. Indeed, pulses using frequency modulation are designed to give a large time × bandwidth product, for reasons explained below.

6.3 RANGE AND DOPPLER ACCURACY—THE UNCERTAINTY RELATION

We saw in Chapter 4 that the output of the matched filter consists of two terms, one due to the correlation of the signal with itself, and a noise term due to the correlation of the signal with the noise. In the absence of noise, the peak of the signal term would give the true time delay corresponding to the target's range. However, the noise term will cause this peak to move around. In fact, if the gradient of the noise term is not zero at the correct time delay, the peak will be displaced. As a result, the measured value of the time delay of the peak can take a range of values. Woodward and Davies[3]

Table 6.1 Fourier transform pairs and values of E, σ_ω and σ_t for some basic pulseshapes

Pulseshape	Fourier transform pairs	E	σ_ω^2	σ_t^2
(a)	$(T+a)\,\text{Sa}\left((T+a)\dfrac{\omega}{2}\right)\text{Sa}\left((T-a)\dfrac{\omega}{2}\right)$	$\frac{2}{3}(2a+T)$	$\dfrac{3}{(2a+T)(T-a)}$	$\dfrac{T^3+2aT^2+3a^2T+4a^3}{10(2a+T)}$
(b)	$2T\,\text{Sa}(T\omega)$	$2T$	∞	$T^2/3$
(c)	$T\,\text{Sa}^2(T\omega/2)$	$2T/3$	$3/T^2$	$T^2/10$

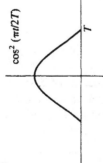

(d) $\cos^2(\pi t/2T)$

$$\frac{T}{1 - (T\omega/\pi)^2}\,\text{Sa}(T\omega)$$

$\dfrac{3T}{4}$

$\dfrac{\pi^2}{3T^2}$

$\dfrac{2T^2}{3\pi^2}\left(\pi^2/2 - 3\right)$

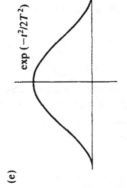

(e) $\exp(-t^2/2T^2)$

$$\sqrt{(2\pi)}\,T\exp[-(T\omega)^2/2]$$

$T\sqrt{\pi}$

$1/(2T^2)$

$T^2/2$

133

succeeded in showing that for large SNR the PDF of the measured delay would be approximately gaussian with a mean value occurring at the true delay and with standard deviation

$$\delta\tau_d = \frac{1}{2\pi\sigma_\omega\sqrt{(2E/N)}} \quad [\text{s}] \quad (6.5)$$

A related result, due to Manasse[4], gives the standard deviation of measurements of doppler frequency as

$$\delta\omega_d = \frac{1}{2\pi\sigma_t\sqrt{(2E/N)}} \quad [\text{Hz}] \quad (6.6)$$

Hence the accuracy of either type of measurement increases with increasing values of the SNR, E/N[†]. Good range accuracy is not dependent on short pulse lengths, but needs large bandwidth; good doppler accuracy needs long pulses.

An immediate question is whether we can *simultaneously* obtain good range and doppler accuracy. Any discussion of this must take account of a fundamental property of Fourier transform pairs: if $u(t) \leftrightarrow U(\omega)$ then

$$\sigma_\omega\sigma_t \geqslant 1/2 \quad (6.7)$$

This relation leads to the famous Heisenberg uncertainty principle of quantum mechanics. Applying it to the measure of accuracy given by Eqs (6.5) and (6.6), we find that

$$\delta\tau_d\,\delta\omega_d = \frac{1}{8\pi^2}\frac{1}{E/N}\frac{1}{\sigma_\omega\sigma_t} \leqslant \frac{1}{4\pi^2}\frac{1}{E/N} \quad (6.8)$$

Hence the product of the accuracies in range and velocity is inversely proportional to the signal-to-noise ratio. This has the important interpretation that both accuracies can be improved simultaneously and without limit by increasing the SNR. For fixed SNR, the product of the accuracies can be decreased by increasing the time × bandwidth product $\sigma_\omega\sigma_t$. This can be achieved by appropriate choice of waveform (see Sec. 6.8).

6.4 RESOLUTION

Accuracy tells us how reliably we can measure the parameters of a single target, but it tells us nothing about our ability to recognize that there are two targets present if their velocities and/or ranges are similar. For this we need the notion of resolution. We have seen in Sec. 4.7 that, after matched

---

[†] These two equations have already been encountered as the range error and doppler frequency error (in Hz) of Eqs (1.19) and (1.22).

filtering, a single stationary target would give rise to an output

$$A_0\rho_u(t - \tau_0) + m(t)$$

where $\rho_u(t)$ is the ACF of the transmitted waveform $u(t)$, and A_0 and τ_0 describe the amplitude and delay effects due to propagation and the target's RCS. If there were two targets present, the linearity of the system would cause the output to be

$$A_0\rho_u(t - \tau_0) + A_1\rho_u(t - \tau_1) + m(t)$$

Figure 6.1 shows the output for two targets with the same RCS for various values of $\tau_0 - \tau_1$, and a signal of the form $u(t) = \exp(-t^2)$. It is clear that the effect of having closely spaced returns is to smear the peak response. For τ_0 close to τ_1, it is difficult to decide whether there is a single or a double peak present in the noisy output. (In fact, for closely spaced targets there will only be a single peak since, near its maximum, each of the two signal terms can be approximated as a quadratic curve. The sum of quadratic curves is again quadratic.) The resolution of the system is a measure of how large the time difference $\tau_0 - \tau_1$ must be before the signal terms give rise to two distinguishable peaks, or meet some other condition by which we can decide if there is more than one target present.

Just as in the case of accuracy, there is no unique criterion by which we can define resolution, since it hinges on what is meant by distinguishability. Many different measures of resolution are therefore in use. Unlike measures of accuracy, they are independent of the noise, but are all defined in terms of properties of the waveform for which we are trying to tell whether there is a single copy present, or two closely spaced copies.

This waveform, after matched filtering, is the ACF $\rho_u(t)$. The ACFs of radar signals have certain general properties, which are useful in analysing the properties of resolution. These include:

1. $E = |\rho_u(0)| \geqslant |\rho_u(t)|$ for all t, i.e. the ACF is maximum at zero lag, and its value there is the signal energy.
2. For finite energy pulses, $\rho_u(t)$ will tend to 0 as t increases, though there may be subpeaks (ambiguity peaks) in addition to the peak at 0.
3. $\rho_u(t) \leftrightarrow |U(\omega)|^2$, i.e. the Fourier transform of the ACF is the energy spectrum of the waveform.

These properties of the ACF have led to such definitions of time resolution as the 3 dB width of $\rho_u(t)$, or the time over which the ACF drops to a value $(1/e)\rho_u(0)$, or the time to the first zero of the ACF. These have their uses for particular types of ACF. A measure of resolution that is more general and is better for analysis is

$$\Delta\tau_d = \frac{\int |\rho_u(t)|^2 \, dt}{\rho_u^2(0)} \qquad [\text{s}] \qquad (6.9)$$

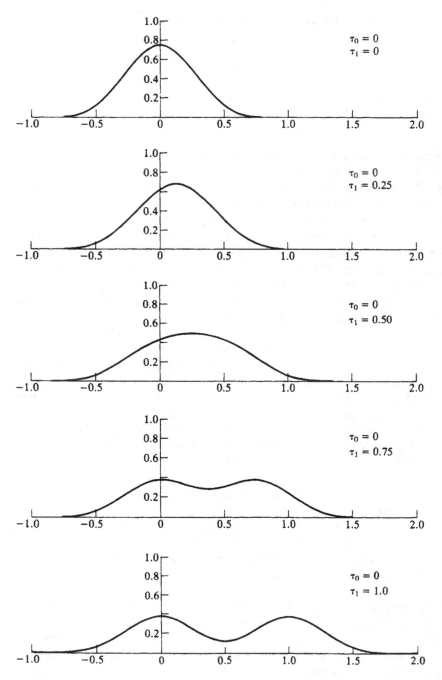

Figure 6.1 The output from the matched filter when two targets with range delays τ_0 and τ_1 are present, for different values of $\tau_0 - \tau_1$. Here τ_0 is set to 0 and τ_1 is marked on each successive plot.

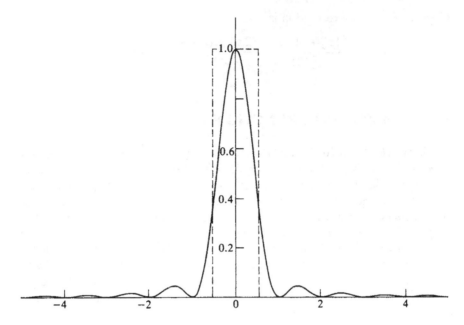

Figure 6.2 The relation of the equivalent-rectangle resolution to the area under the graph of $|\rho_u(t)|^2$ and its value at 0.

This definition has a very simple interpretation (see Fig. 6.2). It is the width of a rectangle of height $\rho_u^2(0)$ that has the same area as is under the curve $|\rho_u(t)|^2$. Hence it is sometimes called the equivalent-rectangle resolution. Using properties 1 and 3 of ACFs and Parseval's theorem, Eq. (6.9) may also be written as

$$\Delta\tau_d = \frac{\int |U(\omega)|^4 \, d\omega}{2\pi E^2} \quad [\text{s}] \tag{6.10}$$

This form of the definition also has a useful and instructive interpretation. If $U(\omega)$ took only the values 1 or 0, and the total length of the intervals in which it took the value 1 was F, then F could be considered as a measure of occupied bandwidth, or what Woodward[2] calls the *frequency span*. The energy of this signal would be $F/2\pi$, and hence for this signal $\Delta\tau_d = 2\pi/F$. This expresses the reciprocal relationship between resolution and bandwidth.

Similar definitions for the frequency resolution are possible. We first need the *frequency-domain ACF*

$$\rho_U(\omega) = \int U(v + \omega)U^*(v) \, dv \tag{6.11}$$

Then the frequency resolution is

$$\Delta\tau_\omega = \frac{\int |\rho_U(\omega)|^2 \, d\omega}{\rho_U^2(0)} = \frac{2\pi \int |u(t)|^4 \, dt}{E^2} \qquad [\text{rad s}^{-1}] \qquad (6.12)$$

6.5 THE AMBIGUITY FUNCTION

Equations (6.9) and (6.10) apply when the targets to be resolved are known to be stationary, while Eq. (6.12) applies when targets are at the same range. However, if we do not have such prior knowledge of the target character- istics then we need to worry about the *combined* effect of a shift in range and frequency. This can be analysed by considering the behaviour of the output of the correlation receiver when the input is doppler-shifted. (This means that the filter is not properly matched to the incoming signal, since it does not replicate its frequency behaviour.) If the transmitted signal is a complex modulation $u(t)$ of a carrier frequency ω_0 rad s^{-1}, and the target velocity causes a doppler shift in frequency of ω_d rad s^{-1}, then the correlation operation will give as output at time t

$$\int_{-\infty}^{\infty} u(s) \exp[j(\omega_0 + \omega_d)s] u^*(s-t) \exp[-j\omega_0(s-t)] \, ds \quad (6.13)$$

which can be rearranged as

$$\exp(j\omega_0 t) \int_{-\infty}^{\infty} u(s) u^*(s-t) \exp(j\omega_d s) \, ds \qquad (6.14)$$

The essential information about resolution is carried by the integral term in Eq. (6.14). Normalizing to a unit energy waveform $u(t)/\sqrt{E}$, this gives the quantity

$$\chi(t, \omega_d) = \frac{1}{E} \int_{-\infty}^{\infty} u(s) u^*(s-t) \exp(j\omega_d s) \, ds \qquad (6.15)$$

which is known as the *ambiguity function* of the transmitted waveform. It has an equivalent frequency-domain expression

$$\chi(t, \omega_d) = \frac{1}{2\pi E} \int_{-\infty}^{\infty} U^*(v) U(v - \omega_d) \exp(jvt) \, dv \qquad (6.16)$$

The cuts across the ambiguity function along the delay (t) and doppler (ω_d) axes are directly related to the time-domain and frequency-domain ACFs (and hence to the time and doppler resolutions), since

$$\chi(t, 0) = \rho_u(t)/E \qquad (6.17)$$

and

$$\chi(0, \omega_d) = \frac{1}{2\pi E} \rho_{\tilde{v}}^*(\omega_d) \tag{6.18}$$

To understand the role of the ambiguity function in a discussion of resolution, we need to consider the combined signal when there are multiple targets present. A target whose response (including doppler shift) is perfectly matched to the receiver will give rise to the output signal $\chi(t, 0)$. (For simplicity, we ignore amplitude effects on the pulse, and assume that the range delay of the returned pulse corresponds to time 0.) Another target of the same RCS but whose range delay and doppler frequency differ from the first target by τ_d and ω_d will give an output $\chi(t - \tau_d, \omega_d)$. Unless $\chi(t, 0)$ and $\chi(t - \tau_d, \omega_d)$ are significantly different, the two targets will be difficult to separate. In particular, if $\chi(0, 0)$ is close to $\chi(-\tau_d, \omega_d)$, it will be hard to recognize the presence of more than one target.

The properties of the ambiguity function will be clearer if we discuss an example. A single rectangular pulse of width T (see Table 6.1b) has an ambiguity function for which

$$|\chi(t, \omega_d)|^2 = 0 \qquad \text{if } |t| > T$$

and

$$|\chi(t, \omega_d)| = \left| \frac{2}{\omega_d} \sin\left(\frac{\omega_d}{2} (T - |t|) \right) \right| \frac{1}{T}$$

$$= \left(1 - \frac{|t|}{T} \right) \left| \text{Sa}\left(\frac{\omega_d}{2} (T - |t|) \right) \right| \qquad \text{if } |t| \leq T \tag{6.19}$$

where $\text{Sa}(x) = \sin(x)/x$. A sketch of $|\chi(t, \omega_d)|$ is given in Fig. 6.3, and Fig. 6.4 shows the cuts along the delay and doppler frequency axes. The important features of this ambiguity function are a central peak of height 1 at $(0, 0)$, which falls off linearly as $|t|$ increases and behaves as a $\sin(x)/x$ function as $|\omega_d|$ increases, surrounded by numerous subsidiary peaks. The way to interpret this figure is that two targets that differ in delay by t and in doppler frequency by ω_d will give nearly the same response and hence will be hard to resolve if $|\chi(t, \omega_d)|$ is nearly equal to $|\chi(0, 0)| = 1$.

Note that there are two conceptually different aspects of resolution involved here. Small values of t or ω_d correspond to points in the central peak that are not far removed from the true values for range and velocity. However, pairs (t, ω_d) lying on one of the subsidiary peaks may be far removed from the correct values of delay and doppler frequency while still giving a comparable response. These correspond to genuine ambiguities in interpretation of the response. Both aspects of resolution are wrapped up in the measures of resolution given by Eqs (6.9) and (6.12).

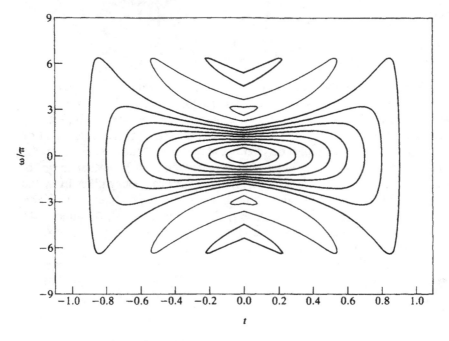

Figure 6.3 The modulus of the ambiguity function of a rectangular pulse ($T = 1.0$).

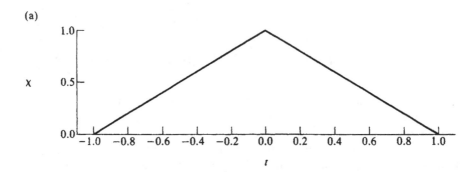

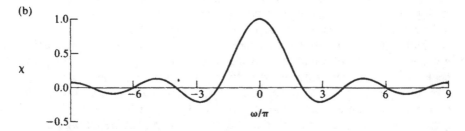

Figure 6.4 Cuts along (a) the delay and (b) the doppler axes of the ambiguity function plot of Fig. 6.3.

These general comments can be made more specific when we investigate the analytic properties of the ambiguity function. It is easy to see that

$$\chi(0,0) = 1 \qquad (6.20)$$

and (using a change of variables in the integral) that the ambiguity function has a rotational symmetry, expressed as

$$|\chi(t, \omega_d)| = |\chi(-t, -\omega_d)| \qquad (6.21)$$

Using the Cauchy–Schwartz inequality (Eq. (4.47)) we find that the modulus of the ambiguity function has its maximum at $(0, 0)$, since

$$|\chi(t, \omega_d)|^2 = \frac{1}{E^2} \left| \int_{-\infty}^{\infty} u(s) u^*(s - t) \exp(j\omega_d s) \, ds \right|^2$$

$$\leqslant \frac{1}{E^2} \int_{-\infty}^{\infty} |u(s)|^2 \, ds \int_{-\infty}^{\infty} |u(s - t) \exp(-j\omega_d s)|^2 \, ds$$

$$= \frac{1}{E^2} \int_{-\infty}^{\infty} |u(s)|^2 \, ds \int_{-\infty}^{\infty} |u(s - t)|^2 \, ds \qquad (6.22)$$

Each of the integrals on the right-hand side is equal to the energy of the signal, so that

$$|\chi(t, \omega_d)| \leqslant 1 \qquad (6.23)$$

Perhaps the most remarkable fact about the ambiguity function is that, if we view $|\chi(t, \omega_d)|^2$ as a surface, then the volume under this surface is always 1, irrespective of the shape of the waveform, i.e.

$$\frac{1}{2\pi} \int_{-\infty}^{\infty} \int_{-\infty}^{\infty} |\chi(t, \omega_d)|^2 \, dt \, d\omega_d = 1 \qquad (6.24)$$

This equation has very important consequences. It tells us that, while we are free to design the waveform $u(t)$ to give a very sharp peak in the ambiguity function at $(0, 0)$, the fixed height of this peak (Eq. (6.20)) implies that the volume of the central peak will then be only a small fraction of 1. Hence more ambiguity must appear away from the central peak, and targets well separated in range and velocity may become indistinguishable.

The squared ambiguity $|\chi(t, \omega_d)|^2$ may be thought of as a quantity of sand of total volume 1. The system designer is free to distribute the sand as he or she sees fit in the (t, ω_d) plane as long as the height at $(0, 0)$ takes the value 1 and the symmetry condition of Eq. (6.21) is met. However, all the sand must be used, so that a narrow peak near $(0, 0)$ can only be achieved by greater ambiguity further from the origin. As long as the ambiguities can be moved to regions of the (t, ω_d) plane where there is reason to believe that targets cannot be present, the limitations imposed by the ambiguity diagram can be escaped. As always, prior knowledge of what is expected can be used to improve performance.

6.6 EXAMPLES OF THE AMBIGUITY FUNCTION

As other examples of types of ambiguities that can occur, we now consider two waveforms, the gaussian pulse and a repeated rectangular pulse. More examples will be found in the later sections of this chapter.

The gaussian pulse

$$u(t) = \exp(-t^2/2T^2)$$

has ambiguity function

$$\chi(t, \omega_d) = \exp(j\omega_d t/2) \exp(-t^2/4T^2) \exp(-T^2\omega_d^2/4) \qquad (6.25)$$

so that $|\chi(t, \omega_d)|^2$ is constant on the elliptical curves

$$\frac{t^2}{4T^2} + \frac{\omega_d^2}{4/T^2} = \text{constant} \qquad (6.26)$$

A contour plot of this ambiguity function is shown in Fig. 6.5. There are no subpeaks, and all the 'ambiguity' arises from the spread of the central peak.

We now deal with a coherent pulsetrain containing M identical rectangular pulses of width T each separated by T_r, where $T_r > 2T$ (see Fig. 6.6). The ambiguity function for this waveform is comparatively

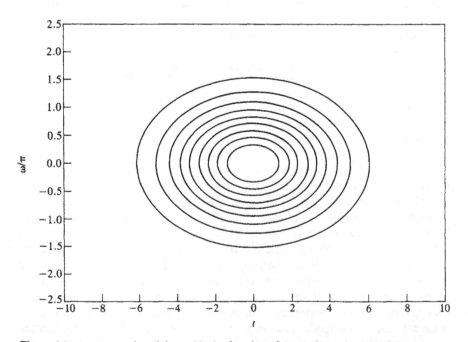

Figure 6.5 A contour plot of the ambiguity function of a gaussian pulse ($T = 2.0$).

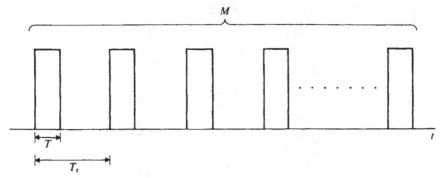

Figure 6.6 A coherent pulsetrain with PRF T_r containing M pulses of width T.

complicated, and has the form

$$|\chi_M(t,\omega_d)| = \frac{1}{M} \sum_{p=-(M-1)}^{M-1} \left| \frac{\sin[T_r(M-|p|)\omega_d/2]}{\sin(T_r\omega_d/2)} \right| |\chi_1(t-pT_r,\omega_d)|$$

$$(6.27)$$

where $\chi_1(t,\omega_d)$ is the ambiguity function of a single pulse, given by Eq. (6.19). A contour plot for the case $M = 3$ is given as Fig. 6.7, and has the 'bed of nails' structure characteristic of the ambiguity diagrams of pulsetrains. The system designer can control both T and T_r in order to ensure that targets of interest only occur near the central peak of this ambiguity diagram, giving enhanced range and doppler accuracy and effective resolution.

The cuts along the delay and doppler axes are given in Fig. 6.8. Along the delay axis Eq. (6.27) takes the form

$$|\chi_M(t,0)| = \sum_{p=-(M-1)}^{M-1} \left(1 - \frac{|p|}{M}\right)\left(1 - \frac{|t-pT_r|}{T}\right) \qquad (6.28)$$

if $|t - pT_r| < T$, and is zero everywhere else. The term in the second parentheses corresponds to a triangular peak of base width $2T$, with its centre at pT_r. The term in the first parentheses corresponds to a triangular weighting. This is shown in Fig. 6.8a. The triangles are spaced at intervals T_r, and the condition $T_r > 2T$ is to prevent these triangles overlapping; if they do so, the expression in Eq. (6.28) becomes more complicated.

Along the doppler axis, Eq. (6.27) has the form

$$|\chi_M(0,\omega_d)| = \frac{1}{M} |\,\text{Sa}\,(T\omega_d/2)| \left| \frac{\sin(MT_r\omega_d/2)}{\sin(T_r\omega_d/2)} \right| \qquad (6.29)$$

Since $MT_r > T$, the Sa term varies less rapidly than the second term (which is periodic with period $2\pi/T_r$), so that we can regard Fig. 6.8b as a modulation of the second term by Sa$(T\omega_d/2)$.

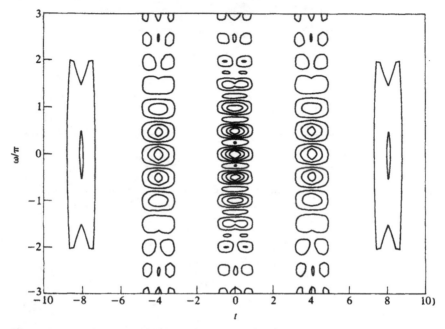

Figure 6.7 A contour plot of the ambiguity function for the pulsetrain shown in Fig. 6.6 ($M = 3$, $T = 1.0$, $T_r = 4.0$).

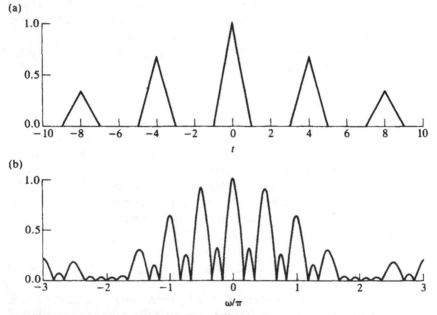

Figure 6.8 Cuts along (a) the delay and (b) the doppler axes of the ambiguity function plot of Fig. 6.7.

6.7 PULSE COMPRESSION

The radar designer's ideal waveform would give good performance as regards the following:

- Target detection
- Range and doppler accuracy
- Range and doppler resolution

These requirements appear to be incompatible. Detectability of targets is dependent on the total energy of the illuminating waveform, which is processed by the matched filter. Good range accuracy and resolution require large energy and high bandwidth, which, for simple pulses, implies short pulses. Short-duration and large-energy pulses require very large peak powers, which may not be available, and short pulses imply poor doppler resolution. Fortunately, there is an escape from these apparently contradictory requirements and the need for large peak powers, by using more sophisticated waveforms.

An essential ingredient of such waveforms is the use of long pulses requiring a reasonable peak power to obtain good doppler resolution and the high energy needed for good detection performance. Range resolution is obtained by designing the waveform shape to give high bandwidth, normally by using frequency modulation. Because the matched filtering 'compresses' the long pulse to an ACF of short duration at the output of the receiver, this type of radar processing design is known as *pulse compression*. Both range and doppler accuracies are good for this form of processing, since the requirements of large bandwidth and pulse duration are both met. The question of resolution is more complicated. If targets are known to be at the same range, then there is good velocity resolution. Similarly, if targets are known to have the same velocity, then there is good range discrimination. If there is no such *a priori* information, and we need to separate targets that are of unknown relative range and velocity, then we cannot escape the mandates of the ambiguity diagram. In order to illustrate this, we consider the effect of linear frequency modulation (FM).

6.8 CHIRP

The simplest form of frequency modulation is linear FM, i.e. a pulse described by

$$u(t) = a(t) \exp(j\pi k t^2) \tag{6.30}$$

where $a(t)$ is a pure amplitude modulation. The instantaneous frequency is found by differentiating the phase, to give

$$\omega(t) = 2\pi k t \qquad [\text{rad s}^{-1}] \tag{6.31}$$

which is clearly linear. The total frequency deviation during the pulse (which we can consider as a reasonable approximation to the bandwidth) is therefore $2\pi kT$. Plots of the in-phase part of the signal and frequency as a function of time for a rectangular pulse with linear FM are shown in Fig. 6.9. This form of pulse is popularly known as *chirp* because of the sound made by a signal of this type at audio frequencies. Lightning strikes also give rise to chirp signals at radio frequencies. Emissions from a lightning strike begin as a very compressed pulse. Different frequencies propagate at slightly different speeds (a phenomenon known as *dispersion*), so that at great distances a long falling tone is heard on a radio receiver. Pulse compression is analogous to receiving this signal and correcting for the dispersion, in order to recover the pulse generated by the lightning strike.

A chirp signal has an ambiguity function of the form

$$|\chi(t, \omega_d)| = \begin{cases} \left|\left(1 - \dfrac{|t|}{T}\right) \text{Sa}\left(\dfrac{(T - |t|)(\omega_d + 2\pi k t)}{2}\right)\right| & \text{if } |t| \leqslant T \\ 0 & \text{otherwise} \end{cases} \tag{6.32}$$

The shape of the ambiguity function is not apparent from this equation, but contours are plotted in Fig. 6.10, and cuts along the delay and doppler axes are shown in Fig. 6.11. The most obvious feature of the ambiguity diagram

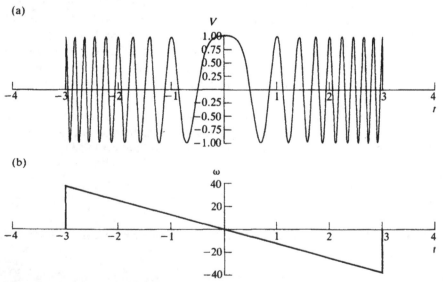

Figure 6.9 (a) The in-phase component of a linear FM signal and (b) the instantaneous frequency of this signal as a function of time.

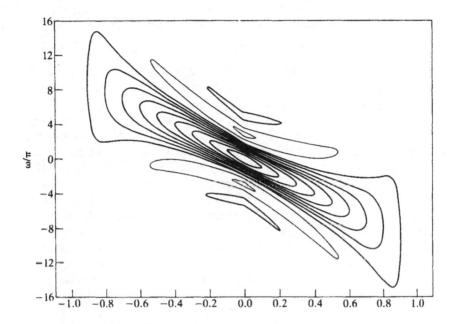

Figure 6.10 A contour plot of the ambiguity function of a linear FM pulse ($T = 1.0, k = 5$).

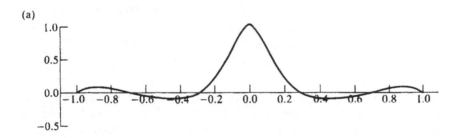

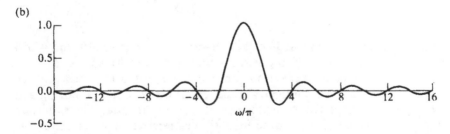

Figure 6.11 Cuts along (a) the delay and (b) the doppler axes of the ambiguity function plot of Fig. 6.10.

is that the principal ambiguities are distributed around the line

$$\omega_d + 2\pi kt = 0 \qquad (6.33)$$

This implies that a misinterpretation of the range of a target will also lead to a misinterpretation of its velocity, since the peak in the ambiguity diagram at a given range has an associated velocity that depends on the range.

The cuts through the ambiguity diagram parallel to the two axes convey no idea of this structure, but confirm our original expectations about the range and doppler resolutions *taken in isolation*. In fact, it can be shown that for a large time × bandwidth product, i.e. $kT \gg 1$, the first null in the cut along the time axis occurs at approximately $1/kT$, which is the reciprocal of the pulse bandwidth. The gain in range resolution by using frequency modulation is clear when Fig. 6.11a is compared with Fig. 6.4a. For the rectangular pulse without frequency modulation, the cut along the delay axis extended to T before becoming 0. If we regard the distance to the first zero along the delay axis as a measure of range resolution, then the resolution has been improved by a factor kT^2, which is known as the *compression ratio*. As we would expect, the compression ratio is directly proportional to k. There is no improvement in the doppler resolution.

A striking example of the ambiguity inherent in a chirp pulse often occurs in synthetic aperture radar (SAR) images. SAR uses a side-looking pulsed radar carried on a moving platform (normally an aircraft or satellite) to generate images of the earth's surface (see Chapter 11). In the simplified configuration of Fig. 6.12, we can see that a stationary scatterer has a doppler frequency relative to the platform given by

$$f_d = \frac{2V \sin \theta}{\lambda} \approx -\frac{2Vx}{\lambda R} \qquad [\text{Hz}] \qquad (6.34)$$

for small θ. The minus sign is present because, in our system of coordinates, if $\theta > 0$, the distance x is negative (see Fig. 6.12). Since $x = Vt$ (taking the scatterer at the origin of the x axis, and measuring time from the instant when the scatterer is broadside to the platform)

$$f_d = -\frac{2V^2}{\lambda R} t \qquad [\text{Hz}] \qquad (6.35)$$

This is a linear FM signal like that described by Eq. (6.31), for which $k = -2V^2/\lambda R$. The SAR processor compresses this signal, and for a stationary scatterer the peak response will occur in the right place in the image relative to other stationary scatterers. If the scatterer is moving and has a velocity component along the line of sight to the radar, the associated doppler shift causes the response from the scatterer to be a cut across the ambiguity diagram parallel to the delay axis, but moved up or down. As we can see from Fig. 6.10, this moves the maximum response to a later or earlier

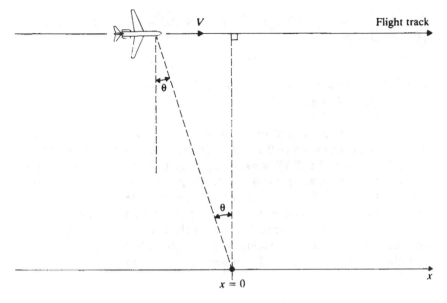

Figure 6.12 A simplified view of the geometry of a SAR system.

time. The time error, using Eq. (6.33), is

$$t_e = -\frac{\omega_d}{2\pi k} = \frac{R\lambda}{2V^2}f_{du} \quad [\text{s}] \qquad (6.36)$$

where we have used the notation f_{du} to indicate the excess doppler frequency uncompensated for in the SAR processing. Since the maximum response occurs at the wrong time, the scatterer will be misplaced in the image. The positional error corresponding to the time error of Eq. (6.36) is

$$Vt_e = \frac{R\lambda}{2V}f_{du} \quad [\text{m}] \qquad (6.37)$$

Because of this, SAR images often show effects such as ships displaced from their wakes, or cars apparently in the middle of fields instead of on the road along which they are travelling. The magnitude of the displacement (which is in the along-track direction, even though caused by cross-track motion) can sometimes be used to estimate the velocities of moving scatterers.

Worked example A ship with speed 10 km h^{-1} is travelling on a bearing 95°. It is imaged by an air-borne X-band SAR travelling due west at 200 km h^{-1} at a range of 40 km. How far will the ship apparently be displaced from its wake in the image?

SOLUTION Since $f_{du} = 2V_r/\lambda$, where V_r is the radial velocity, the displacement is given by

$$d = R(V_r/V) \qquad [m]$$

Here $V_r = -20 \cos 85° \approx -1.74 \text{ km h}^{-1}$, so that the displacement is approximately 348 m.

COMMENTS It is clear that quite small doppler shifts can cause large apparent displacements in the image. In fact, the calculation is complicated by the fact that SAR is sampling the signal. Since the bandwidth of the FM signal used in the along-track processing is typically only a few hundred hertz for an air-borne SAR (see Chapter 11), even modest cross-track velocities can move the scatterer out of the frequency band used in the processing. The scatterer may then be aliased (or signal may be lost). Calculating the ensuing displacement effects requires detailed examination of the way the SAR processing is being carried out.

6.9 PHASE CODING

Another form of pulse compression is one in which a long pulse of duration T is made up of M contiguous subpulses each of length T/M. The subpulses each have their own modulation, which can be in frequency (e.g. Costas[5]) or in phase. We only discuss the latter. A general formula for the ambiguity function of these types of waveforms is given in Skolnik[6] (equation 175 of Chapter 3).

There are numerous schemes for such phase coding, but the simplest are the Barker codes[7]. In these codes, the phase is either 0 or π, so that the transmitted signal is effectively a binary sequence taking the values 1 and -1. The choice of such a sequence is constrained by ambiguity requirements, which means that the ACF of the sequence should have low sidelobes. The Barker codes are those sequences for which the sidelobes at zero doppler do not exceed $1/N$. Only nine such sequences are known, the longest being of length $N = 13$. These are given in Table 6.2, where $+$ represents 1 and $-$ represents -1. A contour plot of the ambiguity function for the Barker code of length $N = 7$ is shown in Fig. 6.13, and the corresponding cut along the axis of zero doppler is shown in Fig. 6.14a. We also give in Table 6.3 the sequence of outputs along the delay axis when the code and the replica it is being correlated with are displaced by an exact multiple of T/M. (In Fig. 6.14a all values of the lag are considered, not just the multiples of T/M.) Though the sidelobes are larger than would be desired, the form of the ambiguity plot is approaching the ideal 'thumb-tack' ('drawing-pin') form, i.e. a sharp central peak surrounded by a comparatively flat plateau. (The

Table 6.2 The Barker codes (here + corresponds to 1 and − to −1)

M												
2	+	+										
2	−	+										
3	+	+	−									
4	+	+	−	+								
4	+	+	+	−								
5	+	+	+	−	+							
7	+	+	+	−	−	+	−					
11	+	+	+	−	−	−	+	−	−	+	−	
13	+	+	+	+	+	−	−	+	+	−	+	−	+

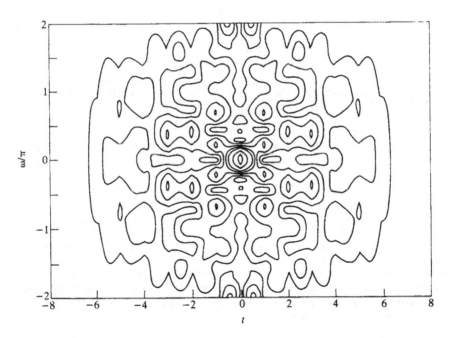

Figure 6.13 A contour plot of the ambiguity function of the Barker code of length 7 ($T = 1.0$, $N = 7$).

plateau must be there, to meet the requirements of the total ambiguity having unit volume.) The cut along the doppler axis is shown in Fig. 6.14b.

A form of phase coding that does not suffer from the restricted length constraints (and associated sidelobe levels) of the Barker codes is known as Frank coding[8]. These codes are of length M^2, and can be thought of as M sub-sequences each of length M. Each sub-sequence starts with zero phase.

(a)

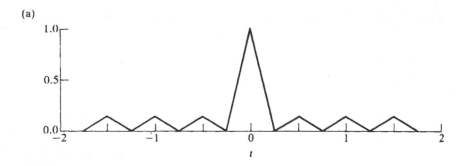

(b)

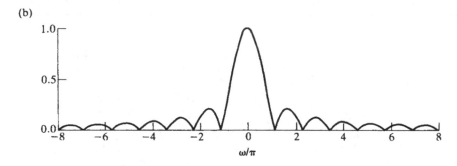

Figure 6.14 Cuts along (a) the delay and (b) the doppler axes of the ambiguity function plot of Fig. 6.13.

Table 6.3 The output sequence for a Barker code of order 7

Code	0	0	0	1	1	1	−1	−1	1	−1	0	0	0
τ_d	−6	−5	−4	−3	−2	−1	0	1	2	3	4	5	6
Output	−1	0	−1	0	−1	0	7	0	−1	0	−1	0	−1

The output is obtained by placing a copy of the code, displaced τ_d places to the right, under the code (zero displacement is when the code and its copy are exactly aligned), multiplying componentwise and summing the products.

In the first sub-sequence, all the phases are 0; in the second, the phase of successive pulses increases by $2\pi/M$; in the third, it increases by $4\pi/M$; and so on (see Fig. 6.15 for the phases of the Frank code of length 16). The average phase change in the pth sub-sequence is $2\pi(p-1)/M$ radians. Since average rate of phase change is a measure of frequency, this means that the frequency of the coded pulse increases linearly with p. Hence we might expect the signal to display some of the properties of a linear FM signal. This is borne out by the contour plot of the ambiguity function of the Frank code of length 16 shown in Fig. 6.16. The central feature of this plot is a diagonal

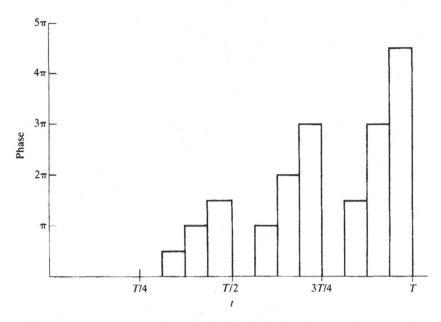

Figure 6.15 The phase changes associated with the Frank code of length 16.

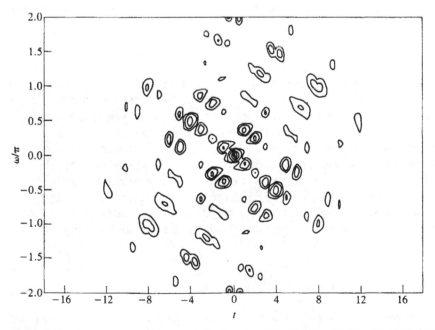

Figure 6.16 A contour plot of the ambiguity function of the Frank code of length 16 ($T = 1.0$, $M = 4$).

(a)

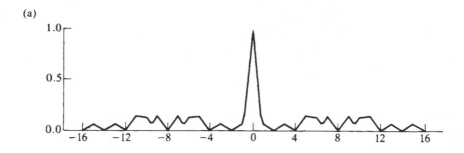

(b)

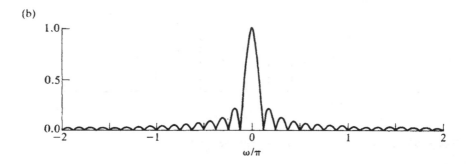

ω/π

Figure 6.17 Cuts along (a) the delay and (b) the doppler axes of the ambiguity function plot of Fig. 6.16.

ridge similar to that in the ambiguity plot of linear FM (Fig. 6.10). There are also parallel ridges (such ridges also occur for linear FM, but at a lower level, and hence are lost by the contour levels used to generate Fig. 6.10).

The cut along the delay axis corresponding to Fig. 6.16 is shown in Fig. 6.17a. The peak sidelobe level is $\sqrt{2}/16$. This demonstrates that the sidelobe levels of the Frank codes do not fall as rapidly as $1/M$, the rate achieved by the Barker codes. However, there is in principle no restriction on the length of a Frank code. For large Frank codes, it can be shown that the peak sidelobe level declines as $1/(\pi\sqrt{M})$, so that arbitrarily low sidelobe levels can be obtained. The cut along the doppler axis is shown in Fig. 6.17b.

6.10 SUMMARY

Detection performance is optimized and range and doppler accuracy are both improved by maximizing the SNR. Since this needs the transmitted pulse to have high energy, long pulses are required unless a very high-power transmitter is available. Long pulses also permit good doppler accuracy, but, for simple pulses, give poor range accuracy. The system designer can avoid this apparent dilemma by using phase-modulated pulses. These can be constructed to give the large bandwidth that, after pulse compression, leads to good range accuracy. Whatever transmitted waveform is chosen, the resolution constraints imposed by the ambiguity function cannot be escaped. All that the designer can do is to attempt to move the significant areas of ambiguity into regions of the range–doppler plane where targets are unlikely to be present.

Key equations

● Effective bandwidth:

$$\sigma_\omega^2 = \frac{\displaystyle\int_\omega^\infty \omega^2 |U(\omega)|^2 \, d\omega}{2\pi E} = \frac{\displaystyle\int_{-\infty}^\infty |u'(t)|^2 \, dt}{E} \qquad [\text{rad}^2 \, \text{s}^{-2}]$$

● Effective pulse duration:

$$\sigma_t^2 = \frac{\displaystyle\int_{-\infty}^\infty t^2 |u(t)|^2 \, dt}{E} = \frac{\displaystyle\int_{-\infty}^\infty |U'(\omega)|^2 \, d\omega}{2\pi E} \qquad [\text{s}^2]$$

● Accuracy of delay measurement:

$$\delta\tau_d = \frac{1}{2\pi\sigma_\omega\sqrt{(2E/N)}} \qquad [\text{s}]$$

● Accuracy of doppler frequency measurement:

$$\delta\omega_d = \frac{1}{2\pi\sigma_t\sqrt{(2E/N)}} \qquad [\text{rad s}^{-1}]$$

● Product of range delay and doppler accuracies:

$$\delta\tau_d \, \delta\omega_d \leqslant \frac{1}{4\pi^2} \frac{1}{E/N} \qquad [\ ]$$

● Equivalent-rectangle time resolution:

$$\Delta\tau_d = \frac{\int |\rho_u(t)|^2 \, dt}{\rho_u^2(0)} = \frac{\int |U(\omega)|^4 \, d\omega}{2\pi E^2} \qquad [\text{s}]$$

● Equivalent-rectangle frequency resolution:

$$\Delta\tau_\omega = \frac{\int |\rho_U(\omega)|^2 \, d\omega}{\rho_U^2(0)} = \frac{2\pi \int |u(t)|^4 \, dt}{E^2} \qquad [\text{rad s}^{-1}]$$

● Ambiguity function:

$$\chi(t, \omega_d) = \frac{1}{E} \int_{-\infty}^{\infty} u(s)u^*(s - t) \exp(j\omega_d s) \, ds$$

$$= \frac{1}{2\pi E} \int_{-\infty}^{\infty} U^*(v)U(v - \omega_d) \exp(jvt) \, dv$$

● Properties of the ambiguity function:

$$\chi(0, 0) = 1$$

$$|\chi(t, \omega_d)| = |\chi(-t, -\omega_d)|$$

$$|\chi(t, \omega_d)| \leqslant 1$$

$$\frac{1}{2\pi} \int_{-\infty}^{\infty} \int_{-\infty}^{\infty} |\chi(t, \omega_d)|^2 \, dt \, d\omega_d = 1$$

6.11 REFERENCES

1. *The Fourier Transform and its Applications*, R. N. Bracewell, McGraw-Hill, New York, 1986.
2. *Probability and Information Theory, with Applications to Radar*, P.M. Woodward, Pergamon Press, Oxford, 1953.
3. A theory of radar information, P.M. Woodward and I. L. Davies, *Phil. Mag.*, **41**, 1001, 1950. [Most of this material is covered in reference 2, but not the details leading to the relation between SNR and accuracy given as Eq. (6.5).]
4. Range and velocity accuracy from radar measurements, R. Manasse, *MIT Lincoln Lab. Report*, 312–326, 1955. [This is the original reference, but there is some discussion in reference 6.]
5. A study of a class of detection waveforms having nearly ideal range–doppler ambiguity properties, J.P. Costas, *Proc. IEEE*, **72**, 996–1009, 1984.
6. *Radar Handbook*, M. Skolnik, McGraw-Hill, New York, 1970.
7. Group synchronizing of binary digital systems, Barker, R.H., in *Communication Theory*, Ed. W. Jackson, Academic Press, New York, pp. 273–287, 1953.
8. Polyphase codes with good nonperiodic correlation properties, R.L. Frank, *IEEE Trans. Information Theory*, **IT-9**, 43–45, 1963.
9. Pulse compression techniques with application to HF probing of the mesosphere, C.A. Gonzales and R.F. Woodman, *Radio Sci.*, **19**, 871–877, 1984.

6.12 PROBLEMS

6.1 What is the equivalent rectangle resolution of the output of a receiver matched to (a) a rectangular and (b) a gaussian pulse? What is the frequency resolution of a gaussian pulse?

6.2 Sketch the output of a receiver matched to rectangular pulses of length T [s], if there are two equal-amplitude targets present separated by range delays of (a) $T/4$, (b) $T/2$, (c) T, (d) $3T/2$ and (e) $2T$ [s].

6.3 Sketch the output from a receiver matched to zero doppler and a rectangular pulse of length T [s] if there are two targets simultaneously present, both with the same RCS and range but one of which has zero doppler and one has doppler frequency $2\pi/T$ (for simplicity, not reality, use $T = 1$).

6.4 Compare the resolutions of the two pulses of problem 4.2.

6.5 A book on radar makes the following statement: 'M pulses, each with duration t_p, can be viewed as a single pulse of duration Mt_p. Since bandwidth B is related to pulselength t by $B = 1/t$, this combined pulse will have a bandwidth f_B/M, where f_B is the bandwidth of a single pulse.'

Consider a radar transmitting a coherent amplitude-modulated pulsetrain $u(t)$ as shown below.

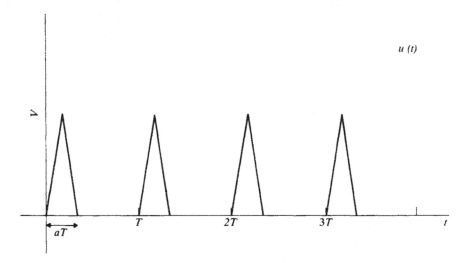

How does the effective bandwidth of this pulsetrain compare with the same quantity for a single pulse? Show that the amplitude of the Fourier transform of this pulsetrain satisfies

$$|U(\omega)| = \left|\frac{\sin(2T\omega)}{\sin(T\omega/2)}\right| |P(\omega)|$$

where $P(\omega)$ is the Fourier transform of a single pulse. Sketch $|P(\omega)|$, $|\sin(2T\omega)/\sin(T\omega/2)|$ and $|U(\omega)|$. Do you agree with the statement that heads this problem? Are there any hidden assumptions in it?

6.6 An air-borne C-band (6 cm) SAR travelling at 300 km h^{-1} carries a 1 m antenna. It produces an image in which a ship is apparently displaced by a distance $R\beta/4$ m in the along-track direction at range R, where β is the beamwidth of the real antenna. What can we infer about the ship's velocity?

6.7 Find the output from the binary codes:
    (a) $1, 1, 1, -1$;
    (b) $1, 1, -1, 1$;
    (c) $1, 1, 1, -1, 0, 0, 0, 0, 1, 1, -1, 1$.

This is an example of a *complementary* code. Sequence (c) is constructed from a pair of shorter binary codes whose sidelobes cancel each other, and which are separated by (at least) the length of the shorter codes. Complementary codes give complete cancellation of the range sidelobes (in the absence of noise) in the vicinity of the main lobe. They are used when very low sidelobes at zero doppler are important, and are widely exploited in mesosphere–stratosphere–troposphere (MST) radars (e.g. Gonzales and Woodman[9]).

Find all the complementary pairs of length 2, and show that there are no complementary pairs of length 3.

SEVEN

SECONDARY SURVEILLANCE RADAR

- What secondary surveillance radar is, and what it does
- Advantages over primary radar
- Problems with secondary surveillance radar
- The future

Secondary surveillance radar is one of the main tools used in air traffic control.

7.1 INTRODUCTION

Secondary surveillance radar (SSR) is not a true radar system at all but a two-way communication system between an interrogator on the ground and transponders fitted to aircraft, which reply automatically. We include SSR here because the system is very similar to radar in the way it operates, suffers from many of the classical radar problems and is widely used throughout the world, often in conjunction with primary surveillance radar.

The origins of SSR lie in the 'identify, friend or foe' (IFF) systems of World War II; a signal was transmitted from the ground towards a suspect aircraft, which was required to reply with the appropriate code or be treated as a foe. Modern SSR works in a similar way; interrogation messages are transmitted on a one-way uplink frequency of 1030 MHz and cooperating aircraft reply on a one-way downlink frequency of 1090 MHz. The replies are fed to a plot extractor, which decodes the aircraft identity and height and passes them on to the air traffic controllers together with the measured range and bearing. A classic work on secondary surveillance radar is Stevens[1].

Secondary surveillance radar cleverly avoids problems with clutter through the use of two frequencies, because the receivers at either end of the link are not in tune with the adjacent transmitter and they do not pick up unwanted echoes. Another advantage of SSR is that the transmitter and antenna gain requirements for the uplink and downlink are much more modest than for a primary surveillance radar operating over the same range—this is because the combined R^2 propagation losses of each one-way link are much lower than the R^4 losses of the two-way radar signal path.

Worked example In the past, the International Civil Aviation Organization (ICAO) specification for SSR has limited the peak effective radiated power (ERP = transmitter power × aerial gain) to 52.5 dB W. What is the uplink range of the system if the transponder needs to receive a signal of -101 dB W for adequate probability of detection of a single pulse? Assume that the transponder antenna on the aircraft has no gain, that propagation losses amount to 2 dB and that a further allowance of 3 dB must be made because the aircraft does not remain at the peak of the azimuth antenna pattern for the entire interrogation time.

What ERP would the primary surveillance radar designed in Chapter 2 need in order to detect the same aircraft if its RCS were 20 dB m^2?

SOLUTION Using the same notation as in Chapter 1, the power density per unit area at the range R of the aircraft is given by

$$\text{Power flux} = \frac{P_t G_t}{4\pi R^2} \quad [\text{W m}^{-2}]$$

This is intercepted by an antenna with an effective area of

$$A_e = \frac{G_r \lambda^2}{4\pi} \quad [\text{m}^2]$$

So the power received by the transponder is

$$P_r = \frac{P_t G_t G_r \lambda^2 L}{(4\pi R)^2} \quad [\text{W}]$$

We can rewrite this to give the maximum operating range R_{max} as

$$R_{max} = \left(\frac{P_t G_t G_r \lambda^2 L}{P_r (4\pi)^2} \right)^{1/2} \quad [\text{m}]$$

Putting in values:

$P_t G_t$	52.5	[dB W]	
G_r	0	[dB]	Nominal omnidirectional antenna
λ^2	-10.7	[dB m^2]	$=0.291$ m
L	-5	[dB]	Total losses
$1/P_r$	101	[dB W^{-1}]	
$1/(4\pi)^2$	-22	[dB]	

R^2_{max}	115.8	[dB m^2]	
R_{max}	616 km		(300 nautical miles)

In practice, the maximum range would probably be less than this because the aircraft would not necessarily be in the centre of the antenna elevation pattern and receiving the full 52.5 dB W illumination.

In comparison, the power needed by the primary radar can be worked out using Eq. (2.11) as follows:

$$P_t G_t = \frac{(\text{SNR}) \times N \times (4\pi)^3 \times R^4}{G_r \times \lambda^2 \times \sigma \times L_s} \quad [\text{W}]$$

$$= +13\,[\text{dB}] - 145.8\,[\text{dB W}] + 33\,[\text{dB}] + 231.6\,[\text{dB m}^4]$$

$$- (+36\,[\text{dB}] - 12.7\,[\text{dB m}^2] + 20\,[\text{dB m}^2] - 5\,[\text{dB}])$$

$$= 93.5\,\text{dB W}$$

Assuming 36 dB as the antenna contribution to the ERP, then the transmitter power would need to be over 500 kW.

COMMENTS The range performance of the SSR that we have just calculated is quite sufficient, for two reasons. First, most aircraft are below the horizon when they are 600 km away; and secondly, they would almost certainly be in a different air traffic control zone.

Primary and secondary radar are often used together (sometimes the antennas are even attached and rotate together); the primary system is used to provide air traffic controllers (ATC) with a 'map' on a plan position indicator (PPI) display of everything moving in the sky in their region. The SSR interrogates each target, usually with the request: 'Who are you and what height are you at?' Cooperative targets, such as all civil airliners and most private and military planes, reply with the information requested, which is then displayed in alphanumeric form at the appropriate place on the PPI display. In today's crowded airlanes, this information is valuable, for after an aircrew have filed their flight plans, including the flight identification number, the aircraft can be tracked automatically along its route without the need for ATC requests for the plane to identify itself. Such requests would be necessary if the only information were from primary radar.

7.2 BASIC PRINCIPLES

The interrogation of an aircraft by SSR involves a pair of pulses modulating the 1030 MHz carrier frequency. The pulses are labelled P_1 and P_3 and the spacing between them determines the information requested from the aircraft (see Fig. 7.1). Different requests are known as 'modes' of operation and, although there are quite a few, mostly military, there are two modes used more frequently than the others because they are common to both civil and military aeroplanes; these are modes 3/A and C:

P_1–P_3 spacing	Request	Mode
8 μs	Identify	3/A
21 μs	Height?	C

Aircraft reply to this interrogation with a train of pulses that are 0.45 μs wide and spaced 1.45 μs apart and which are used to modulate the 1090 MHz downlink carrier. The first and last pulses in this train are always present and are known as framing pulses F_1 and F_2 (Fig. 7.2). In between F_1 and F_2 are 12 pulses, which may or may not be present, depending on the message

Figure 7.1 The interrogation message is determined by the separation of the pulses P_1 and P_3. Pulse P_2 is used for sidelobe suppression.

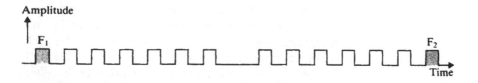

Figure 7.2 The form of the reply from the transponder on the aircraft.

being transmitted, and an extra centre pulse, which is not currently used. The 12 pulses are used as a 12-bit code, which has 4096 possible combinations —enough to give the aircraft identity when in mode 3/A. In mode C, one of the pulses is not used, but the remaining 2048 codes are sufficient to give the aircraft height in steps of 100 feet. When both height and identity are required, the requests are made alternately, which is known as *mode interlacing*. There is plenty of time to interlace modes A and C because, as the SSR antenna rotates, the beam illuminates an aircraft for about 30 ms and during this time its transponder is interrogated about 15 times.

Normally, an airline pilot will select mode A/C and the transponder is programmed to respond to requests from the ground without crew intervention. The aircraft altimeter automatically feeds height information to the transponder. There are also four other switches the aircrew can set, which send out additional messages for unusual situations such as a hijacking, radio communications failure or if ATC are having problems identifying the aircraft.

Secondary surveillance radar is used to locate aeroplanes in a similar way to primary radar. Ranges are measured by the round-trip time of a pulse travelling up to the aircraft and returning, with appropriate allowances for the delays in the equipment on the aircraft and on the ground. The accuracy of the range measurement can be improved by using correlation methods on the entire received pulsetrain, rather than using just one pulse. The aircraft height is known from the information given by its own altimeter, and so the remaining piece of information required to fix its position is the azimuth, as measured at the SSR.

Azimuths are now measured using the monopulse technique described in Chapter 3, and this provides much greater accuracy than earlier methods. The importance of monopulse for SSR is that it provides a measurement of angle on every pulse, so that when a train of pulses is received, these measurements can be combined to form an improved estimate of the azimuth (in general, n independent measurements can be combined to improve the accuracy by a factor of $\sqrt{n}$). A typical SSR antenna is about 8 m wide, giving an azimuth beamwidth of about 2.5°, but after monopulse processing the error in the azimuth of the aircraft is as low as a few minutes of arc. Because the antenna is not required to measure elevation angles (the aircraft height being already known), the avertical size of older SSR antennas has often been only about 0.4 m, giving vertical beamwidth approaching 50°. These long, thin parabolas have given rise to the descriptive term 'hogtroughs'.

Modern SSR antennas are usually flat phased arrays having vertical dimensions of the order of 1.6 m to give improved control over the elevation pattern, and these have become known as LVAs, large vertical apertures. However, they remain several times wider than they are tall, and to some extent still retain the long, thin appearance of the old hogtroughs.

7.3 PROBLEMS WITH SECONDARY SURVEILLANCE RADAR

The advantages of secondary surveillance radar over primary radar are perhaps fairly obvious; less transmitter power is required, all aircraft give the same amplitude response independent of their size and there are no RCS fluctuations to worry about. It is also less expensive and conveniently identifies the aircraft. But SSR is not without problems of its own, although fortunately most of these now have engineering solutions.

One problem is caused by the sidelobes of the ground antenna initiating the transponder response, as shown in Fig. 7.3. The radar system does not know that the echo has come from a sidelobe transmission, and so incorrectly plots the aircraft position as being in the centre of wherever the main beam happens to be pointing. This *sidelobe interrogation* can be resolved by transmitting an extra interrogation pulse P_2 on a separate antenna (this is the mystery pulse shown on Fig. 7.1 and not referred to thus far). This new 'control' antenna is simpler than the main antenna and is designed to have an omnidirectional pattern and lower gain; however, it still has more gain than the sidelobes of the main antenna (Fig. 7.3) because of the rule that the gain integrated all the way round an antenna must come to unity.

On the aircraft, the transponder is programmed to reply to main-beam interrogations, in which case the amplitudes of P_1 and P_3 should be equal and larger than P_2, which has come from the control antenna. However, if

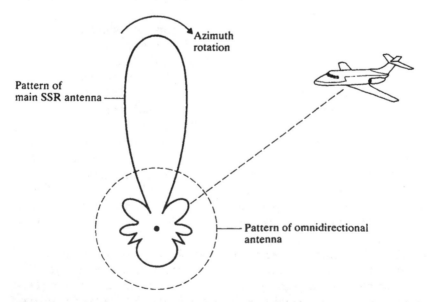

Figure 7.3 SSR main-beam omnidirectional antenna patterns. Aircraft at close range may be interrogated by a sidelobe of the antenna as well as by the main beam.

P_2 exceeds P_1 or P_3, then the request must have come from a sidelobe of the main antenna and the transponder makes no reply.

The idea of using a control beam is an example of good engineering. Rather than spending a great deal of money trying to improve sidelobe performance of the main antenna, the problem can be solved by adding an extra low-budget antenna. Most modern SSR systems do not in fact have a separate control antenna—the control beam is synthesized by the main antenna, so that both main beam and control beam suffer equal multipath fading (see next section), and amplitude comparisons between the two remain valid.

Other problems and solutions involving SSR systems are as follows:

- *Fruit* (false replies unsynchronized in time) occurs when the transponder on an aircraft is triggered by one SSR and the signal is received by another SSR that was not expecting a reply. The fact that fruit replies arrive at the receiver at arbitrary times means that they have random ranges and are relatively easy to filter out.
- *Garbling* is a good descriptive term for overlapping replies from two aircraft. Garbling can occur because a fruit reply happens to occur at the same range as a true reply, in which case the problem quickly resolves itself as the fruit range changes. However, garbling can also occur because two aircraft are at similar ranges (this does not mean that they are in danger of collision, for they may be at different altitudes and azimuths but are at the same radial distance from the radar). Before monopulse, garbling was a serious problem, but now the azimuthal accuracy of monopulse can be used to assign a bearing to each pulse received, thus enabling the two intermixed pulsetrains from the aircraft to be resolved.
- Other difficulties with SSR include *co-channel interference* (because all systems work on the same frequency and may interfere with each other), *capture* when one system monopolizes an aircraft's transponder and causes another system to lose data, and *false replies* in which a transponder may be accidentally interrogated twice because the transmitted signal reaches the aircraft by an additional path such as by reflection from a large building, see reference 2 for more details. A familiar example of radio signals being reflected from buildings is the 'ghost' image sometimes seen on a television picture; the main image is the direct signal, and a weaker signal arrives slighter later after bouncing off a nearby block of flats or other large structure.

The problem of radio signals arriving at the receiver by more than one route and corrupting the information carried is common to most forms of radar and to radio communications in general. It is usually known as *multipath* and frequently involves reflections from the ground as well as buildings.

7.4 MULTIPATH

Figure 7.4 shows a radio signal travelling from a radar system on a tower up to an aircraft; some of the transmitted energy travels by the shortest path and some is reflected from the ground and appears to come from a source beneath the ground. These two radio sources act in the same way as Young's slits in optics, and give sum and difference interference patterns. In fact, a version of the Young's slit experiment using a single source of light and a mirror to generate an apparent second source is called Lloyd's mirror, and is an even better analogy of the multipath case.

When reflected from the ground, the radio signal undergoes a phase change, which is nominally 180°. If the difference in path length between the direct and reflected signal is $\lambda/2$, an interference maximum is formed and the signal arriving at the aircraft is larger than expected. Moments later, however, the path difference might change to λ and almost complete cancellation of the signal occurs. This results in wild fluctuations of amplitude and phase, which are worse at low elevation angles. The problem is the same for systems receiving radio signals and for those transmitting, and it can cause severe problems for tracking radars and SSR systems.

One way of thinking about these interference fringes is to consider them as breaking up the vertical antenna pattern into a series of narrow lobes, as shown in Fig. 7.5. A simple way to demonstrate the reality of these lobes is to use a radar in receive-only mode to watch the sun setting (the sun is quite a powerful radio source). At high elevations, the sun appears as a constant-amplitude radio source, but as it nears the horizon, amplitude fluctuations become apparent, and then finally become severe as the sun sets (the sun is sometimes used in this way to measure the vertical radiation pattern of a radar antenna). An aircraft flying horizontally at low elevation passes through the lobes of the antenna, just as the sun does, and the signal level varies, causing the radar system to keep losing track of the target.

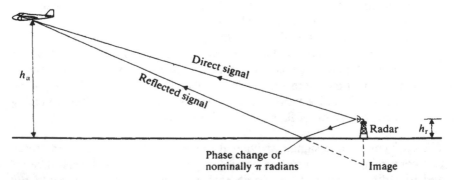

Figure 7.4 Propagation over a plane reflecting surface results in two signal paths that interfere with one another.

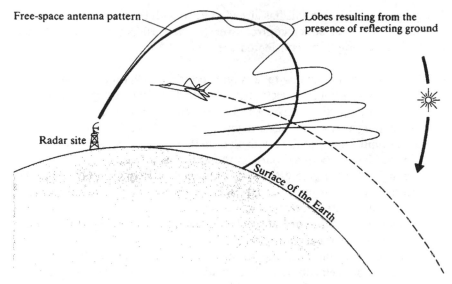

Free-space antenna pattern

Lobes resulting from the presence of reflecting ground

Radar site

Surface of the Earth

Figure 7.5 The presence of a reflecting surface causes the antenna pattern to break up into lobes at low elevation angles.

Perhaps the first attempt at solving the multipath problem might be to try to make use of the time delay between the direct and the reflected signals, but the time difference is usually too small. This can be shown with a simplified calculation using the geometry of Fig. 7.4 with a few assumptions, such as the Earth being flat and the range R of the aircraft being large compared with the height of the radar antenna h_r. A little elementary geometry can be used to show that the difference in range ΔR between the direct and indirect wave is given by

$$\Delta R \simeq 2h_r h_a / R \qquad [\text{m}] \qquad (7.1)$$

where h_a = height of the aircraft. Putting some typical numbers into Eq. (7.1) for example $h_r = 10$ m, $h_a = 1000$ m (~ 3000 feet altitude) and $R = 100\,000$ m (100 km range), gives $\Delta R = 0.2$ m. This is equivalent to a difference in time of 0.7 ns between the reception of the direct and reflected signals, and is far too small in comparison with SSR pulse lengths to form a practical method of separating them. This same difficulty is true of digital communications in general.

Other attempts to deal with multipath include placing scattering structures on the ground to break up the reflected wave and even painting offending structures with radar-absorbing paint, as developed for *Stealth* aircraft applications. The most promising methods of reducing the effect of multipath probably lie in improved antenna design and more advanced signal and data processing. The challenge is to design an antenna with high gain at low elevation angles but which avoids 'looking' at the ground at zero elevation.

The process of achieving this has already begun with the new LVA antennas, which have a much better control over the vertical beam pattern. Other techniques that might be considered are:

- Locking the antenna with a slight positive elevation to reduce the sensitivity to ground reflections, but with the disadvantage that it also reduces the signal-to-noise ratio for low-elevation aircraft.
- Developing an antenna pattern with a deep null that is directed towards the ground—the problem with this method is that the elevation of the ground changes with azimuth as the antenna rotates.
- Arranging for the greatest rate of change of beamshape to lie along the direction of the ground, rather than a null, to give the largest possible difference in output between low-elevation signals.
- Using two antennas, one above the other, and correlating their outputs to separate the multipath signals.
- To use existing antennas and put more effort into rejecting data showing unreasonable behaviour, such as wild variations in range or azimuth, and trying to form tracks with the remaining data. Information from other sensors such as radar, infrared systems and optical tracking systems might be added into the tracking process at this point, especially on military systems.

7.5 MODE S AND THE FUTURE

For all its association with the world of radar, SSR is essentially a digital communications system, and a new mode of operation known as *mode S* is coming into service to exploit this aspect more fully, see Scanlan[3]. Longer pulsetrains are transmitted to increase the information transfer, and the interrogation of aircraft is *selective* in that the ground base *addresses* them one at a time.

The first task of a new SSR system is to find out which of the aircraft it is interrogating are equipped with mode S transponders. This identification is achieved by the use of a fourth interrogation pulse P_4, which can be given a duration of either 0.8 or 1.6 μs. Because pulses P_1 to P_3 are transmitted as before, the older mode A/C transponders continue to reply as normal. A mode S transponder examines the duration of P_4; if it is set to 0.8 μs the transponder makes no reply, but if $P_4 = 1.6\ \mu$s then it replies by giving its own unique address. These addresses are noted by the SSR system, which then schedules the mode S transponders for special interrogation.

The mode S interrogation proper is contained within a long P_6 pulse, which is modulated by differential phaseshift keying (DPSK—a standard digital communications technique) to convey either a 56- or 112-bit word depending on whether the length of P_6 has been set to 16.25 or 30.25 μs.

These bits are used for selectively addressing the mode S transponders. Aircraft fitted with the older transponders do not reply to this new type of transmission because, cleverly, the first two pulses P_1 and P_2 are set to the same amplitude, which older transponders treat as a sidelobe interrogation and therefore ignore.

Mode S transponders reply in a similar manner to the interrogation; a four-pulse preamble is followed by a 56- or 112-bit binary word. Several of these words may be strung together in successive transmissions to enable quite complex messages to flow between aircraft and the ground.

7.6 SUMMARY

Secondary surveillance radar is partly a communication system between aircraft and air traffic controllers on the ground; a limited amount of information (aircraft height and flight identification number) is requested by an interrogator on the ground and automatically supplied by a transponder on the aircraft. In the future, this flow of information will increase.

Secondary surveillance radar also acts as a radar system because the position of the aircraft is found by measuring the range (from the time delay between interrogation and reply) and the azimuth, as measured by an antenna on the ground. Many of the early problems with SSR have now been solved, and the system is in widespread use throughout the world.

7.7 REFERENCES

1. *Secondary Surveillance Radar*, M.C. Stevens, Artech House, Norwood, MA, 1988. [The classic book on modern SSR; it is clear and full of detail.]
2. *Understanding Radar*, H.W. Cole, BSP Professional Books, Oxford, 1985. [Contains more than 60 pages of information on SSR in a very readable form.]
3. *Modern Radar Techniques*, Ed. M.J.B. Scanlan, Collins, Glasgow, 1987. [Chapter 6, also written by H.W. Cole, is dedicated to modern SSR.]

7.8 PROBLEMS

7.1 The Cossor Condor 9600 is a complete ATC system that includes full monopulse SSR. One of the SSR antenna options is the Condor 9642 large vertical aperture, which has a peak gain of 27 dB and a beamwidth of 2.45° at the −3 dB points.

If the power of the Cossor interrogation transmitter is adjusted so that the ERP conforms to the ICAO specification of 52.5 dB W when it is used in conjunction with the 9642 antenna, would you expect an aircraft transponder to receive an adequate signal at the maximum instrumented range of 256 nautical miles? Assume additional propagation losses of 2 dB and an antenna pattern loss of 3.5 dB because the elevation angle of the aircraft places it above the angle of maximum antenna gain.

7.2 For the downlink the ICAO defines the transponder output power as 24 dB W. For an aircraft at a range of 100 nautical miles, what signal strength would the Condor 9600 system receive? Assume 1 dB loss for atmospheric attenuation and a further 7 dB for system and vertical antenna pattern losses. Remember that the downlink frequency is not the same as the uplink frequency used in the previous question.

7.3 If the minimum working signal level of the interrogation system on the ground is −110 dB W, what is the theoretical maximum downlink range of the Condor system? Allow 3 dB in your calculations for azimuth beamwidth loss.

7.4 Why does the downlink apparently give better performance than the uplink if the transmitter is less powerful?

7.5 Estimate roughly the angular uncertainty that might be expected when the Condor system tracks an aircraft with a range of 100 nautical miles. Assume a receiver noise level of −130 dB W.

EIGHT

PROPAGATION ASPECTS

- The radar horizon
- The effect of the atmosphere
- The effect of the ionosphere
- Diffraction effects

How to cope with radio waves not travelling in straight lines in the atmosphere.

8.1 INTRODUCTION

Radars frequently operate through the atmosphere of the earth, often at low elevation angles where most targets occur, but where there are most problems for the radar. Aircraft flying at constant altitude towards a surveillance radar appear over the curvature of the earth and are difficult to detect and track. Similar problems occur with ship surveillance when using maritime radar. When close to the horizontal, radar beams have their greatest path length through the atmosphere, which is itself at its most dense and turbulent. Low-elevation radar beams can also encounter obstacles, such as hills, and become diffracted into the shadow regions behind.

In this chapter we are going to investigate the radio horizon and discover whether the atmosphere, the ionosphere and the terrain significantly affect the propagation of radio waves. Again, elementary optics is useful to describe what is happening.

8.2 THE RADAR HORIZON

Assuming for the time being that radio waves travel in straight lines through the atmosphere, the radio horizon is defined by a line tangent to the surface of the earth, as in Fig. 8.1. If R is the range of a point on the radio horizon (or of a target appearing at zero elevation), h is the height of the point and R_e the radius of the earth, then

$$R^2 = (R_e + h)^2 - R_e^2$$
$$= 2R_e h - h^2 \quad [m^2]$$

It is reasonable to assume that $h^2 \ll 2R_e h$ and therefore

$$R \simeq \sqrt{(2R_e h)} \quad [m] \tag{8.1}$$

Although the earth is not truly spherical, being flattened at the poles and bulging slightly at the equator, there is little error in assuming it to be a

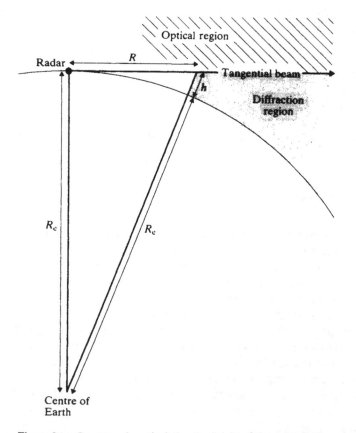

Figure 8.1 Geometry for calculating the height of the radar horizon as a function of range.

sphere with an average value of $R_e = 6378$ km. Many radar engineers have to work with aircraft height in feet and range in nautical miles. (A nautical mile is the distance around the earth's surface corresponding to an angle of one minute measured at the centre of the earth. It equals about 6080 feet or 1.85 km; see Appendix III.) Restating Eq. (8.1) in nautical miles [n. mile], but with h in feet, gives

$$R \simeq \sqrt{(2 \times 3444h/6080)} \qquad [\text{n. mile}]$$

and a rough approximation that can prove useful is

$$R_{\text{n. mile}} \sim \sqrt{h_{\text{feet}}} \qquad [\text{n. mile}] \tag{8.2}$$

If the radar is at an altitude h_{radar}, Eq. (8.1) can be modified as

$$R \simeq \sqrt{(2R_e h_{\text{radar}})} + \sqrt{(2R_e h_{\text{target}})} \qquad [\text{m}] \tag{8.3}$$

Worked example A radar situated at sea level is approached by a missile flying at a height of 10 m, a low-flying aircraft at a height of 100 m and a high-flying bomber at an altitude of 15 000 m. At what ranges will these targets appear over the radar horizon and be detected by the radar, again assuming that radio waves travel in straight lines?

SOLUTION Using Eq. (8.1) and $R_e = 6378$ km, we get the following answers: Sea-skimming missile is detected at a range of 11.3 km. Low-flying fighter is detected at a range of 35.7 km. High-flying bomber is detected at a range of 437 km.

COMMENTS This quick calculation reveals several interesting aspects of radar; low-altitude missiles are very difficult to detect until the last minute or so of their incoming flight. Even then their RCS is often so small that they are hard to identify and track, especially if they are weaving. At the higher altitudes, few civil aircraft are likely to fly above 15 000 m, so there is little need to design air traffic control radars for ranges greater than about 450 km.

Beware, however—the calculations above are pessimistic because radio waves tend to be refracted over the horizon by the earth's atmosphere, thereby extending detection ranges. We will examine the effects of the atmosphere next.

8.3 ATMOSPHERIC EFFECTS

The atmosphere can cause radio waves to be dispersed, attenuated, refracted and retarded. *Dispersion* is small enough to be ignored, except in a few specialist wideband systems.

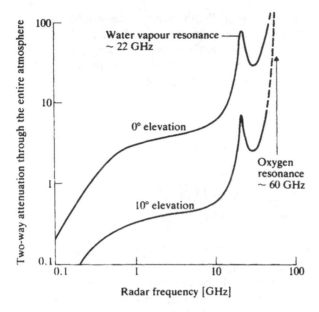

Figure 8.2 Two-way atmospheric absorption through the entire troposphere as a function of radio frequency for two elevation angles.

Attenuation by the atmosphere is a function of frequency and, as Fig. 8.2 shows, it can become quite significant above about 10 GHz, and even below, if long ranges are surveyed at low elevation angles. Atmospheric attenuation decreases with air pressure (and therefore with altitude), and most of the absorption occurs in the troposphere—the lowest, turbulent region of the atmosphere where the weather is found. The attenuation also increases fairly linearly with increasing rainfall rate (usually measured in millimetres per hour, $mm\,h^{-1}$) or increasing snowfall, and radars susceptible to significant atmospheric losses must make use of local weather statistics in their performance predictions.

Refraction by the atmosphere is a serious problem for most radar systems and occurs because of the refractive index profile of the atmosphere. The refractive index of air depends on temperature, pressure and the water vapour content. Of these factors, it is the partial pressure of the water vapour that has the largest influence and the total air pressure that is least critical.

The refractive index of air n is always close to unity; it is about 1.0003 at the earth's surface and falls even nearer to unity with increasing height. The small differences of n from unity are important, so they are deliberately magnified by introducing the *radio refractivity* N defined by

$$N = (n - 1) \times 10^6 \quad [\quad] \tag{8.4}$$

Hence N has a value of about 300 at the surface. The radio refractivity is usually expressed in N units, and can be calculated from:

$$N = 77.6 \left(\frac{p}{T} \right) + 3.73 \times 10^5 \left(\frac{e}{T^2} \right) \quad [\quad] \qquad (8.5)$$

where p = air pressure [millibars], T = temperature [K] and e = partial pressure of water vapour [millibars].

The troposphere changes more quickly in the vertical dimension than horizontally, and the important parameter is therefore the change of refractive index with height, known as the *refractive gradient* dN/dh. Under normal atmospheric conditions this gradient is negative, roughly linear below about 1 km, and causes radio signals to be bent downwards with a radius of curvature ρ given by (see e.g. Meeks[1]):

$$\rho = \left(-\frac{1}{n} \frac{dn}{dh} \cos \phi \right)^{-1} \qquad [m]$$

$$\simeq \left(-\frac{1}{n} \frac{dn}{dh} \right)^{-1} \qquad \text{at low elevation} \qquad (8.6)$$

where ϕ = elevation angle of radar beam.

As Fig. 8.3 shows, the effect of the refractive index decreasing with altitude is to give the target an apparent height greater than its true height. The angle α through which the beam bends can be calculated from the following equation, and is positive for a downward-bending beam[2]:

$$\alpha = \int_{radar}^{target} \cot \phi \, \frac{dn}{n} \qquad [\text{radians}] \qquad (8.7)$$

When radio waves travel in curves, much of the simple geometry used in radar becomes very difficult. A useful trick, much used by radar engineers, is to replace the radius of the earth R_e by a larger number kR_e. This increased

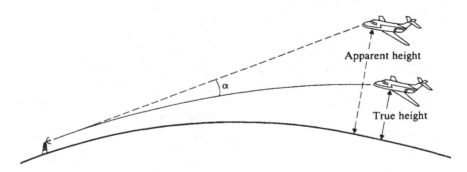

Figure 8.3 Atmospheric refraction causes the apparent height of a target to be greater than its true height.

radius of the earth has the effect of bending the radio waves back up into a straight line over this artificial earth, and allows us to return to our simple geometry provided we remember to replace R_e by kR_e every time it occurs in the calculations. But what value should we choose for k? Geometrically k is given by

$$k = \rho/(\rho - R_e) \qquad [\quad] \qquad (8.8)$$

Combining this transform with Eqs (8.4) and (8.6), and assuming $\cos \phi \simeq 1$ for low elevation angles, gives

$$k = \left(1 + 10^{-9}R_e \frac{dN}{dh}\right)^{-1} \qquad [\quad] \qquad (8.9)$$

using dN/dh in units of N per kilometre (the conventional units) and R_e in m. You might see Eq. (8.9) used with a factor of 10^{-6} instead of 10^{-9}; this happens when R_e is expressed in kilometres.

For a standard atmosphere, $dN/dh = -40\ N/km$, giving $k = 1.255$, but engineering practice has been to use the approximation $k = 4/3$, and the expression 'four-thirds earth' is commonplace. When calculating radar performance there is special paper available for calculating the vertical coverage that is drawn on a four-thirds earth (see Fig. 8.4). If you follow

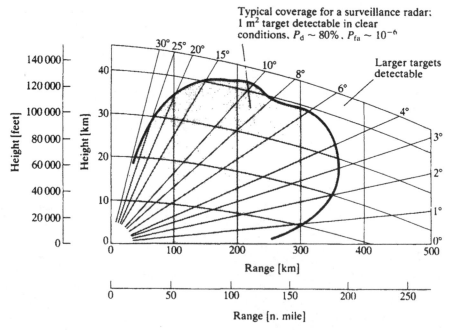

Figure 8.4 A typical graph for plotting radar vertical coverage, using the four-thirds earth model.

the curve of a target flying at an altitude of 10 km, it crosses the horizon at a range of over 400 km; without refraction and the four-thirds earth correction, Eq. (8.1) gives a range of about 357 km.

Worked example Assuming four-thirds earth, at what ranges could the targets described in the worked example in Section 8.2 be detected?

SOLUTION Using Eq. (8.1), but with R_e replaced by $\frac{4}{3}R_e$ we now have: Sea-skimming missile is detected at a range of 13.0 km. Low-flying fighter is detected at a range of 41.2 km. High-flying bomber is detected at a range of 505 km.

For accurate tracking of radar targets, the four-thirds approximation is insufficient and an exponential formula is used for the variation of N with height (see Rotherham in reference 3, for example):

$$N = N_s \exp(-h/H) \qquad [\quad] \tag{8.10}$$

where N_s = surface value of N [   ] and H = scale height of the atmosphere [m, if h is in m]. The CCIR reference atmosphere[4] gives the values $N_s = 315$ and $H = 7.35$ km, and worldwide maps of N_s are to be found in Bean and Dutton[2].

In practice, there can be large day-to-day variations in the refractive index structure of the atmosphere. Sub- or super-refractive layers can occur in which the refractive gradients are respectively less or greater than the standard atmosphere. At low elevation angles (0–2°) radar beams can become trapped in super-refracting ducts close to the earth's surface. When $dN/dh = -157\,N/$km, the radius of curvature becomes equal to R_e, and greater values cause radio signals to bend back to earth (Fig. 8.5).

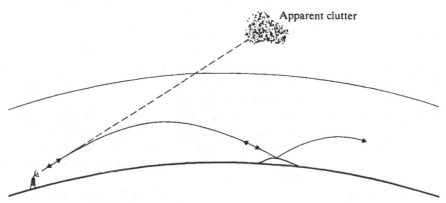

Figure 8.5 Super-refracting ducts can cause radar signals to bend back to earth, creating anomalous clutter problems.

Ducting is usually caused by temperature inversions on land, or by evaporation at sea, and it can permit substantial over-the-horizon propagation when conditions are favourable. This is exploited in communications, but is too lossy and intermittent to form the basis of a reliable over-the-horizon radar system. It does, however, cause some radars to observe ground clutter at greater ranges than usual. Some types of temperature inversion can also cause sub-refraction, so that the radio wave is bent towards the earth less strongly than usual. Sub-refraction can result in various forms of anomalous propagation (anoprop), including the possibility of the radio wave entering an elevated duct formed between layers in the atmosphere having different temperatures.

Atmospheric refraction effects occur at all radar frequencies, but in general they begin to affect systems operating above 500 MHz before affecting those working at longer wavelengths.

Retardation is the effect by which radio waves travel more slowly in air than in free space. Usually the range error that this causes is small enough to be ignored, but there are cases, such as the battlefield radar described later in this chapter, where it becomes necessary to make corrections. The range error depends on the value of n integrated over the path length R, but if we can define some average value $\bar{n}$ then

$$\Delta R = (\bar{n} - 1)R \qquad [\text{m}] \qquad (8.11)$$

Assuming a linear refractivity profile gives

$$\Delta R = 10^{-6} R \left(N_0 + \frac{(h_t - h_r)}{2000} \frac{dN}{dh} \right) \qquad [\text{m}] \qquad (8.12)$$

where N_0 = value of N at the radar site [   ], h_r = height of the radar [m] and h_t = height of the target [m]. (This formula is sometimes expressed with h in kilometres, in which case the divisor 2000 is reduced to 2.)

8.4 DIFFRACTION BY THE TERRAIN

The phenomenon of diffraction was discovered by the Jesuit, Francesco Maria Grimaldi, who described the experiments leading to its discovery in his book *De Lumine*, published in Bologna in 1665. It was Grimaldi who gave the phenomenon the name 'diffraction' from the Latin verb *diffringere = dis + frangere*, meaning 'to break in different directions'. The name is a good description of the effect, which is important at radio wavelengths, because it can sometimes make radar detection possible in the shadow regions behind hills.

There are well established mathematical models and computer programs that describe diffraction over simple obstacles such as knife-edges and cylinders, and the usual engineering approach is then to try to describe real

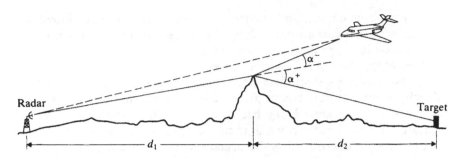

Figure 8.6 The geometry used to evaluate diffraction effects.

terrain in terms of these obstacles. Fortunately, most hills and ridges can be described quite well by single and multiple diffraction edges or by cylinders with knife-edges on top of them. Procedures exist for determining the outcome when there are several different types of these obstacles in the path of the radio wave.

The simplest and most rigorously analysed diffraction obstacle is the knife-edge, shown in Fig. 8.6. Various texts describe the mathematics; see for example Griffiths[5] or Meeks[1] (which includes useful program listings). Although a full mathematical analysis is complicated, the results may be summarized, using the notation shown in Fig. 8.6, as follows:

- Diffraction losses generally increase with frequency f, but for the important case of grazing incidence the diffraction loss is independent of frequency and is about 6 dB. At large diffraction angles, the loss has a $10 \times \log(f)$ dependence.
- The minimum loss occurs when the obstacle is midway between the radar and the target, rather than near either end of the radio path.
- Increasing the diffraction angle α^+ increases the diffraction losses, so increasing the distance between the radar and the obstacle helps to reduce the losses by lowering α^+.
- Rounded hills, modelled as cylinders, behave in a similar manner to knife-edges but with additional losses. Often, though, a hill that is geologically smooth is broken by trees, scrub and terrain irregularities, so that the diffraction edge model remains a good assumption.
- Diffraction from a knife-edge upwards can modify the free-space field by interference, but as α^- increases, this loss quickly falls to zero.

8.5 BATTLEFIELD RADAR SYSTEMS

Radars used on the battlefield for the detection or guidance of weapons are interesting in many respects, not least of which is the sophisticated command

and control communication networks that link them together. However, weapon location radars are particularly interesting because of their sensitivity to variations in the propagation conditions.

One of the most lethal weapons of World War II was the mortar, and radar systems for locating them were soon developed. The task is not so difficult, because mortar bombs are usually fired at a high elevation; they may be detected against a sky background and their flight path can be measured and predicted backwards to find the source from which they came. Unfortunately, modern artillery shells, and rocket barrages fired from mobile launch vehicles, follow flatter trajectories, which makes them harder to detect because of ground clutter. The low trajectory also forces radars to use low-elevation beams, and the geometry is such that small errors in the measurement of the ballistic flight path can cause large errors in the calculation of the weapon site (Fig. 8.7).

Modern artillery duels are deadly affairs fought at ranges of up to a maximum of about 35 km. The objectives of weapon-locating radars are to detect enemy shells or rockets as soon as they rise from the ground, calculate the position of the source and then call for fire by friendly forces on that location. Generally speaking, a gun battery has insufficient fire-power to saturate an area much larger than a football pitch, and an enemy weapon must therefore be located with an accuracy of about 100 m on the ground. Because of the flat trajectories, this ground error can translate to an accuracy of a few metres along the flight path of the shells.

For a microwave weapon-location radar operating at a maximum range of 35 km, the correction in the height of the target due to refraction at low elevation angles can be calculated by integrating Eq. (8.7) from the radar to the target to give the vertical displacement from straight-line propagation. For a standard atmosphere with $dN/dh = -40\ N/\text{km}$, the height correction is about 25 m and is roughly independent of elevation. However, if super-refraction occurred with $dN/dh = -120\ N/\text{km}$, then the correction needed

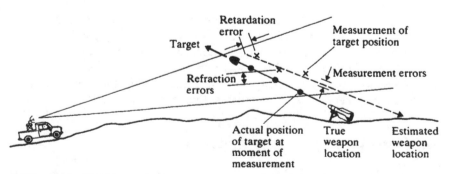

Figure 8.7 Small errors in the measurement of a ballistic trajectory can cause large errors in the location of the source.

would be nearer 75 m and, if a standard atmosphere had been assumed, an error of up to 50 m in the calculated height of the missile would be introduced. The correction to the range due to atmospheric retardation can be calculated from Eq. (8.12) and would be around 10 m in the worst cases. Although this is smaller than the height correction, it remains sufficiently large to cause problems if the wrong atmospheric model is assumed. Failure to locate enemy gun positions accurately can have fatal consequences, not only for the friendly forces but for the radar itself, which will quickly be located (by passive radio monitoring devices) and placed high on the enemy list of target priorities.

There is a need for good meteorological information on a battlefield in order that the flight of projectiles can be predicted accurately and to compensate for the effects of atmospheric refraction on the radar performance. At present, atmospheric data are best supplied by radiosonde balloon systems similar to those used by the meteorological office. Often, though, only surface observations are available, and a standard atmospheric profile is assumed. Even radiosonde data are of limited value because balloons can usually be flown only over friendly territory. Also, balloon instrumentation does not respond to very sharp refractive index changes. New laser, radar and passive sounding systems are being developed to measure temperature, pressure and humidity, and in the future these may be found on the battlefield in support of radar location systems.

8.6 IONOSPHERIC EFFECTS

So far, we have discussed the effects of the neutral atmosphere on radar systems operating at low elevation angles. When radar systems look out into space, or from space down to Earth, the free electrons in the ionosphere—the ionized part of the atmosphere—may play a significant role in radio wave propagation. This can be viewed as a problem if the radar is looking for an undisturbed propagation channel between itself and the target. It can also be turned to advantage as a means of studying the ionosphere from the ground. In fact, ground-based radars play a major role in ionospheric research. Incoherent scatter radar measures the minute accounts of power backscattered by electrons in the ionosphere when a high-power electromagnetic (EM) wave passes through it. Systems such as the European Incoherent Scatter Radar Facility (EISCAT) in Tromso, Sweden, have revolutionized our understanding of ionospheric processes in the last decade. Other systems, such as the PACE radar in Antarctica, and the SABRE and STARE systems in northern Europe, measure the backscatter from irregularities in the ionosphere, and use these to study electric fields and plasma processes.

The ionosphere begins at a height of 60 km and continues up to about 1000 km. It was one of the first targets to be detected by early pulsed radar,

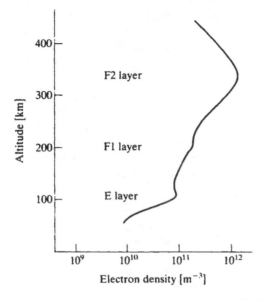

Figure 8.8 The variation of electron density with altitude in the day-side ionosphere.

which, in 1924, revealed the presence of several ionized layers, as shown in Fig. 8.8. The electron content in these layers depends on the solar energy falling on the atmosphere, so has latitudinal, diurnal, seasonal and long-term variations, as well as sudden changes due to outbursts of solar activity. (The structure of these layers is discussed further in Chapter 10.) There are also belts surrounding the magnetic poles in both the northern and southern hemispheres where energetic particles (mainly electrons and protons) bombard the atmosphere. This particle precipitation (and other related processes) gives rise to the visible aurora, so these regions are known as the auroral ovals. It also causes significant modification of ionospheric structure and electron content. These regions can be particularly troublesome for radar systems. The electron content normally varies irregularly, with blobs and holes giving rise to backscatter that could be mistaken for targets. Normally there are also strong electric fields. These cause the irregularities to have velocities and associated doppler shifts that may be comparable to those expected from targets. Hence, extraction of targets from auroral clutter can be a difficult problem.

The effect of the ionosphere on radio waves passing through it is to cause attenuation, refraction and a rotation of the plane of polarization known as Faraday rotation. Reflections from the ionosphere form a special case of refraction, which plays a vital role in skywave over-the-horizon radar. This is discussed separately in Chapter 9. Diffraction effects may also be important when the ionosphere is spatially irregular, such as occurs most of the time at auroral latitudes, and in the post-sunset equatorial region.

Attenuation by the ionosphere increases with increasing ionization, but is inversely proportional to frequency squared. Above a few hundred megahertz, the effect is small. For frequencies exceeding 2 MHz, the attenuation can be estimated very roughly by[6]

$$\text{Two-way loss} = A/f^2 \quad [\quad] \tag{8.13}$$

where f is the radar frequency [Hz] and A [s^{-2}] is a constant, which is of the order of 4×10^{-8} in the daytime and 2×10^{-10} at night. More detailed theory can be found in Hall and Barclay[3], Picquenard[7] and Scanlan[8].

Refraction is the slowing of the wave velocity as it passes through a medium, which causes the direction of travel of a plane wave to alter. The bending of the rays gives rise to the problem you encounter when you try to pick up a coin at the bottom of a pool. If you do not correct for the tilting of the wave-normal to a steeper angle (see Fig. 8.9a), you will touch the pool bottom too far away. For the ionosphere, it is more like having a slab of material that displaces the line of sight to the object (Fig. 8.9b). This aspect of refraction leads to positional errors in the horizontal plane.

The slowing of the wave causes errors in the apparent distance to the target. The refractive index n is given by

$$n^2 \approx 1 \pm 80.6(N_e/f^2) \quad [\quad] \tag{8.14}$$

where N_e is the electron density [electrons m^{-3}] and f is given in hertz. The minus sign is used when we are considering phase velocity, and the plus sign when we are interested in the velocity of propagation of signals. The refractive index tells us by what factor the wave is slowed compared to the speed of light. To find the increase in time for a signal propagating through the ionosphere, we must integrate the refractive index along the ray path. For frequencies in excess of 30 MHz, the ensuing one-way range error is given by

$$\Delta l = \int_s n \, ds \approx \frac{40.3}{f^2} \int_s N_e \, ds \quad [\text{m}] \tag{8.15}$$

where s is distance along the path S from the radar to the target. The integral of electron density along the path is called the *total electron content* (TEC), for obvious reasons. It varies considerably with position and time, but generally does not exceed 5×10^{17} electrons. This value of TEC would give rise to a one-way range error of 900 m at 150 MHz, which is the frequency used, for example, by the TRANSIT positioning system. More problematically, variations in TEC would cause this error to vary between a few metres and hundreds of metres unless they were properly corrected.

Worked example The ray path from a space-based radar operating at 1 GHz must pass through an ionosphere for which the TEC is equally

(a)

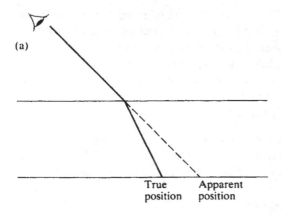

True   Apparent
position position

(b)

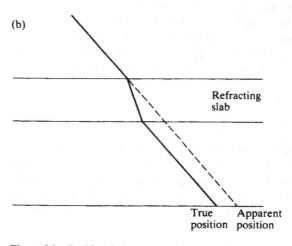

Refracting
slab

True   Apparent
position position

Figure 8.9   Positional errors caused by refraction.

likely to be anywhere in the range 1×10^{16} and 5×10^{17} electrons. What errors would be possible in calculating the true range to a target 600 km away, and what would be the average error?

SOLUTION If f is in gigahertz and TEC is given in units of 10^{16} electrons, the range error equation becomes

$$\Delta l = 0.403 \times (\text{TEC})/f^2 \qquad [\text{m}]$$

In these units $f = 1$ and TEC varies between 1 and 50, so the possible range errors are between 0.4 and 20 m. If all errors are equally likely, the average error is 10.2 m.

Since the refractive index of the ionosphere increases with increasing electron density, a ray entering the ionosphere from the ground is progressively bent further away from its original direction. In the extreme case (which occurs at high frequencies (HF)), the ray path can be bent so much that the ray returns to the ground (if the ray was launched from space, it would return into space). Only rays that are of a high enough frequency to penetrate the peak ionospheric electron density can pass through the ionosphere. (This peak typically occurs at a height of 300–400 km.) This 'reflection' from the ionosphere is used for ionospheric sounding from either the ground or space (the type of radar that carries out this sounding is called an *ionosonde*, and it is discussed in Sec. 10.4). Reflection from the ionosphere is also the principle used by over-the-horizon radar to obtain radar backscatter from the earth's surface at very long ranges (see Chapter 10).

Faraday rotation is caused by a combination of the earth's magnetic field, the electron density and the path length through the ionosphere. The plane of polarization is rotated by an angle

$$\Omega = M \times (\text{TEC})/f^2 \quad [\text{radians}] \quad (8.16)$$

where M is dependent on the geometry of the earth's magnetic field and the wave-normal. A reasonable order of magnitude for M is 0.5 s^{-2}, which indicates that at gigahertz frequencies rotations of at most a few degrees occur (though extreme cases in excess of $80°$ have been recorded). Hence Faraday rotation will cause little problem to, for example, space-borne synthetic aperture radars operating at gigahertz frequencies. However, like attenuation, Faraday rotation has a $1/f^2$ dependence, and will become increasingly important as frequency decreases. Severe effects can be caused at HF.

Attenuation, refraction and Faraday rotation can be thought of as acting on individual ray paths passing through the ionosphere. However, except at the highest frequencies, we need to take account of *diffraction* effects. These occur when the electromagnetic wave passes through an irregular ionosphere. As we have already noted, this is the normal state of affairs in the auroral zones and in the post-sunset equatorial region. Extreme disturbances could be caused by chemical releases into the ionosphere (rocket releases of barium oxide are often used in ionospheric research), or by high-altitude nuclear explosions. If a space-based radar is well above the ionosphere, the transmitted wavefronts will be nearly planar by the time they reach the ionosphere. As the wave passes through an irregular slab of ionosphere, local variation in the refractive index causes phase perturbations. On exit from the ionosphere, the contours of constant phase are no longer planar (see Fig. 8.10). While the wave propagates downwards, interference effects set in, which cause amplitude and phase fluctuations at the ground. An exactly similar process is what causes the stars to twinkle, and the fluctuations are

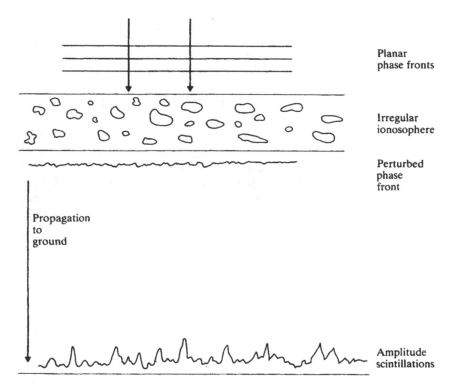

Figure 8.10 The mechanism by which phase perturbations in the ionosphere give rise to scintillations observed on the ground. As the perturbed wave propagates to the ground, interference effects develop, and give rise to large fluctuations in phase and amplitude.

hence known as *scintillations*. On the return path to the satellite, the same type of disturbance will occur.

The amplitude scintillations can be thought of as fluctuations in the RCS of the target, and hence are treated in the same way as for the Swerling cases (though there is still no universal agreement on the correct PDF to describe these scintillations). Whether the received signal varies on a pulse-to-pulse or a scan-to-scan basis depends on the relative motion of the satellite, the target and the ionosphere. Phase scintillations can also have important effects on systems that rely on phase coherence, such as space-borne SARs. If the pulse-to-pulse signal contains a random phase element that changes significantly as the synthetic aperture is being formed, the radar performance may be degraded. Progressively more serious effects include displacement of the beam (leading to geometric errors), increased sidelobe levels (leading to loss of contrast in the image) and destruction of the focus (leading to complete loss of the image). Figure 8.11 shows all these effects for simulated SAR data as the operating wavelength increases. At a wavelength of 0.83 cm, the synthetic

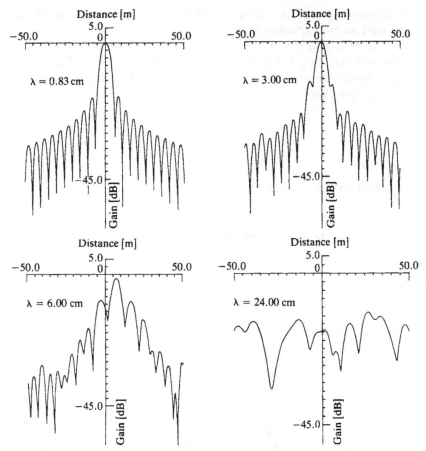

Figure 8.11 Simulated distortions of the synthetic aperture gain pattern caused by ionospheric irregularities. The different patterns correspond to increasing wavelength (and hence increasing synthetic aperture length; see Chapter 11), as marked on each pattern. The ionosphere was the same in all cases.

beam is hardly disturbed. By the time the wavelength has increased to 24 cm, the beam has been almost completely destroyed.

8.7 SUMMARY

At low elevation angles, the propagation of radio waves is degraded by both atmospheric effects and scattering by the terrain. For many general-purpose surveillance requirements, the four-thirds earth approximation is a sufficiently good correction for atmospheric refraction (the strongest effect) and the other problems are ignored. When precision target tracking is required, all atmospheric effects have to be considered, and appropriate corrections made, based on local climatic information.

The electron content of the ionosphere also causes attenuation and refraction effects, and a rotation of the plane of polarization. Radio waves passing through the ionosphere may suffer a form of scintillation that can affect space-borne earth-imaging radars.

Key equations

- Range of a point on radio horizon:

$$R \simeq \sqrt{(2R_e h)} \qquad [\text{m}]$$

which can be restated in nautical miles as:

$$R_{\text{n. mile}} \sim \sqrt{h_{\text{feet}}} \qquad [\text{n. mile}]$$

- Radio refractivity:

$$N = (n - 1) \times 10^6 \qquad [\quad]$$

which can be calculated from:

$$N = 77.6 \left(\frac{p}{T} \right) + 3.73 \times 10^5 \left(\frac{e}{T^2} \right) \qquad [\quad]$$

- Effective earth radius factor:

$$k = \left(1 + 10^{-9} R_e \frac{dN}{dh} \right)^{-1} \qquad [\quad]$$

- Range error due to atmospheric effects:

$$\Delta R = 10^{-6} R \left(N_0 + \frac{(h_t - h_r)}{2000} \frac{dN}{dh} \right) \qquad [\text{m}]$$

- Attenuation by the ionosphere:

$$\text{Two-way loss} = A/f^2 \qquad [\quad]$$

- Refractive index of the ionosphere:

$$n^2 \approx 1 \pm 80.6 N_e / f^2 \qquad [\quad]$$

- One-way range error through the ionosphere:

$$\Delta l = 0.403 \times (\text{TEC}) / f^2 \qquad [\text{m}]$$

- Faraday rotation:

$$\Omega = M \times (\text{TEC}) / f^2 \qquad [\text{radians}]$$

8.8 REFERENCES

There are many good books on the subject of propagation because this is a subject of importance in communications as well as in radar. The following selection is suggested as a useful introduction to the subject.

1. *Radar Propagation at Low Altitudes*, M. L. Meeks, Artech House, MA, 1982. [A very readable monograph complete with program listings and a good bibliography.]
2. *Radio Meteorology*, B. R. Bean and E. J. Dutton, Dover Publications, New York, 1968. [This is the standard text on large-scale atmospheric variations in radio refractive index.]
3. *Radiowave Propagation*, Ed. M. P. M. Hall and L. W. Barclay, Peter Peregrinus for the IEE, Stevenage, Herts, 1989. [Originally derived from course notes accompanying an IEE school on radio wave propagation, this book forms an indispensable guide to the whole subject.]
4. *Reference Atmosphere for Refraction*, Recommendations and Reports of the CCIR, CCIR Rec369-3, V, ITU, Geneva, 1986.
5. *Radio Wave Propagation and Antennas*, J. Griffiths, Prentice-Hall, Englewood Cliffs, NJ, 1987. [A particularly easy to read guide to propagation.]
6. *Radar Handbook*, Ed. M. I. Skolnik, McGraw-Hill, New York, 1970.
7. *Radio Wave Propagation*, A. Picquenard, Macmillan, London, 1974.
8. *Modern Radar Techniques*, Ed. M. J. B. Scanlan, Collins, Glasgow, 1987.

8.9 PROBLEMS

8.1 At what range could a target flying at a height of 1 km be detected by a radar at sea level, assuming (a) no refraction, (b) four-thirds Earth, (c) a refractive gradient of $-20 \, N/km$ and (d) a refractive gradient of $-120 \, N/km$?

8.2 In the cases (c) and (d) in problem 8.1, what range corrections would be needed to allow for atmospheric retardation? Assume $N_0 = 300$ at the radar site.

8.3 (a) In the next chapter it is shown that the power received by a weather radar, when precipitation fills the beam, is proportional to $1/R^2$. If the radar were badly sited such that many of the low-elevation observations were made at grazing incidence to the local horizon, how would the performance be impaired when scanning for light rain nominally detectable at 4.5 km in the clear?

(b) How would the detection range of a point target be affected at the same elevation?

8.4 A space-based radar operating on the two frequencies 200 and 400 MHz measures the range delay to a target as 44 and 14 ms respectively. What is the true range to the target?

NINE

RADAR STUDIES OF THE ATMOSPHERE

- Why atmospheric radars are useful
- Angels and echoes from clear air
- Special-purpose research radars

Our atmosphere is a major environmental issue; radar is one of the key research tools.

9.1 INTRODUCTION

The atmosphere would not appear to be a very promising radar target at first sight, and it is perhaps surprising to discover that there are a considerable number of radars dedicated to observing and investigating the atmosphere at all altitudes. There are three main areas in which atmospheric radars are used, and these are now outlined.

Hazard avoidance Most airliners carry a weather radar in the nose to look ahead and give a warning of severe weather on a display inside the cockpit. Pilots may then take avoiding action within the limits set by their flight lane. Airports prone to severe storms are now also installing ground-based weather radars to give warnings of severe down-draughts, which are a danger to aircraft taking off and landing. At present, strong down-draughts during thunderstorms cause about one aircraft crash every 18 months in the USA alone.

Forecasting Radar maps of rainfall are a familiar sight on televised weather forecasts, but the radars supplying this information also provide other data useful for forecasting. The main data source for weather forecasts remains satellite images, aircraft measurements and the radiosonde balloon flights, which are equipped with instruments to measure temperature, pressure and humidity. However, radar continues to make increased contributions to weather forecasting, and there are plans to replace balloons with a network of stratospheric–tropospheric (ST) radars, which are described in Sec. 9.3. (The balloons themselves are tracked by the type of radar described in problem 3.1.)

Research Balloons rarely reach altitudes in excess of 25–30 km and most of what we know about the upper atmosphere comes from radar and lidar observations. Radar is important in lower-atmosphere research, too. Although balloon networks can provide a synoptic picture of the weather, they cannot observe the evolution of small-scale atmospheric features revealed by continuous radar surveillance.

9.2 SCATTERING MECHANISMS

In this section we look at the different types of scattering mechanisms that cause radars to receive echoes from apparently clear air, and at how these mechanisms can be exploited to investigate the structure and dynamics of the atmosphere.

Radar echoes from the clear atmosphere have interested scientists since the mid-1930s, although during World War II they were regarded as something of a nuisance (see Atlas[1]). After the war, the increasing use of radars, and the strength of some types of echo, led to serious research into understanding the possible scattering mechanisms. By the 1960s, a new generation of high-power, high-frequency radars had been developed especially for the detailed study of weather processes in the lower atmosphere. These systems were often X-band (9.4 GHz), but sometimes S-band (2.8 GHz) and UHF (near 400 MHz). The 1970s saw a major shift to VHF frequencies and the use of large, vertically pointing, phased array antenna systems.

The turbulent lower atmosphere in which we live is called the troposphere, and extends from ground level up to roughly 11 km. Within this region there are many scattering sources including rain, snow, hail, clouds, birds, insect clouds, debris raised by fires, steps in temperature, clear-air turbulence and fog (at millimetre wavelengths). Understanding these scatterers is important to radar engineers for two reasons: first, they are a source of clutter for conventional radar, and secondly, they form a mechanism by which the atmosphere may be studied. The more prevalent scatterers, and their RCS, are now considered.

Hydrometeors

This is the meteorological term for scattering particles (rain, ice particles, etc.). They cause the Rayleigh scattering that we encountered in Chapter 2 because the radar wavelength λ is usually greater than the diameter D of the particles. The RCS of a single droplet is

$$\sigma_{\text{particle}} = C\pi^5 D^6 / \lambda^4 \quad [\text{m}^2] \tag{9.1}$$

where C is a dimensionless constant that depends on the dielectric constant of the particles. For water, C is near 1, whereas for ice particles it is about 0.2. To calculate the RCS of a rain cloud, we must sum over the volume resolution cell of the radar, to give

$$\sigma_{\text{cloud}} = \frac{C\pi^5}{\lambda^4} \sum_1^N D_i^6 \quad [\text{m}^2] \tag{9.2}$$

where N = number of scatterers in the volume illuminated [ ]. The sixth-power dependence on the particle diameter means that heavy rain, which usually contains large droplets, gives a much stronger backscatter echo than light rain. This effect is easily observed on the weather radar colour intensity maps shown on television weather forecasts when storms are present.

In practice, Eq. (9.2) is evaluated by defining $\sum D_i^6$ per unit volume as a radar reflectivity factor Z (see Battan[2], Nathanson[3] or Skolnik[4]). This volume reflectivity is empirically related to the precipitation rate r [mm h^{-1}] by

$$Z = \sum_1^N D_i^6 / \text{unit volume} \quad [\text{mm}^6\,\text{m}^{-3}]$$
$$= ar^b \tag{9.3}$$

The millimetre units creep in because of the meteorological conventions of describing drop size in mm; do not blame the radar engineers! There have been many different experimental determinations of a and b, but some commonly accepted values are presented in Table 9.1. Note that, at low

Table 9.1 Empirically determined constants relating the rainfall rate r to the radar reflectivity factor Z

Precipitation	a	b
Rain	200	1.60
Thunderstorm	486	1.37
Ice crystal	500	1.66
Wet snowflakes ($t > 0°C$)	2000	2.0
Dry snowflakes ($t < 0°C$)	1050	2.0

frequencies (in the Rayleigh region), Z is not frequency-dependent, and by building Z into the radar equation we can develop a formula that will tell us the power received from precipitation by radars with operating frequencies up to X-band. If the volume of the resolution cell V_{res} is given by

$$V_{res} = (R \, \Delta\theta)(R \, \Delta\phi)(c\tau/2) \qquad [m^3] \qquad (9.4)$$

where $\tau =$ pulse duration $[s]$, then the RCS of the resolution cell is given by

$$\sigma = V_{res} \sum \sigma_{particle}/\text{unit volume}$$

$$= \frac{C\pi^5}{\lambda^4} ZR^2 \, \Delta\theta \, \Delta\phi \, \frac{c\tau}{2} \qquad [m^2] \qquad (9.5)$$

and the radar equation becomes (with any appropriate units also shown)

$$[W] \; [\;] \; [\;] \; [\;] \; [\;] \quad [m^3] \quad [rad] \, [rad] \, [m \, s^{-1}] \, [s] \, [\;] \, [\;]$$

$$\frac{[W]}{P_r} = \frac{P_t \quad G_t \quad G_r \quad C \quad \pi^2 \; \overbrace{(10^{-18}Z)} \quad \Delta\theta \quad \Delta\phi \quad c \quad \tau \; L_{sys} \; L_{atmos}}{64 \quad \lambda^2 \quad R^2 \quad 2} \quad [W]$$

$$[\;] \, [m^2] \, [m^2] \, [\;]$$

$$(9.6)$$

with Z in $mm^6 \, m^{-3}$. The propagation loss L_{atmos} may have to include the effects of attenuation through the rain cloud itself. Collecting the constants together (in units of m s^{-1}), and assuming the constant C to be unity, gives

$$P_r = \frac{2.3 \times 10^{-11} P_t G_t G_r Z \, \Delta\theta \, \Delta\phi \, \tau L_{sys} L_{atmos}}{\lambda^2 R^2} \qquad [W] \qquad (9.7)$$

The constant 2.3×10^{-11} has dimensions of speed because it includes the velocity of light c. Occasionally, variations of this formula are used to take into account the elliptical beamshape. Sometimes V_{res} is reduced by a factor of $\pi/4$, and Skolnik[4] suggests a further reduction of $1/(2 \ln 2)$ because the effective two-way illuminated volume is less than the one-way value suggested by the half-power beamwidths used in Eq. (9.4). At long ranges, radar resolution cells are large and are unlikely to be completely filled by precipitation. In this case, a beam-filling factor must also be introduced.

Radars designed specifically to investigate cloud physics sometimes operate at frequencies higher than X-band to increase the RCS of the hydrometers. To surveillance radars, rain represents a form of clutter that is particularly unwelcome because it has a doppler shift imparted by the motion of the cloud, as shown in Fig. 9.1. For this reason, surveillance is carried out at relatively low frequencies, usually L-band, to achieve long ranges.

Worked example In the UK, the Rutherford Appleton Laboratory operates an S-band ($\lambda = 9.75$ cm) high-resolution radar for the study of

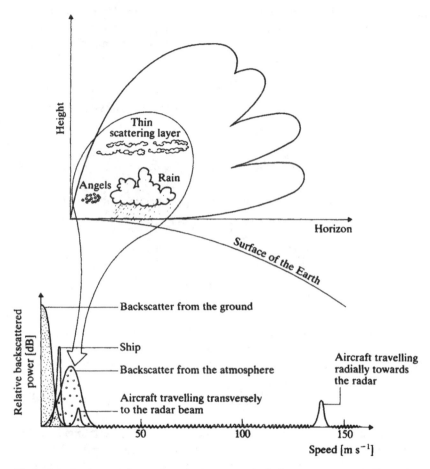

Figure 9.1 Backscatter from rain and other atmospheric features amount to a form of moving clutter that can mask slowly moving targets.

cloud physics. The antenna is a 25 m diameter parabolic dish giving a beamwidth of 0.25° (see reference 1). If the peak power is 560 kW and the pulse duration is 1.0 μs, what echo power would you expect to receive from heavy rain ($r = 4$ mm h^{-1}), if the rain entirely filled a resolution cell, at a range of 50 km?

SOLUTION We have $Z = 200(4)^{1.6} = 32.6$ [dB mm^6 m^{-3}] and then using Eq. (9.7)

$$\begin{aligned}
\text{Constant} &= -106.4 \, \text{dB m s}^{-1} &&\equiv 2.3 \times 10^{-11} \, \text{m s}^{-1} \\
P_t &= 57.5 \, \text{dB W} \\
G_t G_r &= 116.4 \, \text{dB}
\end{aligned}$$

$$
\begin{aligned}
Z &= 32.6 \,\text{dB mm}^6\,\text{m}^{-3} \\
\Delta\theta \,\Delta\phi &= -47.2 \,\text{dB rad}^2 \\
\tau &= -60.0 \,\text{dB s} \\
1/\lambda^2 &= 20.2 \,\text{dB m}^{-2} \\
1/R^2 &= -94.0 \,\text{dB m}^{-2} \qquad (\text{Range} = 50 \,\text{km})
\end{aligned}
$$

$$
P_r = -80.9 \,\text{dB W}
$$

The received power is well above the expected noise level for a 1 MHz bandwidth, even allowing for additional system and propagation losses.

Research radars, such as the Rutherford Appleton instrument, make provision for measuring the polarization of the echo, which gives information on the shape of the hydrometeors. A surveillance radar, on the other hand, would almost certainly select circular polarization during heavy rain; this is because a simple scattering object like a raindrop reverses the direction of polarization on reflection and if left-hand circular is transmitted, then the echo returns as right-hand circular and is ignored by the system. Complex scattering objects such as aircraft, where multiple reflections take place, generally return a mixture of polarizations and so may be detected through the rain.

As well as acting as a radar target, precipitation also absorbs radio waves and causes an attenuation that rises rapidly with frequency. For example at 1 GHz, rain falling at 25 mm/h causes almost no attenuation but at 10 GHz the two-way attenuation is around 0.5 dB/km. More details can be found in Meeks[5].

Angels

This intriguing name was given to the echoes of discrete targets that returned from an apparently empty sky to trouble early radar operators. Birds are probably the main cause, but clouds of insects can also cause angels and, because they drift in the wind, they can cause clutter problems similar to rain clouds.

The RCS of a bird generally lies in the resonance region and large variations are observed, but a rough rule is that a bird has the same echoing area as a plastic bag filled with water having the same weight. Typical RCS values are -40 dB m^2 for a small bird and -20 dB m^2 for a larger bird such as a gull. Flocks of birds appear as much larger, moving targets. Radar can be used to study bird migration, and to identify birds by wing beat frequency, altitude and speed; see for example Eastwood[6].

Insects have an RCS of -50 dB m^2 or lower, but, again, concentrations of insects within the volume resolution cell of the radar can cause sufficient scattering to create the angel effect.

Clear-Air Turbulence

Echoes from genuinely clear air were originally quite unexpected and difficult to explain, although it was guessed that weak fluctuations in the radio refractive index of the air must be responsible. It is now generally understood that echoes from clear air arise from two main types of scattering mechanism: one is *volume scattering* from turbulence and the other is *specular* (mirror-like) reflections from thin layers.

The atmosphere is everywhere turbulent but there exist regions of greater intensity such as convective cells or 'thermals'. Volume scattering from these regions (and also from the whole atmosphere, but fainter) is due to constructive interference from fluctuations in refractive index. Those scatterers with a scale size equal to half the radar wavelength are mainly responsible for the backscattering, in a process similar to the Bragg scattering discussed in more detail in Chapter 11.

One of the hazards of flying is to encounter the sudden turbulence that forms in thin horizontally stratified layers and which sometimes has a mean vertical gradient, or refractive index, much greater than in the surrounding atmosphere. These layers may be only a few tens of metres in thickness and yet they can extend horizontally for tens of kilometres. The radio scattering from these layers causes strong radar echoes in both the troposphere and the lower stratosphere. The stratosphere is the horizontally stable layer above the troposphere and extends in altitude from roughly 12 to 50 km.

The region above the stratosphere is called the mesosphere, and thin layers are found here too. The early Soviet cosmonauts produced some excellent free-hand sketches of mesopheric layers that are visible when viewed edge-ways on against the bright limb of the earth. They have since been confirmed by radar observation. Above 100 km, the atmosphere can no longer be considered entirely neutral, and ionospheric scattering processes dominate.

The refractive index of the neutral atmosphere was described by Eq. (8.5), but in the mesosphere there are sometimes free electrons present, which cause an ionospheric term to be added:

$$n = 1 + 77.6 \times 10^{-6}\left(\frac{p}{T}\right) + 3.73 \times 10^{-1}\left(\frac{e}{T^2}\right) - 40.3\left(\frac{N_e}{f^2}\right) \qquad [\quad]$$

$$(9.8)$$

where N_e = number of free electrons per m^3 [　] and f = radar frequency [Hz]. Equation (9.8) describes the refractive index n, from ground level up to 100 km (to get the radio refractivity N, subtract 1 and multiply by 10^6). Using reasonable values of T, e, p and N_e we find that the wet term (containing e, the partial pressure of water vapour) dominates in the troposphere but above 50 km the ionospheric term begins to determine the refractive index. The usual practice in atmospheric physics is to incorporate refractive index changes into the radar equation by adopting yet another version, based on Eq. (9.6):

$$P_r = \frac{P_t \lambda^2 L_{sys}(C_s^2 + C_r^2)}{16\pi^2 R^2} \qquad [\text{W}] \qquad (9.9)$$

where C_s^2 = contribution due to volume scattering by turbulence [ ] and C_r^2 = contribution due to reflection from layers [ ].

The volume scattering term is given by

$$C_s^2 = \Delta R G \eta \qquad [\ ] \qquad (9.10)$$

where ΔR = range resolution [m], G = antenna gain and η = radar reflectivity of the scattering volume, or RCS per cubic volume [$\text{m}^2 \, \text{m}^{-3} = \text{m}^{-1}$]. The reflectivity of the turbulence is related to the structure constant C_n^2 (a measure of turbulence) by

$$\eta = 0.38 C_n^2 \lambda^{-1/3} \qquad [\text{m}^{-2/3} \, \text{m}^{-1/3} = \text{m}^{-1}] \qquad (9.11)$$

C_n^2 is related to the scale size of turbulence in the atmosphere, which increases with height. Formally C_n^2 is calculated from

$$C_n^2 = 5.26 \times \overline{\Delta n^2} \, L_0^{-2/3} \qquad [\text{m}^{-2/3}] \qquad (9.12)$$

where $\overline{\Delta n^2}$ = mean-square variation in the refractive index [ ] and L_0 = scale size of the turbulent eddies [m].

These rather complicated, but well established, formulae mean that we can relate turbulent refractive index changes to the power received by a radar observing them, for all altitudes in the neutral atmosphere.

Although clear-air echoes may be a source of clutter for surveillance radars, they are increasingly being exploited as a means of investigating waves, winds and turbulence in the Earth's atmosphere, as the next section shows.

9.3 MESOSPHERE–STRATOSPHERE–TROPOSPHERE RADAR

During the 1970s it was realized that VHF radars operating at wavelengths of 1–10 m had considerable potential for studying atmospheric dynamics because these wavelengths are well matched to turbulence scale sizes. These radars, known as mesosphere–stratosphere–troposphere (MST) radars, have several advantages over rocket, balloon and aircraft measurements, including:

- Very good time and height resolution.
- The capability of making observations simultaneously over a wide range of heights.

● The possibility of continuous operation.
● The capability of measuring vertical air motions.

A typical MST radar operates at 50 MHz and uses a high-power transmitting system driving a large antenna array to form several narrow beams. One beam is pointed vertically and the others are tilted a few degrees off-vertical in the compass directions N, E, S and W, as shown in Fig. 9.2. The same antenna array is used for reception in order to gain the maximum possible power × aperture product. The need for such a large radar system lies in the volume scattering mechanism for turbulence, which, despite the careful choice of wavelength, produces relatively weak echoes compared with most other types of radar target.

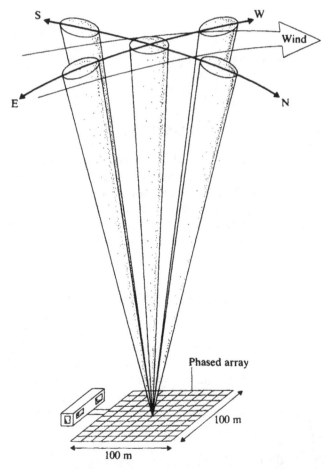

Figure 9.2 Typical MST radar arrangement using several fixed beam positions to measure horizontal and vertical air motions.

The structure constant C_n^2 typically has a value of 10^{-17} $[m^{-2/3}]$ at 11 km, falling by roughly 1 dB per kilometre increase in altitude. This leads to a value for η of 2.1×10^{-18} $[m^{-1}]$, falling by 1 dB per kilometre, giving a very small RCS even when quite large volumes are illuminated.

Worked example A 50 MHz MST radar uses a pulse duration of 1 μs, a peak power of 125 kW and an array of dipoles 100 m × 100 m. Would you expect volume scattering from 11 km altitude to be detected?

SOLUTION If the antenna aperture of 10^4 m^2 has an efficiency factor of 0.5 (mainly due to weighting to reduce the sidelobes) then from $G = 4\pi A_e/\lambda^2$ we estimate the antenna gain to be 1745 (32.4 dB). Assuming $\Delta R = 150$ m, we can calculate C_s^2 to be 5.5×10^{-13}. The power received can be calculated from Eq. (9.9), with $C_r^2 = 0$:

$$
\begin{aligned}
P_t &= 51.0 \, \text{dB W} \\
\lambda^2 &= 15.6 \, \text{dB m}^2 \\
L_{\text{sys}} &= -7.0 \, \text{dB} \\
C_s^2 &= -122.6 \, \text{dB} \\
1/(16\pi^2) &= -22.0 \, \text{dB} \\
1/R^2 &= -80.8 \, \text{dB m}^{-2} \\
\hline
P_r &= -165.8 \, \text{dB W}
\end{aligned}
$$

If the system were internally noise-limited, the level would be -204 dB W Hz^{-1} + 60 dB (1 MHz bandwidth) plus the noise factor of the receiver. In this case the volume scattering echo from a single pulse would be below noise level. In practice, a coherent integration time of about 1 s would be used and the final bandwidth would therefore be 1 Hz. However, external noise dominates internal noise by about 19 dB at this frequency, and so the final noise level is about $-204 + 19 = 185$ dB W, and volume scattering would be detected with a 19 dB SNR.

COMMENTS Roughly speaking, at 11 km the signal strength falls by about 1 dB per kilometre because of the $1/R^2$ factor, and a further 1 dB per kilometre because of the change in C_n^2, giving a total of 2 dB per kilometre. If no useful atmospheric science can be carried out with a SNR of less than 10 dB, these results imply that the radar would be a useful instrument only up to an altitude of about 15 or 16 km (longer coherent integration cannot be used owing to the short timescales of turbulent processes). The presence of any scattering layers, adding a C_r^2 term into the radar equation, would increase the SNR at lower altitudes, but by 20 km, they too are beginning to disappear. There is no escape from the conclusion that higher-altitude atmospheric research requires more powerful transmitters, larger antenna farms and more money.

A MST radar will measure returns from the troposphere and the lower stratosphere, and may sometimes receive echoes from the mesosphere when free-electron scattering occurs, but there is no easy way of investigating the region of the *middle atmosphere* between the two. For this reason, some systems were not designed to try to see beyond the lower stratosphere and, as a result, two distinct types of radar have evolved: the true MST system, and the smaller stratosphere–troposphere (ST) radars.

An example of a powerful modern MST radar is the Japanese MU radar, mentioned in problem 9.6 at the end of the chapter. The MU system uses an active antenna array (see Chapter 15) consisting of 475 solid-state transmitter modules, and Yagi antennas, to develop a total peak transmitted power of 1 MW. At the other end of the scale, there is an increasing use of ST radars to study turbulence and severe wind shears in the lower atmosphere, and in the USA, operational ST systems are to be installed at 47 of the country's busiest airports to warn pilots of potentially dangerous wind conditions. Some of the lower-atmosphere radars are more compact UHF *wind profilers*, which operate in much the same way, but at frequencies around 400 or 900 MHz.

New methods of operating ST and MST radars, and of carrying out the signal and data analysis, are continually being developed, and this field of radar is expanding rapidly.

9.4 METEOR WIND RADAR

Winds in the upper atmosphere between about 70 and 110 km have been studied since the 1950s by *meteor wind radar*. When tiny meteoritic particles enter the earth's atmosphere, they heat up due to friction (like the Space Shuttle during re-entry). The heating causes a meteor to boil away in a process known as ablation, which leaves behind a long, slightly cone-shaped, column of ionization, which is a strong scatterer of lower-frequency (HF and VHF) radio waves (see Fig. 9.3). The ionized trails drift in the neutral wind with a velocity that is easily measured by pulse doppler radar. There are always sufficient numbers of sporadic meteor echoes per hour for even a radar of modest power to be able to measure the larger-scale atmospheric features. At regular times each year, the earth passes through comet debris causing meteor showers that give an increased echo rate for a few days.

Although meteor wind radars do not have the resolution of the newer MST radars, they remain of considerable interest because the meteor trail formation process gives some clues about upper-atmosphere chemistry (recent work suggests that this could also include ozone concentrations). There is also an interest in over-the-horizon communications by forward scatter from meteor trails, and astronomers are interested in learning more about the distribution and sizes of meteoritic material within the solar system.

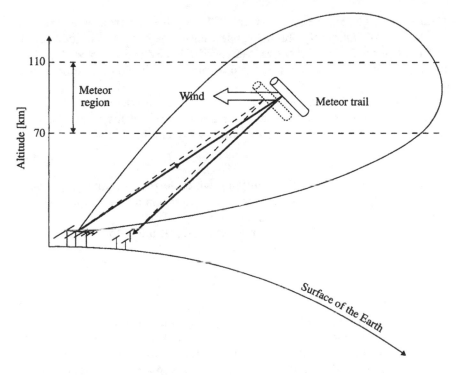

Figure 9.3 Radar scattering from meteor trails can be used to measure winds at high altitudes.

A typical meteor radar operates at about 40 MHz with a 15 dB gain Yagi–Udah or phase array transmitting antenna. The transmitter pulse has a 10 μs duration and a peak power of 200 kW in the larger systems, although useful results can be obtained with 20 kW. The receiving antenna often consists of several low-gain elements arranged as an interferometer to measure the echo angles of arrival, which, combined with the range, are used to determine the height of the echo and wind measurement[7].

Although a full treatment of meteor trail scattering theory is complicated, an estimate of the RCS of a meteor trail can be made in the following way. Although a few of the brightest meteor trails are said to be *overdense* (the electron density is so high that the radio wave cannot penetrate the trail, rather like a tube of copper), most meteor trails are *underdense*, with an RCS given by

$$\sigma_{\text{meteor}}(t) = \sigma_1 \sigma_2 \sigma_3 \exp(-2t/T_{\text{u}}) \qquad [\text{m}^2] \qquad (9.13)$$

Taking each of the terms in turn, the first is the principal echoing area given by

$$\sigma_1 = 2\pi \lambda q^2 r_{\text{e}}^2 R \qquad [\text{m}^2] \qquad (9.14)$$

where λ = radar wavelength, typically 7.5 m; q = electron line density, which for an averagely bright underdense meteor would be about 10^{13} electrons m^{-1}; r_e = effective radius of the electron = 2.8×10^{-15} m; and R = range to closest point on meteor trail, typically 90 km. These values give a typical RCS of 3300 m^2.

The second term is a loss factor to account for interference between the illuminating radiation and re-radiation from the electrons when the trail diameter is of the same order as the radio wavelength. It is given by

$$\sigma_2 = \exp(-8\pi^2 r_i^2 / \lambda^2) \qquad [\ \ ] \qquad (9.15)$$

where r_i = initial radius of the trail [m], which has been found empirically to be about 0.22 m at 80 km altitude, 0.5 m at 90 km and 1.1 m at 100 km. Using 0.5 m for the initial trail radius gives $\sigma_2 = 0.7$.

The third term σ_3, and the exponential factor, allow for the expansion of the trail by diffusion; σ_3 is given by

$$\sigma_3 = \exp\left[-K\left(\frac{R}{2\lambda^3}\right)^{1/2} \right] \qquad [\ \ ] \qquad (9.16)$$

The constant K [m] is related to the way the trail diffuses and the velocity of the meteoritic particle. The value of K is about 0.09 m at 80 km, 0.05 m at 90 km and 0.1 m at 100 km. A typical value for σ_3 is therefore 0.6.

The exponential term gives a characteristic decay to meteor echoes, which is used in some systems as a method of determining the echo height. The time constant for the received power to decay by a factor e^2 is given by

$$T_u = \frac{\lambda^2}{16\pi^2 D} \qquad [s] \qquad (9.17)$$

The rate of diffusion D (known as the ambipolar diffusion constant) can be found from a formula by Greenhow and Neufeld[8]:

$$\log_{10} D = 6.7 \times 10^{-5} h - 5.6 \qquad [m^2 s^{-1}] \qquad (9.18)$$

where h = echo height [m].

Putting these factors together gives the RCS of a typical meteor as 1400 m^2, but decaying with time in an exponential fashion. More details can be found in Sugar[9], who also gives an expression for the RCS of overdense meteors.

The number of sporadic (background) meteors encountered by a radar system can be estimated from

$$\text{Number of meteors with } q > q_0 = \frac{160}{q} \qquad [\text{meteors m}^{-2} \text{s}^{-1}] \qquad (9.19)$$

During meteor showers, the meteor echo rate can increase dramatically. The best known, and most spectacular, meteor showers are the Perseids, around 12 August each year, and the Geminids, around 13 December.

Besides their use in atmospheric physics, these formulae describing meteor rates and RCS can be used to estimate the extent of meteor clutter detected by various types of radar (for example, HF radars, described in the next chapter, experience meteor clutter problems).

9.5 OTHER RADAR STUDIES OF THE ATMOSPHERE

Another method of studying the atmosphere between about 5 and 100 km is *lidar* (light detection and ranging). This technique involves the emission of light pulses in a vertical beam, which are then backscattered by atmospheric gases and particles. The scattering height is determined from the time delay, as with conventional radar. The receiver usually involves some form of frequency selection followed by a photomultiplier detector[10]. The scattering process may be Rayleigh scattering from molecules, resonant scattering from specific atomic species to which the laser has been tuned, or Mie scattering from aerosols and ice crystals in the stratosphere.

Lidar has found applications in measuring stratospheric densities, temperature, composition and the distribution of pollutants. The recent development of coherent lidar means that troposphere and lower-stratosphere wind measurements should soon be possible, perhaps as part of a satellite-borne global wind field mapping system.

Among the variety of other experimental radar soundings of the atmosphere is the radar acoustic sounding system (RASS). This ingenious technique involves the vertical emission of a pulse of sound, which travels upwards at a velocity that depends on the air temperature and density. The wavefront is tracked by radar backscatter from the refractive index discontinuity induced by the sound pulse. The technique can provide useful information at low altitudes, provided the wavefront is not distorted by wind shear.

9.6 SUMMARY

From initially being a nuisance, the scattering of radar signals from the atmosphere has been turned into a useful, and expanding, research technique for the study of atmospheric physics and meteorology. Scattering occurs from discrete sources (rain, birds, etc.) and also from changes in the refractive index of the air, mainly caused by turbulence.

Weather radars, investigating lower-atmosphere cloud physics and precipitation, operate at frequencies in S-band and above, but new VHF phased array radars have emerged as a tool for probing the atmosphere from 1 to 100 km altitude.

Key equations
- Radar reflectivity factor:

$$Z = ar^b \quad [\text{mm}^6 \, \text{m}^{-3}]$$

- Radar cross-section of resolution cell:

$$\sigma = \frac{C\pi^5}{\lambda^4} ZR^2 \, \Delta\theta \, \Delta\phi \, \frac{c\tau}{2} \quad [\text{m}^2]$$

- Radar equation for hydrometeor scattering:

$$P_r = \frac{2.3 \times 10^{-11} P_t G_t G_r Z \, \Delta\theta \, \Delta\phi \, \tau L_{\text{sys}} L_{\text{atmos}}}{\lambda^2 R^2} \quad [\text{W}]$$

- Radar equation for clear-air turbulence:

$$P_r = \frac{P_t \lambda^2 L_{\text{sys}} (C_s^2 + C_r^2)}{16\pi^2 R^2} \quad [\text{W}]$$

- Volume scattering factor due to turbulence:

$$C_s^2 = \Delta R \, G \times 0.38 C_n^2 \lambda^{-1/3} \quad [\quad]$$

9.7 REFERENCES

1. *Radar in Meteorology*, Ed. D. Atlas, Battan Memorial and 40th Anniversary Radar Meteorology Conference, American Meteorology Society, 1990. [A large but fascinating book full of the history of meteorological radars, as well as current thinking.]
2. *Radar Meteorology*, L. J. Battan, University of Chicago Press, Chicago, 1959.
3. *Radar Design Principles*, F. E. Nathanson, McGraw-Hill, New York, 1969.
4. *Introduction to Radar Systems*, M. I. Skolnik, McGraw-Hill, New York, 1985.
5. *Radar Propagation at Low Altitudes*, M. L. Meeks, Artech House, MA, 1982.
6. *Radar Ornithology*, Sir. E. Eastwood, Methuen, London, 1967.
7. *Meteor Science and Engineering*, D. W. R. McKinley, McGraw-Hill, New York, 1961.
8. Turbulence at altitudes of (80–100) km and its effects on long duration meteor echoes, J. S. Greenhow and E. L. Neufeld, *J. Atmos. Terr. Phys.*, 384–392, 1959.
9. Radio propagation by reflection from meteor trails, G. R. Sugar, *Proc. IEEE*, **52**, 116–136, 1964.
10. *Laser Monitoring of the Atmosphere*, Ed. E. D. Hinkley, Springer-Verlag, Berlin, 1976.

Other titles that may prove useful are:

11. *Radar Observation of the Atmosphere*, L. J. Battan, University of Chicago Press, Chicago, 1973. [A pre-MST radar book.]
12. *Doppler Radar and Weather Observations*, R. J. Doviak and D. S. Zrnic, Academic Press, New York, 1984. [A bit mathematical.]
13. *Radar Observations of Clear Air and Clouds*, E. E. Gossard and R. J. Strauch, Elsevier, Amsterdam, 1983.

9.8 PROBLEMS

9.1 The UK Meteorological Office weather radar network has used Siemens–Plessey S-band and C-band radar sensors to measure rainfall. The Siemens–Plessey type 45S sensor features a 3.66 m diameter parabolic antenna giving a 2° pencil beam and a gain of 37 dB at 2.9 GHz. The peak transmitter power is 650 kW and the pulse duration is 2 μs. If the noise level is -137 dB W, what is the SNR of the echo from a rain cloud filling the beam at a range of 20 km, if the rainfall rate is 1 mm h^{-1}?

9.2 What would be the effect of increasing the rainfall rate to 100 mm h^{-1}, in the above problem?

9.3 What would be the dynamic range requirement of a receiver covering rainfall rates of 1/8 to 64 mm h^{-1} and ranges from 1 to 200 km? Could this be reduced by sweeping the gain of the receiver to increase the sensitivity with increasing range? If so, what power law should be used for the swept gain?

9.4 The German SOUSY MST radar system located in the Harz Mountains operates at 53.3 MHz and has an antenna array 70 m in diameter giving a beamwidth of 5° and a gain of 31 dB. The beamformer uses 4-bit phaseshifters (see Chapter 15) to steer the beam anywhere within a 30° cone centred on the vertical. At 7° off-vertical, it is found that the echo power received from a thin layer at an altitude of 11 km is 10 dB lower than the measurement in the vertical beam. Could this be explained by the geometry and the longer range to the layer in the off-vertical beam?

9.5 The peak transmitter power of the SOUSY radar is 600 kW. What is the strength of an echo scattered from turbulence at 11 km, assuming a 2 μs pulse duration and 7 dB system losses?

9.6 The Japanese MU MST radar system operates at 46.5 MHz, uses a circular antenna array 103 m in diameter and a peak transmitter power of 1 MW. How much larger would the SNR be than that of the SOUSY system, assuming the other parameters are identical?

9.7 If the UK MST radar (near Aberystwyth, Wales) uses a coherent integration of 512 pulses, and is set to have an unambiguous range of 24 km and a total observation time of 10.5 s, what is the width of the doppler spectrum in the fast Fourier transform?

9.8 If the UK MST radar uses a vertical beam, and beams steered to a maximum of 12° off-vertical, would an aircraft travelling at 300 m s^{-1} cause any problems?

9.9 Can an MST radar be used as a meteor radar? Would you expect to detect many meteors in the vertical beam?

TEN

OVER-THE-HORIZON RADAR

- The two types of over-the-horizon radar
- Surface-wave radar
- Skywave radar
- Over-the-horizon radar equation

Over-the-horizon radars are even less perfect than microwave systems, but the rewards for seeing over the horizon are worth pursuing.

10.1 INTRODUCTION

It has been known since the experiments of Guglielmo Marconi in 1901 that radio waves could propagate beyond the horizon because of the signals he successfully transmitted from Poldhu in Cornwall, UK, to St John's, Newfoundland. Investigations into the cause of this propagation soon revealed that the solution to Maxwell's equations for a wave at a plane interface between two media gives a *space wave* (free-space propagation) and a *surface wave* (a wave guided along the interface). With the discovery of the earth's ionosphere in the 1920s, it was realized that there was also a third possible mode of propagation, the *ionospheric wave*, which turned out to be the explanation of Marconi's transatlantic communications.

The propagation of HF (3–30 MHz) radio waves over great distances has always been exploited in communications, and sometimes frequencies lower than HF are used, as listeners to long-wave radio will know. During World War II the UK air defence radar 'Chain Home', operating on 20–30 MHz, was occasionally troubled by '*n*th-time-around' clutter created when the radio signal was scattered by the ionosphere and travelled unusually long distances. Under these conditions the normal operating PRF of 25 Hz was reduced to 12.5 Hz (details in Neal[1]). In other countries, similar discoveries were made, and many radar engineers began thinking of turning this unwanted propagation to advantage.

From the early experiments with long-range propagation, two kinds of HF radar or 'over-the-horizon' (OTH) radar have been developed, known as surface-wave (or groundwave) radar and skywave radar, making use of the surface-wave and the ionospheric modes of propagation respectively. Surface-wave systems were first operated in the early 1950s and effective skywave systems a little later. The reason for the slow development of OTH radar compared to more conventional systems is not a reflection on the ability or the imagination of the engineers involved but rather the constraints of the technology available at the time.

Surface-wave radar uses the surface-wave propagation mode to look over the immediate horizon, and it may be used to survey ranges up to a maximum of certainly no more than 400 km. It is most useful as a local area defence system and as a method of collecting good-quality wave and tidal information over a restricted area of ocean. Although bistatic systems have been operated successfully, surface-wave radar is regarded as being a predominantly monostatic technique with a relatively low capital cost.

Skywave radars, on the other hand, are almost always large, bistatic and very expensive. These radars make use of the ionosphere to scatter radio waves very long distances beyond the horizon, sometimes in forward scatter mode to a receiver beyond the target. The minimum range is about 1000 km and the maximum useful range is around 4000 km. Skywave radar is thus more suited to the defence and remote ocean sensing needs of countries of such continental proportions as the USA, the former USSR and Australia.

10.2 SURFACE-WAVE RADAR

The principle of surface-wave or groundwave (gw) radar is that a surface radio propagation mode can be utilized to make a radar signal follow the surface of the sea as it disappears over the horizon. The method works only for vertically polarized antennas in contact with salty conducting water; it cannot be used over land, on freshwater lakes, or where fresh water dilutes the sea, such as in the Baltic or the Nile Delta.

The propagation of radio waves along surfaces has been analysed extensively in the past; the theory is complex and is, for all practical purposes, impossible to solve manually[2]. However, intuitively we would expect to find such propagation because, although the sea is a good conductor, it is not perfect and it supports a small horizontal electric field induced by

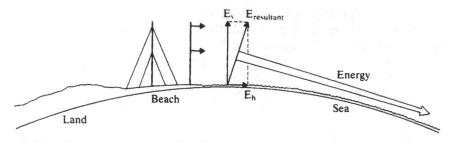

Figure 10.1 Surface-wave propagation.

the transmitted signal (see Fig. 10.1). The vector sum of the large vertical transmitted signal and a small induced horizontal field causes the resultant wavefront to tilt over such that the Poynting vector (the direction of energy flow) enters the sea at the Brewster angle[†]. One way to imagine this is to picture the lower part of the wavefront having a slightly lower velocity due to the water, dragging behind, and so bending the beam downwards. This propagation mode has various names, but it is possibly best known as the Norton surface wave.

The block diagram of a typical gw radar is shown in Fig. 10.2. 'Floodlight' illumination of a sector of sea is provided by a log-periodic transmitting antenna, which has vertically radiating elements of varying length to enable it to operate over a wide bandwidth. The receiving antenna is a 100 m array of monopoles parallel to the coast, which are used to form a number of beams simultaneously. If the system were installed in a confined area such as on an oil-rig or ship, the transmitting antenna would almost certainly have to be a monopole as well.

The transmitter is usually a wideband linear amplifier chosen from a range of commercial HF communication equipment. Unlike pulse transmitters, which can only deliver their peak power in short bursts, communications transmitters are *continuously rated*, meaning that they can deliver their specified power output continuously. The SNR of an HF radar system can thus be improved by finding waveforms that extend the percentage of time that the transmitter is on.

---

† The Brewster angle, named after a distinguished Scottish physicist of the last century, Sir David Brewster, is best known in elementary optics. A beam of light striking a glass block at a steep angle of incidence will enter and be refracted, whereas a beam striking at a very oblique angle will be reflected from the surface. In between these two cases lies the Brewster angle, at which the light runs along the interface between the glass and the air. In the case of gw radar, the air/sea boundary acts in a similar way to the air/glass interface, and the radio wave travels along the sea surface.

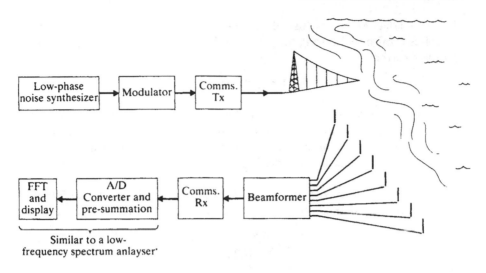

Figure 10.2   A typical surface-wave radar installation.

The modulation used in HF radar may be a pulse, but frequency-modulated interrupted continuous wave (FMICW) is also used, as this makes communications transmitters work harder. Continuous-wave (CW) modulation gets the highest mean power out of a transmitter, but it gives no range information because it occupies an almost infinitely narrow bandwidth. Frequency-modulated continuous wave (FMCW) involves sweeping the frequency of the CW tone to gain bandwidth and therefore range resolution. FMICW interrupts the transmission process (with a roughly half-on, half-off ratio) so that a co-sited receiver can receive echoes during the 'off' time. Bistatic systems, with a receiver some distance from the transmitter, can make use of FMCW operation to transmit and receive continuously.

Often, the receiver is also a wideband communications receiver, suitably modified to accept the particular modulation chosen. The two outputs of the receiver (one is needed to find the doppler shift and the other to determine its sign) are digitized by an A/D converter and stored in the memory of a computer. For each range gate, the samples are added together coherently for a period that might vary from a few seconds (in the case of missile or aircraft detection) up to about 200 s for oceanographic observations. Beyond 200 s integration, the echoes from the sea add together less coherently. The number of samples averaged in each range gate, before further processing, is known as the *coherent integration, predetection integration* or *pre-summation* figure *n*. After this integration, the reduced number of samples is analysed by a fast Fourier transform (FFT) to produce the doppler spectrum.

If the FFT has M points (usually a power of 2), we can relate the number of pulses transmitted during the integration time t_i to the number processed by the receiver as

$$(\text{PRF}) \times t_i = nM \qquad [\quad] \qquad (10.1)$$

Worked example An HF gw radar is required to operate up to a maximum range of 300 km and must be able to integrate data up to 200 s. If the array processor can handle FFTs up to a maximum size of 1024 points, what pre-summation is needed?

SOLUTION The maximum range of 300 km suggests that a PRF of 500 Hz be used. The maximum integration n that may be needed is given by

$$n = \frac{500 \times 200}{1024} = 97.7$$

COMMENTS In practice, these values would not be chosen. It is not only easier to carry out pre-summation in powers of 2, but also preferable, in order that n can be adjusted to compensate for changes in the FFT length M without requiring changes to the PRF or integration time. Also, 500 Hz would be an unwise choice of PRF because of harmonic interference from 50 Hz European powerlines, and the unambiguous range of 300 km could give rise to second-time-around echoes from the F region of the ionosphere. An HF radar engineer would probably choose

$$325 \text{ Hz} \times 201.6 \text{ s} = 64 \times 1024 \text{ point FFT}$$

In this way the integration time is easily changed by powers of 2 while keeping the unambiguous range constant at 461 km, which is beyond E and F region ionospheric problems.

In the case of FMCW or FMICW modulation, there is a primary FFT, to sort the echoes into the appropriate range bins on the basis of their frequency, followed by a second FFT for each range bin to develop the doppler spectrum.

The extensive reliance of HF radar on digital processing, often carried out in array processors, is the reason for its late development, although some early systems used analogue filters (and analogue integration) to produce a crude form of doppler spectrum.

The range of frequencies used by HF gw radar varies from about 3 to 15 MHz. At the lower end of the band, propagation over the sea is really very good (at 150 km range the extra loss in addition to R^4 is about 7 dB) and the state of the sea makes little difference to this loss. However, there are problems with the size of antennas, with low doppler shifts and with unwanted reflections from the ionosphere. Propagation at higher frequencies

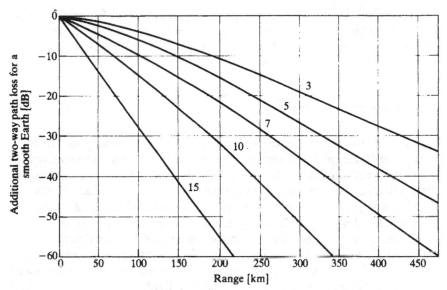

Figure 10.3 The additional two-way loss for propagation over a smooth surface: four-thirds Earth, 4 ohm^{-1} m^{-1} conductivity. Frequency [MHz] is shown next to each curve.

deteriorates rapidly; at 15 MHz the additional loss for a range of 150 km is 42 dB and a rough sea increases these losses. Typical values for the two-way loss over a smooth sea are shown in Fig. 10.3. Computer programs have been developed that predict these losses for any range, frequency, sea state and radar or target height.

One of the main problems with HF radar is the difficulty in finding a sufficiently wide bandwidth clear of interference and communications traffic. Channels are often allocated 3 kHz at a time and it is arduous trying to operate with a bandwidth greater than about 10 kHz. Pulse durations are thus of the order of 100 μs and range cells are 15 km wide; this is larger than is desirable for both military and oceanographic applications.

Dynamic range is also a problem for surface-wave radar. As the example in Chapter 2 showed, dynamic ranges of up to 90 dB can be required if small targets are to be detected against a background of sea clutter. The huge echo from the sea at a low doppler shift, and the small echo from an aircraft or ship at a doppler frequency only a few hertz away, is an extreme test of the linearity of a receiver. Indeed, one form of test used in receiver design involves the injection of two closely spaced tones. The receiver must be linear over a wide dynamic range, otherwise sum and difference frequencies would appear in the spectrum as spurious signals, which would cause false alarms. The linearity, dynamic range and wideband requiredments of OTH radar is the reason why many systems are based around good-quality HF communications equipment.

Despite all the problems of HF gw radar, the rewards for being able to stay in the safety and comfort of the shore, and yet monitor the sea and sky beyond the horizon, are such that there remains considerable interest in this form of radar.

10.3 SKYWAVE RADAR

Skywave (sw) radar makes use of scattering from the ionosphere to look down on a 'footprint' well beyond the horizon. The main applications lie in detecting ballistic missile launches and tracking military and civilian air targets. Because wind direction is relatively easy to extract from sea clutter, storm tracking is also possible. Most skywave radars also have a significant capability for detecting surface targets and for sea sensing. Cruise missiles are not easily detected because they have a small RCS when viewed at long wavelengths. The possible exception to this is submarine-launched cruise missiles, which leave the water vertically and for a few moments may present a large RCS before they settle into horizontal flight.

Unlike gw radar, skywave systems do not need to be near the sea and are perhaps best located inland, where they are safe from storms, salt spray and enemy attack. Skywave radars also tend to use up a lot of ground space, and it is often easier and cheaper to find suitable sites inland, especially since the receiver needs a radio-quiet location. The receivers and transmitters are almost always separated, often by as much as 100 km, thus permitting the use of FMCW modulation to maximize the mean power output of the transmitters.

A typical arrangement for a sw radar is shown in Fig. 10.4. A number of continuously rated high-power transmitters amplify an FMCW waveform and drive a total power of about 500 kW into a 200 m long array. The type of transmitting antenna is chosen to provide a good impedance match to the power amplifiers over a wide bandwidth and to have a beamwidth somewhat wider than the receiving antenna. There are two reasons for using a relatively wide transmitting beam: first, the dwell time may be increased if several narrow receive beams are used simultaneously to survey the area illuminated by the transmitter array (Fig. 10.5); secondly, for a given antenna gain product, the clutter cell size is smaller when the two antenna gains are unequal.

Continuous scanning is not usually possible with sw radar because a finite amount of time is needed to change the beamformer of the transmitting antenna to a new position. It may also be necessary to retune to a new frequency to illuminate a different range, in the same way that HF communications engineers have to. For surveillance of a wide area, a *step scan* technique is used, shown in Fig. 10.5, in which the radar illuminates each sector for several seconds before moving on to the next. As is usual

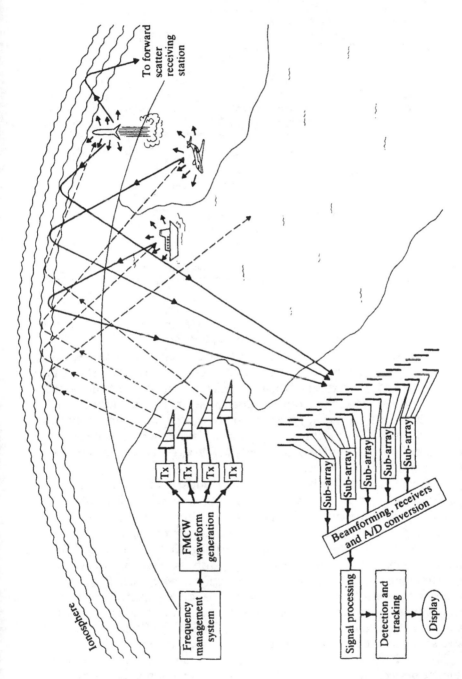

Figure 10.4 Typical skywave radar installation.

213

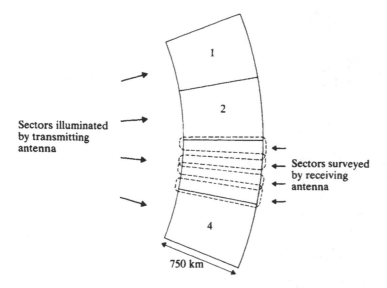

Figure 10.5 Step scanning is used to survey a wide area. Each sector is illuminated for a few seconds and surveyed by four narrow receiving beams, before moving on to the next sector.

with surveillance, the objective is to scan a wide area and return to the starting sector before a target can pass through it undetected.

The transmitting array elements are typically vertically polarized log-periodic, although the North American CONUS-B system uses short, fat dipoles canted 45° off-vertical ('fattening' a dipole increases its bandwidth). The elevation pattern of the array depends on the impedance of the ground, which in turn can be affected by the weather, so usually the ground is levelled and some form of conducting wire mesh is laid down to stabilize the impedance. The improved ground conductivity also helps to predict and control the antenna sidelobes.

The receiving antenna is usually a long (greater than 1 km) array of monopoles, sometimes with a backscreen to reduce the backlobe and occasionally with the monopoles fattened to improve the match at low frequencies. Often, antenna losses at low frequency are not important because the system is *externally noise-limited*, meaning that the external noise (galactic, atmospheric, man-made) arriving with the signal is greater than the internal noise of the receiver. In these circumstances, some losses at the antenna are permissible because noise and signal are attenuated equally and the SNR remains unaffected. But the losses must not be allowed to become too severe—in HF engineering it is regarded as something of a crime to permit losses so large that a system becomes *internally noise-limited*.

The receiving array is connected to a beamformer. This may be an analogue device, but more often these days this function is performed as a digital process using one receiver plus A/D converter for each antenna element, or for each *sub-array* group of elements. Analogue beamformers steer the beam by introducing a frequency-independent time delay (see Chapter 15); digital beamforming usually involves a phase delay that is equivalent to the time delay at one specific frequency f (the operating frequency). For this reason, phase delay beam steering has a restricted bandwidth B, given roughly by

$$B/f \, [\%] \leqslant \Delta\theta \, [\text{degrees}] \qquad (10.2)$$

where $\Delta\theta$ = antenna beamwidth. This bandwidth restriction also implies a limit to the range resolution ΔR and a similar working rule of thumb is given by

$$\Delta R \geqslant D \sin \theta \qquad [\text{m}] \qquad (10.3)$$

where D = length of the antenna array [m] and θ = beam angle off-bore-sight [degrees].

After the beamforming and receivers, the signal and data processing is based extensively on the mapping of the doppler spectrum into range/azimuth resolution cells. These cells are searched using an adaptive thresholding process, and any targets detected are tracked making use of the doppler measurements to compensate for the relatively poor range and azimuth information. The azimuth resolution is poor because even a 1° beam diverges to 35 km at a range of 2000 km. The range resolution is determined by the FM frequency sweep, which in practice is limited to about 100 kHz, corresponding to a range resolution of 1.5 km. Higher bandwidths would run into problems with the array (see Eq. (10.2)) and the frequency-dependent nature of ionospheric propagation, and it is not uncommon to find lower bandwidths being used.

Ionospheric propagation over long ranges is a lossy process and large power × aperture products are needed by sw radars. One method of comparing these radars is to examine their power × aperture products $P_t G_t$ or $P_t G_t G_r$ [dB W], but sometimes the product of the transmitter power, the total antenna gain and the coherent integration time is used; this is known as the figure of merit or FOM [dB J]. Table 10.1 sets out these parameters for the best known systems, as far as they can be determined from published information. Descriptions of current military systems are to be found in *Jane's Radar*[3].

10.4 SKYWAVE PROPAGATION AND FREQUENCY MANAGEMENT

In Chapter 8 we discussed the effect of the ionosphere on the propagation of radio waves. The skywave radar engineer seeks to capitalize on these

Table 10.1 Relative performance of the best known skywave radars

Radar		Power							Resolution (typical)		
Country	Name or type of system	P_t [dB W]	G_t [dB]	G_r [dB]	$P_tG_tG_r$ [dB W]	t [dB s]	FOM [dB J]		$\Delta\theta$ [degrees]	ΔR [km]
Australia	Jindalee-B	52	21	32	105	17	122		0.5	20
Canada/USA	RADC/SARA	42	10	22	74	17	91		3	3
China	OTHB	50	18	26	94	6	100		5	15
France	Valensole	24	20	26	70	25	95		1	22.5
UK	Monostatic	30	21	21	72	17	89		4	75
	Bistatic	40	14	22	76	4	80		5	75
USA	CONUS-B	61	23	28	112	3	115		3	20
	Madre	27	28	22	77	20	97		1	7.5
	ROTH	53	21	34	108	11?	119		0.5	?
	Sea-echo	51	24	24	99	14	113		2	15
	WARF	43	20	30	93	11	104		0.5	0.8
USSR	'Woodpecker'	61	?	?	?	?	?		?	?

effects, particularly refraction, to look beyond the horizon. Of the three ionized regions in the ionosphere, D, E and F, only the E and F regions turn out to be useful for sw radar. The free-electron content of the D region (70–90 km altitude) is insufficient to scatter HF radio waves and the layer acts as an unwanted absorber of radio energy. This absorption occurs because the electrons, while trying to oscillate with the incident radio wave, experience many collisions with neutral air molecules.

The E region is a narrow layer of ionization at about 110 km altitude. The ionization is uneven, but can be very intense at times (owing to auroral effects and intense patches known as sporadic E), and it is essentially a daytime phenomenon. The maximum electron density of approximately 10^{11} [electrons m^{-3}] occurs near midday and corresponds to a critical frequency (called the f_oE) of 2.8 MHz, meaning that this is the highest frequency that will be reflected at vertical incidence. Frequencies higher than f_oE pass straight through the E layer if travelling vertically but can be scattered if they are transmitted at oblique incidence.

The F layer often divides into two layers, F1 and F2, during daytime (see Fig. 10.6). The lower F1 layer lies between 130 and 210 km altitude and has a maximum ionization of about 2×10^{11} [electrons m^{-3}] at noon, giving $f_oF_1 = 4$ MHz. The F2 layer usually has the strongest electron density at about 10^{12} [electrons m^{-3}] during the day ($f_oF_2 = 9$ MHz), and this

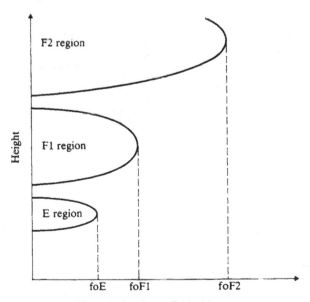

Figure 10.6 The E and F regions of the ionosphere.

layer persists during the night with the ionization falling to around 5×10^{10} [electrons m^{-3}] [$f_oF_2 = 2$ MHz).

The bottom of the ionosphere is not smooth, but has a roughness comparable with the scale sizes of the hills and valleys on the earth's surface, with the added complication that the roughness is moving and changing with the background neutral wind and electric fields. All three ionospheric layers are subject to various disturbances, irregularities and anomalies, and the electron densities may vary with time of day, season, location and solar activity. The ionosphere is thus an uncertain and ever-changing medium of propagation, and there may be many paths by which a signal can pass from the transmitter to the target and back to the receiver—this can cause a single target to appear at many apparent ranges (see Fig. 10.7). There is also a limited band of frequencies that can be used to illuminate a target at any given range and, as with HF communications, the band required may be congested with radio traffic and high background noise levels.

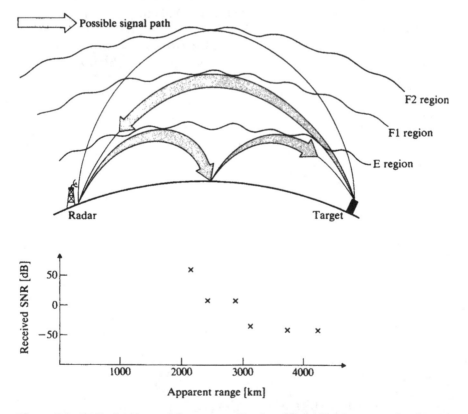

Figure 10.7 Radio signals may follow any combination of E and F region outward and return paths to give several apparent ranges and values of SNR for a signal target.

The answer to the problems of using the ionosphere for OTH radar lies in a process known as *frequency management*. Extensive mathematical modelling and continuous observations of the ionosphere are used to select the best frequency band and then wideband *'look-ahead'* or *channel occupancy monitoring receivers* are used to find a channel free of interference.

Over the years, quite realistic mathematical models of the ionosphere have been constructed (as computer software) to include both local data-bases of observations built up over several solar cycles and maps of the mean ionosphere produced by the CCIR (International Radio Consultative Committee). Statistically, these models provide a good prediction of expected conditions, but on any given day the vagaries of the ionosphere are such that the prediction can be substantially in error. These errors can be minimized by *ionospheric sounding*, which measures the current state of the ionosphere as an aid to selecting the most appropriate model. Occasionally, part of the radar system itself is used for oblique sounding, but it is more usual to use a *vertical sounder*, or *ionosonde*, a small pulsed radar system that transmits upwards and sweeps in frequency to locate the critical frequencies of the E and F layers.

The most common form of display for sounding data is an ionogram, a plot of *virtual height* (not always equal to the true height because of the slowing of radio waves in the ionosphere) against frequency, as shown in Fig. 10.8. After some calculation, Fig. 10.8 can be replotted as true height against electron density to produce a form suitable for ray tracing programs to predict where signals transmitted on any given frequency will end up. This is the information needed by sw radars to choose the best operating frequency.

There are practical methods of ensuring that the frequency management program is working as predicted. The radar operator can ensure that easily

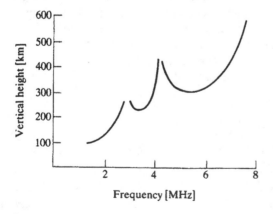

Figure 10.8 A typical ionogram.

recognized targets (cities, coastlines, islands) are detected and appear in the correct locations. Also known targets and transponding devices can be placed within the coverage area to give confidence in the terrain illumination and some feedback on the ionospheric absorption.

Ionospheric modelling, frequency management and calibration form a major part of skywave radar operation and contribute to the expense of these large surveillance radars.

10.5 THE OVER-THE-HORIZON RADAR EQUATION

The skywave radar equation is generally expressed (see Kolosov[4], for example) as

$$\text{SNR} = \frac{P_t G_t G_r \sigma \lambda^2 t_i l_s l_{it} l_{ir}}{(4\pi)^3 R_t^2 R_r^2 F_a k T_0} \qquad [\quad] \qquad (10.4)$$

where P_t = mean transmitted power [W]; $G_t G_r$ = antenna gains relative to an isotropic radiator in free space [   ]; σ = target cross-section, not free-space but as measured, i.e. including any surface reflection effects [m^2]; λ = radar wavelength [m]; t_i = coherent integration time [s]; l_s = system loss factor [   ]; l_{it}, l_{ir} = additional path losses on outward and return paths due to D and E region absorption [   ]; R_t, R_r = outward and return distances to target [m]; and $F_a k T_0$ = apparent external noise level [W Hz^{-1}], with F_a the effective antenna noise factor, usually expressed in dB above kT_0, i.e. above -204 dB W Hz^{-1}. The bandwidth is assumed to be $1/t_i$.

Except during auroral absorption the loss factors l_{it} and l_{ir} may amount to only a few decibels during the day and are near zero at night (details of how to calculate these losses may be found in reference 4). System losses are more severe and are due to a variety of causes related to the antenna design and signal processing. Typical system losses are listed in Table 10.2 and, while they may vary from one system to another, they seldom fall below 10 dB.

The effective antenna noise factor F_a (the result of external noise) increases with increasing wavelength, and over the range 3–15 MHz a realistic value to choose for back-of-the-envelope performance estimates is

$$F_a \simeq 60 - 2f_{\text{MHz}} \qquad [\text{dB}] \qquad (10.5)$$

so that the noise level in a bandwidth B [dB Hz] is given by

$$N_{\text{ext}} = 60 - 2f_{\text{MHz}} - 204 + B \qquad [\text{dB W}] \qquad (10.6)$$

More detailed formulae, representing different noise environments in the USA, are to be found in Skolnik[5]. The classic work in this field is the CCIR Report 322[6], which, although published in 1963, remains a very useful database of global atmospheric noise levels.

Table 10.2 Typical over-the-horizon radar system losses

Loss	Typical value
Transmitter feeder and voltage standing-wave ratio losses	1.0 dB
Transmit antenna ground and resistive losses	0.5 dB
Transmit antenna weighting loss	0.8 dB
Target not in centre of the transmit beam	
Azimuth	0.9 dB
Elevation	0.5 dB
Target not in centre of the receive beam	
Azimuth	0.9 dB
Elevation	0.5 dB
Receive antenna weighting loss	1.1 dB
Receive antenna ground loss	1.0 dB
Degradation by internal noise sources and intermodulation products	0.5 dB
Eclipsing losses due de-ramping (only correctly timed at one range)	
—pulse radars use protection gating giving a similar eclipsing loss	0.9 dB
Receiver filter and digitization losses	0.2 dB
Range weighting loss over one frequency modulation sweep	1.2 dB
Target not in centre of range bin	0.5 dB
Fast Fourier transform weighting loss in doppler processing	2.2 dB
Target not in centre of fast Fourier transform doppler bin	0.4 dB
Range and doppler smearing due to target movement during dwell time	0.6 dB
Loss due to constant false-alarm rate thresholding process	1.3 dB
Total system loss L_s for over-the-horizon radar	15.0 dB

Worked example Could the CONUS-B system detect an aeroplane with a 15 m wingspan at a range of 2000 km?

SOLUTION The size of the aircraft places it in the resonance region at HF wavelengths and the greatest RCS would be observed when the operating frequency was such that the wings formed a short-circuited half-wave dipole. In this case the frequency would be 10 MHz and the RCS is $0.86\lambda^2$ (Table 2.2), giving $\sigma = 28.9$ dB m^2. The radar equation would now look something like:

$$
\begin{aligned}
P_t G_t G_r t_i &= \quad 115.0 \text{ dB J} & \text{FOM, see Table 10.1} \\
\sigma &= \quad 28.9 \text{ dB m}^2 \\
\lambda^2 &= \quad 29.5 \text{ dB m}^2 \\
l_s &= \quad -15.0 \text{ dB} & \text{Typical, see Table 10.2} \\
l_{it}, l_{ir} &= \quad -6.0 \text{ dB} & \text{May be worse at times} \\
1/(4\pi)^3 &= \quad -33.0 \text{ dB} \\
1/(R_t^2 R_r^2) &= -252.0 \text{ dB m}^{-4} \\
1/(F_a k T_0) &= +164.0 \text{ dB W}^{-1} \text{ Hz} \\
\hline
\text{SNR} &= \quad 31.4 \text{ dB}
\end{aligned}
$$

Under these conditions the aircraft would be detected. The CONUS-B data processing software can probably detect air targets with a SNR as low as 6 dB..

COMMENTS In practice, the interference between signals on different ionospheric paths would cause the apparent RCS to fluctuate, and at times the aircraft track might be lost. At night, with only weak F layer propagation paths available, sw radars move to lower frequencies (together with much of the HF communications traffic) and the RCS of the aircraft would fall, making target detection more difficult.

The ground-wave variant of the radar equation differs from the sw version because the presence of the conducting ground modifies the antenna gains and creates the ground-wave effect (see Hall and Barclay[2] and Shearman in Scanlan[7]). A frequently used version of the gw radar equation is

$$\text{SNR} = \frac{P_t G_t' G_r' \sigma' t_i l_s 16 l_p^2}{(4\pi)^3 R^4 F_a k T_0} \quad [\quad] \tag{10.7}$$

where $G_t' G_r'$ = equivalent free-space gain of the antenna relative to isotropic (the actual gain of the antenna is height-dependent, but for an antenna at ground level the gain would be measured as $4G'$); σ' = free-space RCS of the target $[m^2]$; and l_p = additional loss over R^4, including sea-state losses.

It is desirable to produce a common gw and sw form of the radar equation to avoid the type of confusion that can occur when a mixed mode of operation is used, such as a target illuminated by a distant sw radar but detected by a nearby gw receiver. The propagation factor $(4\pi R/\lambda)^2$ can be replaced by the one-way transmission factor f (usually provided by computer modelling of the gw or sw path) to derive the following form of the radar equation for use in all OTH situations;

$$\text{SNR} = \frac{4\pi P_t G_t G_r \sigma m t_i l_s}{F_a k T_0 (f h \lambda)^2} \quad [\quad] \tag{10.8}$$

where P_t = mean transmitted power [W]; $G_t G_r$ = antenna gains as measured *in situ*, relative to an isotropic antenna in free space [ ]; σ = free-space RCS $[m^2]$; m = RCS modification factor [ ] (for gw, $m = 1$; but for sw, m can lie between 0 and 16 depending on the propagation paths, the mean sw value being 4); f = one-way propagation parameter, reducing to $(4\pi R/\lambda)^2$ at short ranges [ ]; and h = target height factor [ ] (for targets on the ground, $h = 1$; for elevated targets, $h = 2$). In this definition of the radar equation, the backscatter RCS of the fully developed sea surface should be taken as -23 dB m^2/m^2.

10.6 PROBLEMS WITH HIGH-FREQUENCY RADAR

Besides the propagation difficulties, there are four other problem areas encountered when operating both gw and sw HF radar. First, there are problems associated with licensing powerful HF transmitters and with getting the frequency allocations clear of other traffic that are wide enough to give good range resolution. These difficulties should not be underestimated.

Secondly, there are often siting difficulties. It can be difficult to find sufficiently large and level radar sites that meet the requirements of being radio-quiet at the receiving end and of presenting no sensitive environmental issues at the transmitter location.

Thirdly, both gw and sw radar experience ionospheric clutter (unwanted backscatter from the moving ionosphere) and meteor clutter, which can at times swamp the presence of wanted echoes. The extent of the meteor echo problem can be evaluated using the equations given in Chapter 9. Skywave radars can also suffer significant polarization losses and focusing/defocusing due to ionospheric effects.

Lastly, military HF radars are susceptible to deliberate jamming and only modest powers are required to disable them, even from relatively long ranges. However, OTH radars are usually classed as early warning devices, rather than as accurate tracking radars, and so they probably fulfil their warning role to some extent if they report that electronic warfare has been initiated.

10.7 SUMMARY

There are two forms of OTH radar, surface-wave and skywave. Surface-wave systems are relatively inexpensive and have found applications in sea sensing (see next chapter), for the defence of localized areas against low-flying missiles and to some extent for monitoring ship traffic.

Skywave radars are used to monitor very large areas of land and sea to search for air targets, ballistic missiles during launch phase and some types of surface target. They also have remote sensing capabilities, especially storm tracking and ocean wave monitoring. These radars are large, powerful, expensive and require sophisticated frequency management systems in order to operate via the ever-changing ionosphere.

Key equations

● For pre-summation and fast Fourier transformation:

$$(\text{PRF}) \times t_i = nM \qquad [\quad]$$

- Restriction for bandwidth in phase delay beam steering:

$$B/f\,[\%] \leqslant \Delta\theta\,[\text{degrees}]$$

- Approximation for effective antenna noise factor:

$$F_a \simeq 60 - 2f_{\text{MHz}} \quad [\text{dB}]$$

- Maximum radar cross-section of a short-circuited dipole (such as aircraft wings) or a vertical monopole shorted to the surface (such as a missile rising out of the sea):

$$\text{RCS} = 0.86\lambda^2 \quad [\text{m}^2]$$

- The radar equation in a form suitable for use in all over-the-horizon situations:

$$\text{SNR} = \frac{4\pi P_t G_t G_r \sigma m t_i l_s}{F_a k T_0 (f h \lambda)^2} \quad [\quad]$$

10.8 REFERENCES

1. CH—the first operational radar, B. T. Neal, *GEC J. Res.*, **3**(2), 73 – 83, 1986.
2. *Radiowave Propagation*, Ed. M. P. M. Hall and L. W. Barclay, Peter Peregrinus for the IEE, Stevenage, Herts, 1989. [There are chapters on both gw and sw propagation.]
3. *Jane's Radar and Electronic Warfare Systems 1989–90*, Jane's Information Group, London, 1989.
4. *Over-the-Horizon Radar*, A. A. Kolosov *et al.*, Artech House, MA, 1987. [This book attempts to lay out the foundations of OTH radar engineering, but the coverage is patchy and biased towards single-hop skywave backscatter systems. The WARF radar is extensively cited.]
5. *Radar Handbook*, Ed. M. I. Skolnik, McGraw-Hill, New York, 1990. [Chapter 24 is dedicated to HF OTH radar.]
6. *World Distribution and Characteristics of Atmospheric Radio Noise*, CCIR Report 322, 1963.
7. *Modern Radar Techniques*, Ed. M. J. B. Scanlan, Collins, Glasgow, 1987. [Chapter 5 is a useful review of skywave propagation, clutter problems and HF remote sensing by Professor E. D. R. Shearman.]
8. Real-time sea-state surveillance with skywave radar, T. M. Georges, J. W. Maresca, J. P. Riley and C. T. Carlson, *IEEE J. Ocean Eng.*, **OE-8**(2), 97–103, 1983.

10.9 FURTHER READING

Over-the-Horizon Radar, Y. A. Mischenko, Zagorizontnaya Radiolokatsiya, Military Publishing House, Ministry of Defence, Moscow, 1972. [The principles of skywave radar are introduced, a number of operational systems examined and some interesting military applications discussed.]
Introduction to Radar Systems, M. I. Skolnik, McGraw-Hill, New York, 1985. [Headrick and Skolnik made one of the first skywave detections of ships in 1974. Consequently, the section in Chapter 14 is an authoritative review.]

10.10 PROBLEMS

10.1 SRI International have used the Wide Aperture Research Facility (WARF) in California to explore the benefits of high range and azimuth resolution for skywave OTH radar[8]. A unique feature of WARF is a 2.5 km long receiving aperture. The system operates over the range 6–30 MHz but, for all the problems below, assume a typical operating frequency of 15 MHz:

    (a) What is the azimuth beamwidth?

    (b) The elevation beamwidth is 36° between half-power points. What is the antenna gain?

    (c) Two 10 kW transmitters feed an antenna of gain 20 dB. What is the total power × aperture product $P_t G_t G_r$?

    (d) What is the figure of merit when an integration time of 12.8 s is used?

10.2 (a) If the transmitter in problem 10.1 is of the FMCW type and sweeps from 15.000 to 15.050 MHz, what is the area of water surveyed by the radar at a range of 1500 km?

    (b) Using the definition in Eq. (10.8) of the reflectivity of sea water as -23 dB m^2 m^{-2}, what would be the RCS of the sea echo?

10.3 What noise level would you expect to find in the problem above?

10.4 What SNR should be received for the sea echo in the problems above? Assume the propagation term $(f h \lambda)$ is 10 dB worse than $(4\pi R)^2/\lambda$.

10.5 An aircraft is detected in the same resolution cell as the sea echo in the problem above. If the tracker reveals the aircraft to be approaching the radar with a doppler shift of 15 Hz and a SNR of 23 dB, what can you deduce about the type of aircraft?

ELEVEN

RADAR REMOTE SENSING

- Measuring waves and currents
- Microwave scatterometry
- Microwave altimetry
- Synthetic aperture radar and its applications

The way radio waves scatter from the earth's surface can be used to investigate large-scale features of land and sea.

11.1 HIGH-FREQUENCY RADAR SCATTERING FROM THE SEA

Random, noise-like processes are often interesting, but the surface of the sea has a particular fascination. People stare at it for hours. For a radar engineer, the challenge is first to find out how radio waves are scattered from the sea and then to turn the problem round and ask, 'How can we determine what is happening at sea from our radar observations?' In some ways, this type of inverse problem is like being given an answer and having to find the question.

What is an ocean wave? We must answer this question before we can make any progress with the problem. Wave motion is carried by particles of water exhibiting circular motion as the wave travels past. This is easily demonstrated by floating a cork, or small piece of wood, on the sea and watching it move as a wave travels by. At the surface of the sea these circular motions can have quite large amplitudes, but they quickly die away with

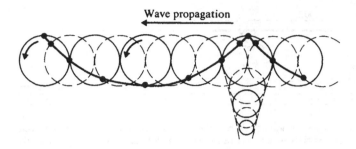

Figure 11.1 The trochoidal shape of sea waves generated by the circular motion of water.

depth (see Fig. 11.1). Submarines soon avoid the effects of wave motion when they submerge.

Figure 11.1 shows why the shape of the wave itself is not a sinusoid but is a rather flat-bottomed/pointed-crest type of shape known as a trochoid. The height of a sea wave is small compared to its length; a typical wave steepness is 1/18 for open ocean sites, with 1/10 rarely exceeded.

Waves are generated by the wind either locally or by distant storms. These latter waves, known as swell, have long wavelengths, which enable them to travel great distances with low attenuation. Locally driven waves are more sophisticated. The wind blows over the water and begins to form short-wavelength waves, but as these build up, and processes become non-linear, the energy is transferred into longer waves with larger amplitudes. For a constant wind speed, an equilibrium is reached with the largest waves having a speed close to that of the wind. If the wind strengthens, then longer and larger waves will be generated, giving the type of spectrum shown in Fig. 11.2.

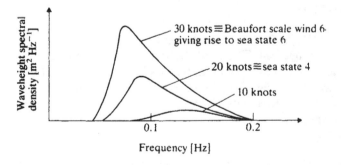

Figure 11.2 A typical waveheight spectrum. This is a *non-directional spectrum* because it gives no information about wave direction. In fact, the largest waves tend to be aligned with the driving wind force but the smaller waves are omnidirectional, giving rise to the complex 'choppy' look of the sea surface.

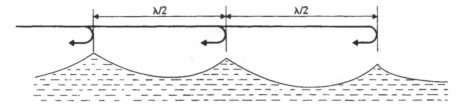

Figure 11.3 When the sea wavelength equals half the radar wavelength, the backscatter from one crest has a path length shorter by a whole radio wavelength than the backscatter from the next crest.

High-frequency (HF) radar, with its capability for making over-the-horizon measurements of a wide area, would seem to be an ideal way of investigating ocean waves from the safety of dry land; but how does the chaotic spectrum of sea waves appear when viewed with the long-wavelength 'eyes' of an HF radar?

The spectrum of Fig. 11.2 shows that all waveheights (and wavelengths) are present up to some maximum value. Any of these waves which are travelling directly towards, or away from, the radar will give rise to backscatter, but those having a length equal to half the radar wavelength ($\lambda/2$) will give a much stronger echo because the reflected radio waves add together coherently as illustrated in Fig. 11.3. This selectivity in the echoing process is known as Bragg resonant scatter, after W.L. Bragg, who first proposed this mechanism to explain the scattering of X-rays from crystals. Bragg scattering is equally important for microwave systems.

Worked example A narrow-beam groundwave HF radar, operating at 10 MHz, has a 15 km range cell size. How much stronger is the echo arising from Bragg resonant scattering than the contribution from waves of similar wavelengths?

SOLUTION At 10 MHz, $\lambda/2 = 15$ m and therefore 1000 sea waves will be present in each range cell. The contribution from n coherent terms is proportional to n^2, whereas incoherent terms add up proportionally to n. The echo power from the Bragg resonant waves will therefore be $(1000)^2/1000 = 1000$ times, or 30 dB, greater than from non-resonant waves of similar amplitude.

COMMENT Since most radar systems use range cells many times larger than the radar wavelength, this enhancement due to Bragg scattering becomes the dominant mechanism irrespective of the radar frequency used.

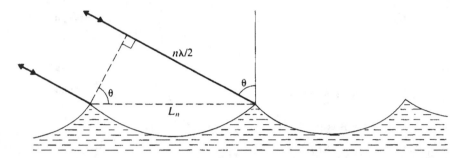

Figure 11.4 Bragg resonance occurs when the path difference between the echo from two wave crests is a multiple of $\lambda/2$.

We now work out the doppler shift of the Bragg resonant sea waves, as observed by the radar. To do this, three pieces of information are needed. The first of these is the *Bragg formula*, which should be slightly more general than the case shown in Fig. 11.3 because the radio wave could be incident on the sea surface at an oblique angle. The full case is shown in Fig. 11.4 and the following equation:

$$L_n = n\lambda/(2 \sin \theta) \quad [\text{m}] \tag{11.1}$$

and for groundwave radar $\sin \theta = 1$. Note the possibility that signals backscattered from the sea waves of wavelength $2 \times \lambda/2$, $3 \times \lambda/2$, etc., will also add coherently at the receiver.

The second piece of information needed is the *dispersion relation* for sea waves, which relates the phase velocity V_n to the sea wavelength L_n:

$$V_n = \sqrt{(gL_n/2\pi)} \quad [\text{m s}^{-1}] \tag{11.2}$$

where g is the acceleration due to gravity.

Finally we need the *doppler formula* of Eq. (1.20), derived in Chapter 1:

$$f_n = \pm(2V_n \sin \theta)/\lambda \quad [\text{Hz}]$$

Putting these together (an exercise left to the reader!) gives

$$f_n = \pm\sqrt{\left(\frac{ngf_{\text{radar}} \sin \theta}{\pi c}\right)} \quad [\text{Hz}] \tag{11.3}$$

At first Eq. (11.3) appears to be a surprising result, because the doppler shift from the waves has a square-root dependence on the radar frequency $f_{\text{radar}} = c/\lambda$, whereas for a 'hard' target (such as a ship or aircraft) f_n is directly related to f_{radar}. On further consideration, it is apparent that changing the radar frequency causes a different set of sea waves to be selected by the Bragg process and these waves have a velocity with a square-root dependence in the dispersion relation. This distinction between waves and hard targets

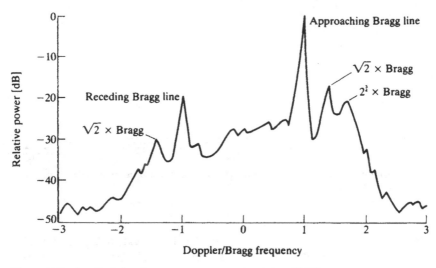

Figure 11.5 A typical doppler power spectrum recorded at 8 MHz.

is a useful property when tracking ships with HF radar. If a ship is travelling at the same speed as the dominant Bragg waves, it usually cannot be detected because the sea echo is much larger than that of the ship. However, a change in radar frequency will cause the two echoes to be separated in the doppler spectrum.

When $n = 1$ in Eq. (11.3) we call this the 'first-order' doppler spectrum and it is the most conspicuous of the signals in the radar output. The features of a doppler spectrum can be observed by attaching a spectrum analyser to an HF radar and using an integration time of around 100 s to display the spectral output. A typical example is shown in Fig. 11.5, where the first-order spectrum consists of the two prominent 'Bragg lines' that represent the velocities of the advancing and receding Bragg resonant waves.

11.2 MEASURING OCEAN CURRENTS

A surprising amount of information can be derived from an inspection of the Bragg lines in a doppler spectrum such as that shown in Fig. 11.5. The two lines are not generally of equal amplitude and the difference is related to the wind direction, such that a wind blowing predominantly towards the radar will cause the approaching Bragg line to be greater than the receding line. The ratio of the Bragg lines is sometimes exploited by skywave radars in order to plot wind maps over large areas of ocean, especially when there is danger from a hurricane.

The effect of a tidal current in the sea is to shift the whole of the doppler spectrum by an amount equal to the radial component of the drift. In practice, current measurements are often made by measuring the displacement of just the two Bragg lines. A single radar can only measure the radial component of the current and it is common practice to use two separate radars, with overlapping beams, in order to derive the full current vector information.

There are now several commercial HF radar systems for measuring tidal currents. Even a short-range system, measuring only coastal currents, can have many applications because of the need to understand more about such problems as coastal erosion, sandbank movements, sewerage distribution, warm effluent distribution (from power stations), docking large container ships and stress on dykes during tidal surges.

The measurement of the Bragg line displacement represents a weighted average of the current over the top layers of the ocean. The radio wave itself does not propagate far into the water (the attenuation with depth being about 60 dB per metre at 3 MHz and 189 dB per metre at 30 MHz). However, the orbital motion of the water particles, shown in Fig. 11.1, means that the radar is sensitive to subsurface currents at depths up to 4 m at 3 MHz and 0.4 m at 30 MHz.

Surface currents measured remotely by HF radar agree well with *in situ* measurements made by current meters and drift measurements and are accepted as being reliable. HF radar also has the advantages of being able to survey a wide area of sea simultaneously and of being unaffected by bad weather and rough seas. It is sometimes possible to gain further insight into small-scale surface current structures by examining any broadening of the Bragg lines or by using several frequencies simultaneously.

11.3 MEASURING WAVES

The extraction of wave information from a doppler power spectrum is quite difficult and depends upon the mechanisms that generate the remainder of the spectrum after the Bragg lines have been explained. There are essentially two classes of second-order processes at work.

The first mechanism arises because the waves are not sinusoids but have the trochoidal form shown in Fig. 11.1. This shape may be represented as a fundamental sine wave plus harmonics all travelling at the same speed. Bragg resonant scatter can take place not only with the principal wave ($n = 1$ in Eq. (11.3)) but also with the harmonics of longer waves ($n = 2, 3, \ldots$). The speed of these waves corresponds to $\sqrt{2}, \sqrt{3}, \ldots$ times the velocity of the first-order Bragg line waves. In practice, the first of these harmonics is quite frequently observed in the doppler power spectrum; for example, such harmonics can be seen in Fig. 11.5.

The second type of process is an electromagnetic effect in which radio waves are scattered back to the radar from two sea waves that are travelling at right angles to each other, reminiscent of the way a corner reflector works. This echo is strongest when the radar bore-sight bisects the 'corner' and gives rise to a feature in the doppler spectrum that is at $2^{3/4}$ times the Bragg line velocity. This effect can just be discerned on the positive side of Fig. 11.5. The remainder of this figure is a continuum of echoes that fall away either side of both Bragg lines in a way reminiscent of a sea wave spectrum such as that shown in Fig. 11.2. Indeed, the continuum is generated by sea waves of all sizes, present in the proportions given by the sea spectrum.

The forward problem is not too difficult; given a knowledge of the wind and sea-state conditions, it is possible to predict fairly reliably the doppler power spectrum that an HF radar will record. The inverse problem is much harder because the formula relating an observed doppler spectrum to an unknown spectrum of waves on the sea surface is a non-linear integral equation. This type of equation is notoriously difficult to invert. In some ways it is like trying to evaluate a polynomial such as $y = a + bx + cx^2 + \cdots$. It is easy to calculate y, given the values on the right-hand side, but try going the other way and solving for $x, a, b, c, \ldots$ for a given value of y. There are obviously many solutions that will add up to the same value of y.

For the HF radar operator there are some added difficulties:

1. The sea-wave spectrum is two-dimensional, whereas observations from a single radar are one-dimensional.
2. Radar echoes are contaminated by noise and interference.
3. The sea spectrum is an average description of the sea surface, but any given observation will see a random surface that will not be exactly represented by the mean.
4. The whole equation is ill-posed in the sense that it contains singularities and ambiguities and because large changes in the sea conditions may cause only small changes in the observations.

Despite all these problems, mathematicians have worked towards finding methods of extracting wave information from HF radar observations and useful techniques are now emerging. There are plenty of incentives to succeed because of the interest in accurate long-term wave measurements. For example, the offshore oil industry would like to be able to predict the largest wave likely to occur in a 50- or 100-year period. Oil-rigs having platforms lower than this extreme waveheight are in danger of being swamped, but building rigs excessively high increases the cost by millions of pounds for each extra metre of height added.

Although waves have been measured for many years by tethered buoys and ship-borne systems, these methods are not without their own problems, and they do not provide the very wide instantaneous coverage and long-term prospects offered by radar remote sensing.

11.4 THE FUTURE OF HIGH-FREQUENCY REMOTE SENSING

Bistatic Operation

Most HF sea sensing systems now use at least two radars to observe a single patch of water in order to increase the amount of information on the two-dimensional sea wave spectrum. If two radars are operated together, further information can be derived by also operating them bistatically such that transmissions by one are received by the other. This arrangement gains a third view of the sea spectrum along a line that bisects the angle between the two radar beams. In future, this may help to increase the accuracy with which wave information can be obtained from HF radar observations.

Sea Ice

Detecting the presence and movement of sea ice is possible using HF radar, and to some extent the thickness of ice may also be estimated. The penetration of radio waves into sea ice is somewhat variable because ice is a complicated substance whose dielectric properties depend on such things as the temperature, the age of the ice and the amount of brine trapped within it.

Skywave mapping of the Greenland ice cap shows that thin ice has a reflectivity similar to that of sea water, and the contour of reflectivity 10 dB lower corresponds to an ice thickness of 1000 m. Groundwave observations show pack ice reflectivities between 2 and 10 dB greater than the reflectivity of the sea, and icebergs appear to be 12 or 13 dB greater.

Ice is a 'hard' target, and drift speeds may be inferred directly from the doppler shift—just as for a ship or an aircraft—although long integration times may be necessary to detect slow ice movements. An alternative approach is to plant transponders on the ice to increase the accuracy of the velocity measurements by imposing a small calibrated frequency shift on the echo. This frequency shift makes it easier to distinguish the ice movement from echoes due to 'stationary' targets such as land, stationary ice, second-order sea clutter and transmitter–receiver breakthrough.

Worked example If the doppler power spectrum shown in Fig. 11.5 were obtained from an HF groundwave radar operating at 8 MHz and an echo from a transponder antenna were just visible as a spike at the centre of the spectrum, what would be the improvement in signal/sea clutter ratio if the antenna impedance were modulated such that half the signal appeared at a doppler shift of 0.85 Hz?

SOLUTION Substituting 8 MHz into Eq. (11.3) shows that the Bragg lines in Fig. 11.5 must have a doppler shift of $f_1 = 0.29$ Hz ($\sin \theta = 1$

for groundwave radar). A doppler shift of 0.85 Hz is close to three times the Bragg line frequency and it would move the transponder echo from the centre of the spectrum to near the edge. At the centre the second-order sea echo is at a power (relative to the peak of the advancing Bragg line) of about -27 dB. Near the edge of the spectrum, the sea clutter has fallen to about -46 dB, which is 19 dB lower. Only half the transponder echo appears in this sideband, giving a signal/clutter gain of $19 - 3 = 16$ dB.

At high latitudes, the mapping of ice movement and the tracking of icebergs by HF radar is likely to remain a subject of considerable scientific, and possibly commercial, interest.

Ship Tracking

A necessary part of wave sensing is the removal of 'ship clutter' appearing as spikes in the doppler spectrum and which may be confused with second-order sea echo features. Conversely, when tracking ships, it is necessary to remove sea clutter. There seems to be a good case for developing integrated systems with two-stage data processing. First, recorded data would be scanned for the presence of ship echoes, which would then be tracked through the recordings, removed and the information transferred to the appropriate users. Secondly, the ocean wave and current information could then be derived from the cleaned-up data, and distributed to interested parties.

Besides the obvious military interest in ship detection and tracking, there are also civil applications for the control of shipping. This already takes place in small, densely populated shipping areas using microwave radar systems, but there is no long-range control of shipping equivalent to air traffic surveillance.

With the wide range of applications for HF radar remote sensing and the possibility of combining many of these functions into a single system, it is likely that this will remain an active and interesting field of radar research for many years.

11.5 MICROWAVE SCATTEROMETRY

High-frequency (HF) radar may offer continuous monitoring of sea conditions in a local area, but it can never hope to provide the global coverage available from satellite-borne systems. Satellite- and air-borne remote sensing frequently involves the use of microwave radars, and one of the simplest and yet most useful of these instruments is the scatterometer.

Microwave scatterometers are radars that look down at the earth's surface from above, normally at oblique angles of incidence between 20° and

60°. Their purpose is to make accurate and precise measurements of the radar cross-section σ^0 of extended targets. The targets may be natural, such as the ocean surface, snow, vegetation, etc., or man-made, such as tarmac or concrete runways. Such targets are of interest in their own right for the purposes of remote sensing, or can be considered as the clutter background against which hard target detection is carried out. Scatterometers typically operate in the gigahertz part of the radio spectrum, and can be mounted on ground-based platforms, such as towers or cranes, or on air-borne or space-borne platforms. They have been used for extensive studies of the variation of σ^0 with frequency, polarization and incidence for many types of target[1].

In order to make accurate measurements of σ^0, scatterometers need to be carefully calibrated. Methods of doing this are discussed by Ulaby *et al.*[2]. In order to achieve precision, they need to counteract the severe fading that is characteristic of observations of the RCS of clutter. Single measurements may give a very imprecise estimate of the mean RCS. In order to get round this problem, scatterometers must be designed to average or integrate many independent measurements. The independent samples may be achieved using temporal, spatial or frequency sampling.

For a scatterometer carried on a moving platform, the doppler spectrum is normally used to obtain independent samples. This is a comparatively complicated topic, and the reader interested in a full treatment would be advised to consult volume 2 of Ulaby *et al.*[2]. However, we can easily explain the basic idea. Let us assume that, as a result of platform motion, doppler frequencies of bandwidth B are returned to the scatterometer antenna (we have already discussed how this would occur in Chapter 6, in the context of SAR). To obtain the greatest possible resolution from the system, the full bandwidth would be used (this is what a SAR does). Another option is to divide the available bandwidth into M non-overlapping frequency bands of equal width. Each sub-band can be used to perform a measurement of the RCS. This gives M independent (because the frequency bands do not overlap) measurements of the RCS, which can be added to achieve greater precision in the estimate. Since the bandwidth used per measurement is decreased by a factor M, the resolution is degraded by the same factor.

Worked example When we are using square-law detection, and the clutter is due to many independent scatterers per resolution cell, the power returned in a single measurement will be exponentially distributed, i.e.

$$p_I(I) = (1/\bar{I})\,e^{-I/\bar{I}} \qquad \text{for } I \geqslant 0$$

Here $\bar{I}$ is the average power we want to estimate. We saw in Chapter 4 that, for this distribution, the standard deviation and mean both have the value $\bar{I}$, and that single measurements are likely to give an unreliable estimate of $\bar{I}$.

If we average M independent measurements from this distribution, the mean is unchanged, but the standard deviation is reduced to $\bar{I}/\sqrt{M}$. The PDF of the average is a chi-squared distribution with $2M$ degrees of freedom, which is tabulated in most statistics books. For M of the order of 10 or greater, this distribution is approximately gaussian with mean $\bar{I}$ and standard deviation $\bar{I}/\sqrt{M}$. How many samples must we average in order to have a 95 per cent probability of estimating the mean power $\bar{I}$ to within $\bar{I}/10$?

To answer this, we need to write the spread of values about the mean in terms of multiples of the standard deviation. Clearly $\bar{I}/10$ is equivalent to $\sqrt{M}/10$ standard deviations from the mean. From Fig. 11.6, we can see that we need each of the shaded areas in the gaussian

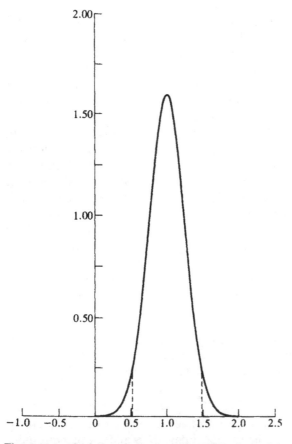

Figure 11.6 The approximate gaussian PDF obtained by averaging measurements from exponential clutter; the shaded areas must have combined area less than 5 per cent to give 95 per cent chance of estimating the mean within a tenth of its value.

PDF to be of area 0.025. Using Table 4.1 and the associated figure, we can see that this occurs when $\Phi(t) = 0.475$, so that the distance from the mean is 1.96 standard deviations. Hence $\sqrt{M} = 19.6$, or we need 384 independent samples. Such large sample sizes are normal if very precise measurements are needed in exponential fading.

Scatterometry is likely to play a major role in current efforts to understand the ocean–atmosphere interactions that play a large part in determining our weather and climate. This is because the RCS of the ocean is dependent on the surface wind velocity (as well as the incidence angle and polarization of the incident wave). Several different formulae have been derived to describe this relationship, all of them empirically derived from observations that are themselves usually hampered by an inadequately detailed description of the wind over the area of water illuminated by the radar.

The relationship between wind speed and measured RCS was first demonstrated using air-borne scatterometers operated by NASA and the US Naval Research Laboratory. Its operational potential was established by observations from the Skylab and Seasat satellites (see Ulaby *et al.*[2] for references and a general account of these developments). Wind vector scatterometery requires radar measurements made in two or more directions to determine the wind vector. Seasat carried a four-beam scatterometer with the beams oriented fore and aft looking at an angle of 45° to the track on either side of the track. This geometry led to a 180° ambiguity in the estimated wind direction. Nonetheless, it was possible to produce global wind field maps over the earth's oceans using measurements from the Seasat scatterometer. A three-beam scatterometer operating at 5.3 GHz is one of the main instruments carried on the European Space Agency's ERS-1 satellite, which was launched on 17 July 1991. The beams are oriented broadside and at 45° to the track (see Fig. 11.7 overleaf) in order to avoid ambiguities. This will provide surface wind measurements in a continuous strip 500 km wide parallel to the satellite track with 50 km spatial resolution. Such measurements can be incorporated into models of atmospheric circulation, and will form an important part of the World Ocean Circulation Experiment.

11.6 RADAR ALTIMETRY

Another source of information on ocean dynamics from satellites is provided by altimeters. These are radar systems designed to produce very precise measurements of the two-way delay suffered by a radar pulse as it propagates vertically to and from the earth's surface. Combined with a knowledge of the satellite orbit, this permits very precise measurements (of the order of 10 cm or less) of surface heights, as long as the surface can be regarded as

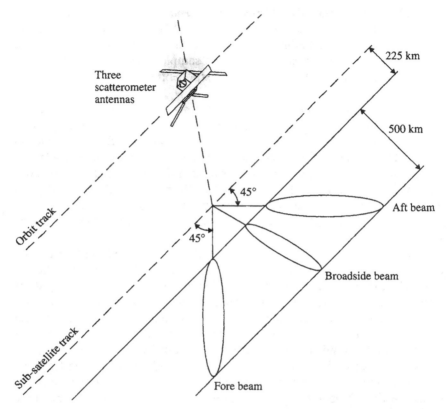

Figure 11.7 The geometry of the space-borne scatterometer carried on the ERS-1 satellite.

approximately planar within the radar footprint. This is the case over the ocean and the earth's great ice shelves. One such device was carried on the Seasat satellite. It provided remarkable images showing that the ocean surface has a topography reflecting the trenches and submerged mountain ranges in the deep oceans. Large-scale currents such as the Gulf Stream can also be extracted from the data because they give rise to deviations from the mean height.

An altimeter operates by emitting a pulse vertically downwards. A *pulse-limited* altimeter uses a very short pulse, and the sequence of events that follow is as outlined below:

1. The leading edge of the pulse strikes the surface, and a return signal begins to propagate back to the receiver.
2. The illuminated area grows like an expanding disc, until the trailing edge of the pulse reaches the ground. As the illuminated area grows rapidly in size, so does the returned power.

3. The illuminated area now assumes the form of an annulus, which propagates outwards along the surface, while preserving its area and hence the returned power. The power received at the earth's surface begins to diminish when the leading edge of this spreading annulus reaches the edge of the main beam, defined by the antenna pattern of the transmit antenna. This in turn leads to a decline in the returned power.

This sequence and the associated returned power is illustrated in Fig. 11.8.

The ERS-1 satellite carries an altimeter that operates at 13.8 GHz, using an antenna of 1.2 m diameter. In order to avoid the generation of very short pulses and the associated rapid processing, a slightly different operating strategy is used. A chirped pulse is transmitted, and the analysis of timing and shape is carried out in the frequency domain. The essential principle is unchanged. Because the altimeter operates on a single frequency, the correction for atmospheric delay (see Chapter 8) must make use of ionospheric models or ancillary data.

The returned waveform in fact yields more information than just the time delay to the surface. Over the ocean, which is anything but a flat surface

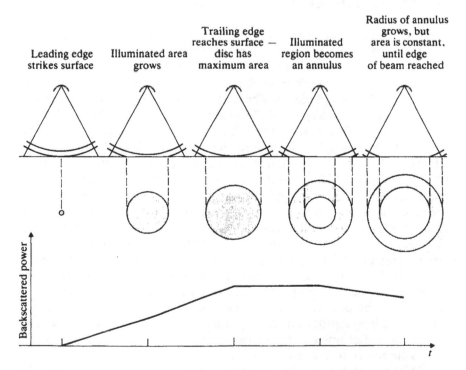

Figure 11.8 The sequence of events that occurs for a single pulse of a pulse-limited altimeter system, and the expected form of the return at the satellite.

on the length scales the altimeter is measuring, the midpoint of the slope of the leading edge of the return corresponds to the mean distance to the sea surface. This is effectively an indication of the *geoid*, on which the earth's gravitational potential is constant. The slope of the leading edge of the return is determined by the spread of surface heights in the altimeter footprint. This gives information on the variance of surface height, i.e. waveheight. The total returned power depends on the small-scale roughness of the ocean surface, which depends on the wind speed; this is the dependence that a scatterometer exploits.

Altimetry can also be used over the ice shelves. This aspect of altimeter operation is less well understood, because of the possibility of penetration of the surface by the radar wave and because we do not have a simple description of the surface, as we do for ocean waves. Nonetheless, this is an active area of research and experiment using the ERS-1 altimeter, because the dynamics of the ice shelves are a critical indicator of the effect of global warming. They also play an important part in the amount of energy absorbed by the earth from the sun, and in earth–atmosphere interactions.

11.7 SYNTHETIC APERTURE RADAR

Perhaps the most sophisticated radar technique (certainly the most technically demanding) used in remote sensing is SAR. The major attractions of SAR are that it can provide high-resolution images of extensive areas of the earth's surface from a platform operating at long ranges, irrespective of weather conditions or darkness. The resistance to weather conditions derives from the use of wavelengths of the order of centimetres, with the X-band (3 cm), C-band (6 cm) and L-band (24 cm) being favoured. Some systems use shorter wavelengths (down into the millimetre bands); these will be adversely affected by precipitation and cloud, for the reasons discussed in Chapter 8. P-band (68 cm) SAR is also in operation. A few reservations are needed here. For air-borne SAR operating at longer ranges (see Fig. 11.9), the propagation path may have to traverse extended regions of rain in the troposphere. The effects discussed in Chapter 8 can then affect performance. For space-borne SAR, the ray paths through the troposphere are much shorter (see Fig. 11.9), and only the most extreme tropical rainfall could have any effect, and even then only at the shorter wavelengths. However, depending on the orbit height, the path lengths through the ionosphere may be long. In the post-sunset equatorial regions and in the auroral zones, ionospheric irregularities can destroy the phase coherence essential to SAR performance. Examples of the consequences for the antenna pattern of a SAR were shown in Fig. 8.11.

This all-weather capability makes SAR a most attractive tool for environmental monitoring in regions affected by clouds or darkness and for

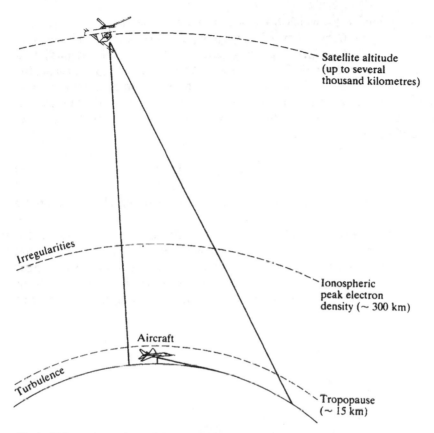

Figure 11.9 A comparison of the propagation paths of air-borne and satellite-borne SARs.

military reconnaissance. An essential element of monitoring changes in the earth's environment is the ability to observe them on a regular and reliable basis. Over many parts of the globe this is not possible using optical and infrared radiation. For large parts of the year the polar regions are in darkness, but observing ice formation and break-up is an important element of safe operation at these latitudes. SAR not only has the ability to operate at these times of the year, but also has sufficient resolution to detect ice leads and edges and icebergs. Ice monitoring is a major aim of the Canadian Radarsat project, which will place an orbital SAR in space in 1995. Other parts of the globe of critical concern are often inaccessible to optical remote sensing. An outstanding example is provided by the tropical rainforests, which are normally affected by cloud, or by smoke if significant amounts of forest burning are taking place. Natural disasters such as floods rarely occur during good weather. The ERS-1 satellite carries a C-band SAR, which will be used to investigate the utility of SAR in monitoring such events.

The principle by which SAR obtains high spatial resolution was first pointed out by Carl Wiley of Goodyear in 1957, and is in fact very straightforward. Consider a moving platform (aircraft or satellite) that carries a pulsed radar pointing sideways to its motion. At a given range, the antenna illuminates a strip of scatterers; as it moves forwards, new scatterers enter the beam, and others leave it (see Fig. 11.10). For a single scatterer, on each pulse of the radar, the two-way propagation to and from the scatterer causes the phase to change by

$$\psi_l = 2 \times \frac{2\pi}{\lambda} s_l \qquad [\text{rad}] \qquad (11.4)$$

Here s_l is the distance between the radar and the scatterer when the lth pulse is emitted.

Buried in the sequence of returns during which the scatterer is illuminated by the antenna is this phase history, which is *unique* to a scatterer in that position. By storing the sequence of returns and correcting for this particular phase history (by subtracting ψ_l from the phase of the lth pulse), the returns from the scatterer can all be brought into phase. Adding all the corrected

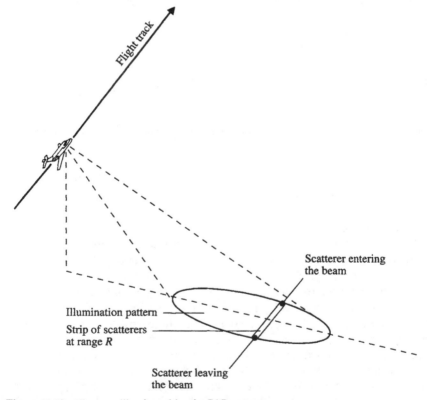

Figure 11.10 The area illuminated by the SAR antenna.

pulses together will cause the returns from the scatterer to add up constructively (see Fig. 11.11). For scatterers with different phase histories, this process will be *destructive*, and they will contribute far less to the summation than the scatterer to which the phase correction is 'tuned'. Scatterers that are close together in the along-track direction will have similar phase histories, and they will have comparable weightings in the summation. The separation that causes one of them to suffer significant destructive interference when the processing is tuned to the position of the other is a measure of the system resolution. From Fig. 11.11 we can see immediately that the SAR simply operates like a big lens or antenna array. The only real difference is that the individual rays through the lens are gathered sequentially rather than simultaneously. In doing this, a large-aperture antenna is synthesized.

Large apertures lead to good angular resolution, and this improvement

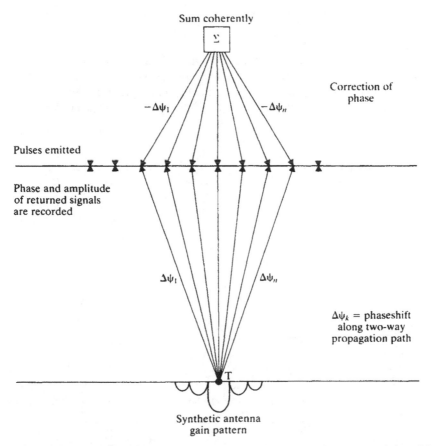

Figure 11.11 The SAR principle from the viewpoint of correction of a sequence of phase delays. The beam pattern of the synthesized antenna is indicated at the bottom of the diagram.

in resolution is the whole purpose of SAR processing. Though the principle is simple, its implementation is very demanding. The treatment below avoids many of the complexities, but contains the central ideas of the implementation. A fuller treatment will be found in Ulaby *et al.*[2]. We can work out the gain in resolution using ideas we have already met. First, let us find the length of the synthesized antenna, using the simplified geometry of Fig. 11.12a. At range R [m], the synthetic antenna has length D [m] given by the distance travelled by the radar platform while a stationary target remains within the main lobe of the real antenna. Thinking of the real beam as sweeping over the target, this means that

$$D = R\beta \qquad [\text{m}] \qquad (11.5)$$

where β is the beamwidth of the real antenna. Since this is given by

$$\beta = \lambda/d \qquad [\text{rad}] \qquad (11.6)$$

where λ is the radar wavelength and d is the real antenna length, we can write

$$D = R\lambda/d \qquad [\text{m}] \qquad (11.7)$$

The key point of this relation is that the synthetic antenna is proportional to range. Longer ranges automatically have longer synthetic antennas. We would now like to use again the relation in Eq. (11.6) between wavelength and antenna length to find the beamwidth of the synthetic antenna. We need

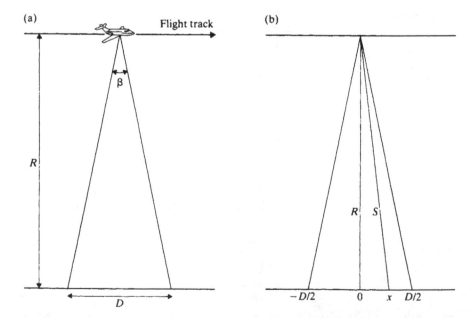

Figure 11.12 A simplified form of the geometry of a SAR system.

to be a little careful, however, because this relation is valid for *one-way* propagation, whereas in SAR we have *two-way* propagation before the synthetic antenna is formed[†]. Taking this into account improves the angular resolution of the SAR by a factor 2, so that it becomes

$$\beta_s = \frac{\lambda}{2D} = \frac{d}{2R} \qquad \text{[rad]} \qquad (11.8)$$

The angular beamwidth of the SAR is inversely proportional to range. But the spatial resolution on the ground will be the product of the angular resolution and the range. Hence the along-track resolution of the SAR is

$$r_a = R\beta_s = d/2 \qquad \text{[m]} \qquad (11.9)$$

This is a key equation. It tells us that the ground resolution of the SAR in the along-track direction is *independent* of the range and also of the wavelength. It also illustrates the curious property that, in order to obtain better resolution we should *shorten* the antenna, in complete contrast to what we expect for real antennas. This property arises because shorter real antennas give rise to longer synthetic antennas (see Eq. (11.7)). In fact, other operating characteristics, such as the width of the usable swath, impose constraints on the minimum length of the antenna[2] (see problem 11.5) and hence on the along-track resolution. Equations (11.8) and (11.9) should be compared with what happens in a conventional radar for which β is given by Eq. (11.6), so that the along-track resolution at range R is $R\beta = R\lambda/d$.

Worked example Compare the azimuthal resolution of a conventional radar antenna and a focused SAR for a radar frequency of 3 GHz and an antenna size of 10 m. Calculate the results for ranges of 10, 100 and 1000 km.

SOLUTION At 3 GHz, the wavelength is 0.1 m. For the focused SAR, the resolution is independent of range at $d/2 = 5$ m. The resolution of an ordinary antenna varies from 100 m at 10 km range to 10 km at a range of 1000 km.

The discussion above only relates to the along-track direction. Cross-track resolution is obtained by transmitting a chirped pulse and pulse compression. In fact, the along-track processing can also be thought of as linear FM with pulse compression. To see this, consider how the range to a stationary scatterer varies as it passes through the beam. Using Fig. 11.12b we can write

$$S^2 = R^2 + x^2 = R^2(1 + x^2/R^2) \qquad \text{[m}^2\text{]} \qquad (11.10)$$

[†] If you are familiar with antenna theory, you will know that Eq. (11.6) is valid in the *far field* of the antenna. For SAR, scatterers are in the far field of the real antenna but in the near field of the synthetic antenna. That causes the quadratic phase variation discussed below.

Since $x \ll R$ in all practical applications of SAR,

$$S = R(1 + x^2/R^2)^{1/2} \approx R + x^2/(2R) \qquad \text{[m]} \qquad (11.11)$$

The phase change along the two-way propagation path is then

$$2 \times \frac{2\pi}{\lambda} S = \frac{4\pi R}{\lambda} + \frac{2\pi x^2}{\lambda R} \qquad \text{[rad]} \qquad (11.12)$$

The first term on the right is a phase delay, which is constant for all scatterers at a fixed range and can be ignored. The second is a phase delay, which is quadratic with distance through the beam. If the platform is travelling with uniform velocity V then $x = Vt$, and the time dependence of phase is

$$\psi(t) = -\frac{2\pi V^2}{\lambda R} t^2 \qquad \text{[rad]} \qquad (11.13)$$

(The minus sign occurs because we must subtract the phase delay to get the phase; the absolute phase reference is unimportant, and we have set it to 0.) Frequency is the time derivative of phase, so this is a linear FM signal; the instantaneous frequency is

$$\omega(t) = -\frac{4\pi V^2}{\lambda R} t \qquad \text{[rad s}^{-1}\text{]} \qquad (11.14)$$

We arrived at exactly the same result from a doppler frequency point of view in Chapter 6 (see Eq. (6.35)).

From Eq. (11.9), we can see that SARs should have spatial resolutions of the order of metres using practical antennas. (For example, the C-band antenna on ERS-1 is of length 10 m, giving a theoretical resolution of 5 m. When a more precise defiinition of resolution is used, and the antenna and processing weighting is taken into account, the quoted resolution is 6 m.) What price is being paid for this resolution? There are two principal costs.

The first of these is that each pixel represents only a single sample from what is, for many types of surface, an extended clutter-type target giving exponential fading. This is the *speckle* phenomenon, which causes major problems in SAR image interpretation. Many systems choose to sacrifice the highest resolution possible from the system in order to reduce the effect of speckle. They do this by averaging the detected powers from several neighbouring pixels (either along- or cross-track, or both). This causes the resolution to be degraded; averaging M pixels in either of the directions degrades the resolution by the same factor in that direction. The gain in accuracy (and reduced variability) of the measured RCS comes about because the pixel distribution changes from an exponential to a chi-squared distribution with $2M$ degrees of freedom, where M is the total number of pixels being averaged for each measurement. As we already remarked when discussing scatterometry, this distribution tends to normality when M

becomes larger; the mean μ is preserved, and the standard deviation becomes $\mu/\sqrt{M}$. This process of averaging powers (or, in some cases, amplitudes) is known as *multi-looking*. Note that the averaging of pixels can be built into the processing; the formation of the full-resolution pixels may never occur. This is, in fact, what a scatterometer does; a scatterometer and a SAR are very closely related instruments. (On ERS-1 they are integrated in what is called the Active Microwave Instrument; while sharing electronics, they use different antennas because of the different operational requirements of the scatterometer.)

The second cost is hidden in Eq. (11.5). To obtain the best possible resolution at range R, the SAR must save and process all the returns in the synthetic aperture of length D. The number of these returns is determined by the PRF. This is, in turn, determined by the Shannon–Whittaker sampling theorem: the sample rate must be at least twice the bandwidth (in Hz). From Eq. (11.14), we can see that the frequency sweep due to a scatterer is determined by the time it is illuminated by the real antenna, i.e. the time during which the platform travels a distance D. This is simply D/V. Hence the total frequency sweep is given by

$$\Delta f = \frac{2VD}{\lambda R} = \frac{2V}{d} \qquad [\text{Hz}] \qquad (11.15)$$

For a broadside beam the frequency sweep is symmetrical about 0 Hz, so that the frequency changes from V/d to $-V/d$ Hz. This gives a bandwidth V/d, which forces the condition

$$\text{PRF} \geqslant 2V/d \qquad [\text{s}^{-1}] \qquad (11.16)$$

Since the time between samples is $1/\text{PRF}$, this is equivalent to the condition that the distance moved by the platform between samples must not exceed $d/2$, i.e. the radar must emit at least two pulses in the time the platform moves a distance equal to the real antenna length.

This means that, to carry out the processing at range R, we must save the returns from $2D/d$ pulses for each pixel we generate, i.e. $2R\lambda/d^2$ pulses are needed. Every one of these pulses must be phase-corrected. This implies a complex multiplication (four real multiplications) for each pulse, using the I and Q channels of the returned signal. Therefore, each image pixel at range R requires $8R\lambda/d^2$ real multiplications. Since new samples are produced at the rate $2V/d$ per second at the minimum sampling rate, a real-time processor would require $16RV\lambda/d^3$ real multiplications per second at range R. The total number of operations must also take into account the number of range gates M_R. If we ignore the variation with range in order to get an order-of-magnitude estimate of the task, we arrive at the need for $M_R RV\lambda/d^3$ multiplications per second for real-time imaging, even without the processing corrections that are normal in most modern processors. Note the dependence on λ; longer wavelengths yield longer synthetic antennas, which give a greater processing load. Note also that the process is described in the time domain, but can also be carried out in the frequency domain.

> **Worked example** The UK Royal Signals and Radar Establishment has operated an X-band SAR for many years. Taking the antenna length as 1 m, let us find the processing requirement for real-time operation if the aircraft is flying at 200 km h^{-1} and producing SAR images at an average range of 60 km, with 1000 range gates.
>
> We have all the information needed; the processing must be able to carry out
>
> $$1000 \times 60\,000 \times \frac{200\,000}{3600} \times 0.03 \approx 10^8 \text{ multiplications per second}$$

The discussion above has assumed that the processing in the cross-track direction does not add significantly to the processing cost. This is true for many air-borne systems, which use surface acoustic wave (SAW) devices to perform the compression of the chirped pulse on reception. For satellite systems and the modern generation of air-borne systems, the whole process is carried out digitally. This increases the estimate of the processing required by three orders of magnitude or more. As a result, the processing requirement derived in the worked example is modest by the standards of current systems. (As an example, the data rate of the ERS-1 SAR is 105×10^6 bits per second. This amount of data cannot be stored on board, and the SAR can only operate if it is within line of sight of a receiving station. All the processing is carried out on the ground.) Consequently, most SAR processors work several orders of magnitude slower than real time, and for satellite SAR real-time processing is considered very much state of the art. Even most air-borne systems do not operate in real time. This occurs because high image quality relies on a number of corrections being applied in the processing. Uncorrected images are often produced as a quick-look product, with a precision product being produced later (and slower). In fact, real-time processing was normal in most early systems, using optical processing elements (see Kovaly[3] for a survey of the early development of SAR). However, these do not offer the flexibility and control of image quality possible with digital methods, and almost all modern systems use digital technology.

Despite these drawbacks, SAR offers our only possibility of reliable high resolution monitoring of many parts of the earth's surface. This has caused major efforts to understand the interaction of the transmitted waves with the surface, which gives rise to the observed backscatter. Nowhere is this more true than over the ocean. The nature of the imaging mechanisms by which SAR reveals the presence of ocean waves has been a subject of heated debate ever since the L-band SAR carried by Seasat revealed the immense wealth of structure in images of the ocean surface. This led to questions about how SAR systems can be optimized for oceanographic purposes. There is also military interest in SAR for applications such as ship tracking and the detection of submarines by locating the way they disturb large-scale underwater wave structures.

Large-scale ocean waves do not give rise to direct backscatter at the centimetre wavelengths used by SAR. The observed variations in backscatter caused by such waves are thought to be due to a modulation they impose on the shorter wavelengths in the sea surface spectrum. SAR is sensitive to these shorter wavelengths by, for example, the Bragg scatter mechanism.

Three processes seem to be important here, although the relative contributions of each may depend on the circumstances of both the sea and the radar:

1. The slope at different parts of a large wave may be important in altering the backscatter from ripple waves. This tilt modulation mechanism is shown in Fig. 11.13.
2. The short ripple waves observed by the SAR are modulated by a hydrodynamic interaction with the longer waves. When the phase velocity of a ripple wave is in the same direction as the drift velocity, the ripple wavelength increases and the amplitude decreases. When the phase and drift velocities are opposed, the ripple wavelength is reduced but the amplitude increases. The orbital motion of the water in large wavelengths causes a periodic modulation of the ripple waves, which may be detected by the SAR (see Fig. 11.14).
3. The velocity perturbations of the Bragg ripple waves imposed by the orbital motion of the water in large waves may cause the SAR image to go in and out of focus, thus revealing the presence of the larger wave. This is known as velocity bunching and is shown in Fig. 11.15.

A difficulty in reconstructing an accurate representation of the ocean surface from its SAR image arises from the sensitivity of SAR to the doppler

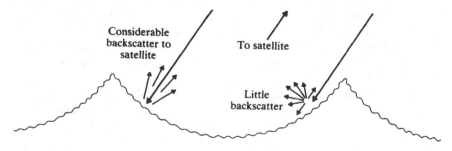

Figure 11.13 Tilt modulation of the backscattered signal.

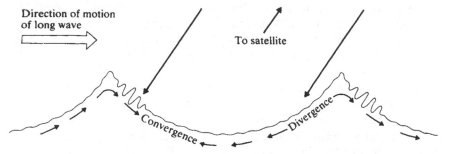

Figure 11.14 Long-wave modulation of Bragg waves by hydrodynamic interaction.

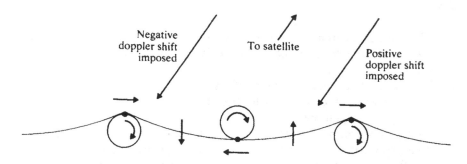

Figure 11.15 Velocity modulation of small waves by the circulation of water in larger waves.

frequency of the scatterers and is related to this third form of modulation. As we saw in Chapter 6, scatterers with a velocity along the line of sight to the radar will be misplaced in the image, unless corrective steps are taken. Since the ocean surface is continuously in motion, this effect must be accounted for in interpreting the data. There is also the problem of speckle. Nonetheless, both theoretical and experimental studies indicate that SAR data can be inverted to recover the ocean wave spectrum, as long as the sea state is not too high. Many interesting questions about the way SAR images waves remain. It is hoped that the major experimental and calibration efforts associated with the ERS-1 SAR will help to resolve some of these questions, and will allow us to assess the viability of SAR as an ocean imaging tool.

The use of SAR over land presents different problems. Much effort has been devoted to investigating the possible utility of SAR for monitoring vegetation, such as agricultural crops. To do this successfully, the SAR needs to operate at resolutions that are significantly less than the typical sizes of fields. This is to allow averaging to remove the speckle. Very accurate estimates of RCS are required because of the comparatively small dynamic range (of the order of 10 dB) of RCS values encountered across different crop types. It has been realized that effective use of SAR for this application is likely to require multifrequency and/multipolarization systems. The ability to gather and relate measurements gathered at different times also seems essential. Most of the work in this field has been carried out using air-borne radars. The use of SAR in forestry has also attracted much attention. This has been accelerated by the focusing of the world's attention on the importance of the great forests for the long-term habitability of the earth. The effects of acid rain, felling and burning are having drastic consequences on the extent and structure of forests throughout the world. An important task for the SAR carried on ERS-1 is to provide information on these processes.

Operating SAR in mountainous regions gives rise to another peculiarity of SAR imaging, known as *lay-over*. Because the SAR positions objects in the cross-track direction by the time of propagation to them, the tops of mountains are apparently moved towards the radar compared to their bottoms, in the two-dimensional image (see Fig. 11.16). In the extreme, as the slope steepens, the top may be closer to the radar than the bottom. This will cause their respective positions in the image to be reversed compared to the normal two-dimensional map projection. The slopes on the lee-side of mountains compared to the radar are also likely to be in radar shadow (Fig. 11.16). Such effects hamper the use of SAR in studies of snow mapping, but can be mitigated by the use of digital terrain models in combination with the SAR data.

The role of SAR in ice mapping has already been mentioned. It seems clear that SAR has a lot to offer in this application. This will include the gathering of information on ice age and thickness, ice type and ice dynamics. Together with location of ice leads and icebergs, this information should help to make operations at higher latitudes safer.

SAR techniques have also been dramatically successful at imaging the planets from orbiting spacecraft. First among these are the images of Venus taken by the Pioneer and Venera 15 and 16 spacecraft. Venus is entirely covered by cloud and nothing can be seen of the surface using ordinary optical methods—indeed, the speed and direction of rotation of the planet were not known until radar revealed it to be spinning in a retrograde fashion. The Pioneer images showed that the surface topography is not unlike that of earth, but with any oceans now gone and with what appears to be an out-of-control greenhouse effect, giving high temperatures and very strong winds. The Magellan spacecraft has now produced much higher-resolution images of Venus, which show a remarkable amount of detail, and should tell us more about the differences between the two planets.

SAR-type processing can also be carried out when the radar is stationary but the target is moving. This technique, known as inverse SAR, is useful for imaging objects, and has been used in radar astronomy to map planets by making use of the differential doppler shifts across the planetary surface,

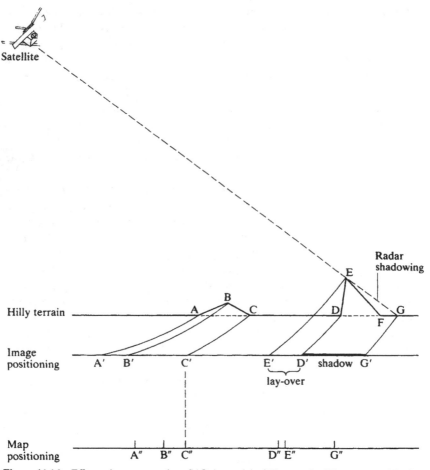

Figure 11.16 Effects that occur when SAR is used in hilly terrain. The map positioning of points has been displayed so that point C coincides in the image and map. The distortion due to terrain effects is obvious, as well as the more extreme effects of lay-over and shadowing.

as viewed from earth. Inverse SAR is a powerful tool in the development of target recognition algorithms because the doppler stability of an object (such as the way a ship rolls) may help to identify it.

The survey of uses for SAR set out above does not come close to summarizing the activity and potentials in this field. We indicate some of the more recent developments in Chapter 15, but much more will be found in Ulaby *et al.*[2] and Elachi[4].

11.8 SUMMARY

The scattering behaviour of the earth's surface at radar wavelengths can provide useful information about many natural processes. Since most applications of remote sensing require large areas to be surveyed, HF radars provide the only useful ground-based systems. These are used for sea sensing. Extensive coverage can be provided by air-borne platforms. Scatterometers mounted on helicopters or aircraft, and air-borne side-looking radars and SARs are employed around the world for a wide range of applications. Perhaps the most exciting prospect is the new generation of space-borne radar instruments, which promise to be a major source of information on global-scale processes. Weather conditions have little effect on these instruments at the range of wavelengths employed (though ionospheric effects have to be allowed for in altimetry and SAR operation). This will allow reliable gathering of surface information, which cannot be guaranteed at optical wavelengths. Radar instruments can also provide information, such as the global wind field over the oceans, that is not available by other means. Much of this information, such as the global wind field or mean sea height, can be gathered at comparatively low spatial resolutions by scatterometry or altimetry (though high-resolution altimetry is attracting much current interest, and we can expect significant progress in the range of problems to which altimetry can be applied). For applications requiring high spatial resolution, SAR provides a possible answer, at the expense of increased system complexity (and cost!) and problems of data interpretation in the presence of speckle. With the launch of ERS-1 and other space-based radars throughout the 1990s, this decade should see major advances in understanding these techniques and their application to monitoring the earth's environment.

Key equations

● The Bragg scattering condition:

$$L_n = n\lambda/(2 \sin \theta) \qquad [\text{m}]$$

● The length of the synthesized antenna:

$$D = R\lambda/d \qquad [\text{m}]$$

● Along-track resolution of a SAR:

$$r_a = d/2 \qquad [\text{m}]$$

● Along-track phase history of a scatterer in the SAR beam:

$$\psi(t) = -\frac{2\pi V^2}{\lambda R} t^2 \qquad [\text{radians}]$$

● Associated linear frequency modulation:

$$\omega(t) = -\frac{4\pi V^2}{\lambda R} t \qquad [\text{radians s}^{-1}]$$

● Total frequency sweep:

$$\Delta f = \frac{2VD}{\lambda R} = \frac{2V}{d} \quad [\text{Hz}]$$

● Sampling condition:

$$\text{PRF} \geqslant 2V/d \quad [\text{Hz}]$$

11.9 REFERENCES

1. *Radar Reflectivity of Land and Sea*, M. W. Long, Lexington Books, Lexington, Mass., 1975.
2. *Microwave Remote Sensing: Active and Passive*, F. T. Ulaby, R. K. Moore and A. K. Fung, vols 1–3, Addison-Wesley, Reading, Mass., 1981.
3. *Synthetic Aperture Radar*, J. J. Kovaly, Artech House, MA, 1976.
4. *Spaceborne Radar Remote Sensing: Applications and Techniques*, C. Elachi, IEEE Press, New York, 1987.

11.10 FURTHER READING

Remote sensing of sea state by radar, D. E. Barrick, chapter 12 in *Remote Sensing of the Troposphere*, Ed. V. E. Derr, NOAA Environmental Research Laboratories, Boulder, Colorado, 1972.
Radio science and oceanography, E. D. R. Shearman, *Radio Sci.*, 18(3), 299–320, 1983. [A good starting place for beginners.]
Extraction of sea state from HF radar: mathematical theory and modelling, B. J. Lipa and D. E. Barrick, *Radio Sci.*, 21(1), 81–100, 1986.
Theory of SAR ocean wave imaging, S. Rotheram, in *Satellite Microwave Remote Sensing*, Ellis Horwood, Chichester, 1983.

11.10 PROBLEMS

11.1 A 15 MHz HF radar is arranged to measure tidal currents across the entrance to a harbour. The dominant Bragg line is observed to have a doppler shift of 0.7 Hz, just as a large ship is seen approaching the harbour. Do you think it is important to convey the result of your observation to the ship?

11.2 A scatterometer mounted on a tower, as shown in the diagram, uses frequency modulation of bandwidth B [Hz] to give a maximum possible time resolution $1/B$ [s]. Show that this gives a slant range resolution

$$r_s = c/(2B)$$

Each of these resolution cells corresponds to an independent measurement of the RCS of the ground (see diagram). Show that the number of independent samples available for averaging is

$$N \approx \frac{2B}{c} h\beta \sec\theta \tan\theta$$

where β is the beamwidth and θ is as shown in the diagram. Plot the variation of N as θ increases from 10° to 70°.

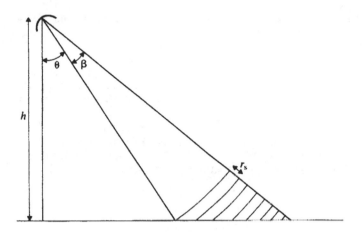

11.3 Assuming a beamwidth of at most a few degrees, show that the annulus generated by a pulse-limited altimeter has approximate area $2\pi Rc\tau$, where the satellite is at altitude R, c is the speed of light and τ is the pulse length. Show also that the annulus forms while the leading edge of the pulse is still in the main beam if the condition

$$\tau < R\beta^2/(8c)$$

is met.

The ERS-1 altimeter operates with a 1.2 m diameter dish at a frequency of 13.8 GHz; the satellite has a nominal altitude of 785 km. If it were operating in pulse-limited form, show that the pulse length would need to be less than 0.11 μs. For a pulse length of 0.1 μs, a loss factor of 50 per cent and a transmitted power of 500 W, what would the signal-to-noise ratio be at the satellite if the surface has an RCS of 5 dB m^2 m^{-2}?

11.4 Compare the azimuth processing requirements of (a) an air-borne SAR travelling at 200 km h^{-1} with a 1 m antenna, operating at a range of 50 km, and (b) a space-borne SAR travelling at 7.5 km s^{-1} with a 10 m antenna, operating at a range of 1000 km, if real-time operation is required. Do the comparison for the cases where both platforms are carrying a C-band (6 cm) and an L-band (24 cm) radar.

11.5 The returns from each pulse of a SAR come from a *swath* defined by the maximum and minimum slant ranges (see diagram). Returns from successive pulses must not overlap on

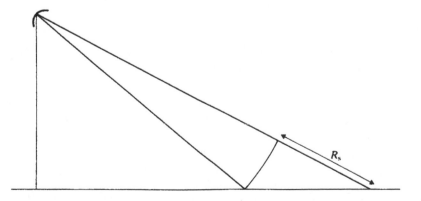

reception. Show that, for a SAR platform whose speed is V and whose antenna has an along-track dimension d, this leads to the two conditions (using Eq. (11.16)

(a) $\mathrm{PRF} \leqslant c/(2R_s)$

(b) $R_s \leqslant cd/(4V)$

What is the widest possible swath that can be imaged by an air-borne SAR if it is carrying a 2 m antenna and travelling at 300 km h^{-1}? What is the best possible resolution in the along-track direction available to a space-borne SAR travelling at 7.5 km s^{-1} if it is designed to produce images from a slant swath 100 km wide?

11.6 The ERS-1 satellite operates at incidence angles varying from 20.1° to 25.9°, with a mid-swath incidence angle of 23°. Are lay-over and shadowing likely to be major problems? Compare this with the effects likely in SAR images gathered by a long-range air-borne SAR operating at an altitude of 10 km and with coverage from 50 to 100 km ground range.

TWELVE

GROUND-PROBING RADAR

- Applications
- Propagation in the ground
- Carrier-free radar
- Wideband antennas

Short-range radar is proving to have many applications for probing the ground and other structures, such as walls.

12.1 INTRODUCTION

Ground-probing radar, also known as *subsurface radar*, is becoming an important subject. The number of applications for ground-probing radar is growing and the technology is beginning to be employed in other radar fields; for example, *impulsive* or *carrier-free radar*, which is in widespread use for ground probing, is now being developed by the military for its anti-Stealth capabilities (see Chapter 14).

The development of ground-probing radar has been driven by its applications, and these are so widespread that the subject has become a fertile area for the creative electronics engineer. Many applications now lie outside ground probing, as the following list of uses indicates:

- *Buried objects:* detecting landmines; locating pipes and cables.
- *Civil engineering:* finding voids and structural flaws; inspection of reinforcing bars; measurement of wall thickness; tunnel and mineshaft location.

- *Security:* locating objects hidden in walls or floors.
- *Archaeology:* grave and burial-mound investigation; locating foundations of buried buildings; investigating interiors of pyramids.
- *Geophysics:* examining structure and strata; surveying subsurface of the Moon and Mars.
- *Earth resources:* coal and peat reserves mapping; locating water tables in dry climates.
- *Ice mapping:* of ice caps and glaciers.

There are many other ways of imaging the subsurface and the interior of structures; these include seismology, ultrasound, magnetometry, impedance imaging and low-frequency induction methods. However, ground-probing radar systems have the advantage of being relatively lightweight, mobile, well focused and easily configured to detect particular types of target. The disadvantage of ground-probing radar is that good radio propagation is only achieved in dry or low-conductivity materials, and the penetration into wet rock or soil may be restricted to a few metres.

Over the years, a great many articles have been published on ground-probing radar in a wide variety of journals and conference proceedings. Fortunately, in 1988, a special issue of the *IEE Proceedings–F* was devoted to this topic[1], and the introductory article presents a comprehensive review that is a useful starting place for those wishing to discover more.

12.2 DESIGNING GROUND-PROBING RADAR SYSTEMS

Subsurface radar design is constrained by several factors; the propagation conditions, the depth (range) resolution, the need for any near-surface (very short-range) measurements and the antenna requirements.

Propagation in the ground is limited by the dielectric losses, which are well documented for different types of ground. The attenuation in most types of ground rises with frequency, and also with water content, because water has a relative permittivity ε_r of about 80 whereas most dry rocks and soils have values of ε_r in the range 2 to 6.

The solution of Maxwell's equations for propagation in the ground shows that radio waves are exponentially damped with depth. The attenuation expressed in dB therefore becomes a linear function with depth, and propagation losses can conveniently be expressed in dB m^{-1}. Typical values for one-way attenuation in different materials are shown in Fig. 12.1.

We learned in Sec. 1.5 that the two dominating factors in the radar equation are the R^4 losses and the noise level; few radar systems can tolerate R^4 losses of more than 200 dB, and so the right-hand scale of Fig. 12.1 instantly gives us a feeling for the sort of depths that can be plumbed using this technique. Geological ground probing is restricted to frequencies below

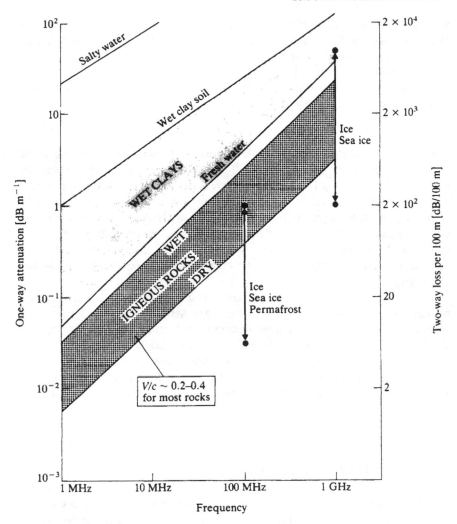

Figure 12.1 Propagation losses in various materials.

100 MHz and buried object location to below 1 GHz. Few systems operate above 1 GHz and antenna constraints effectively prevent operation below 1 MHz.

The range/depth resolution is determined by the bandwidth, but this is not always the bandwidth of the transmitter (which in the case of impulsive radar may be inherently very wide). One limiting factor is the transfer function of the antenna and, although wideband antenna designs are used, the transmitter pulseshape has usually suffered some distortion before it enters the ground. A second limitation is the nature of the ground, which acts like a low-pass filter by attenuating the higher frequencies. The phase velocity v of radio waves in the ground is usually well below the speed of light c and values of v/c are found in the range 0.2–0.4 for many types of rock.

In defining the resolution, these effects can be avoided if the effective bandwidth B_{eff} of the system is defined in terms of the received power spectrum. In this case the resolution is given by

$$\Delta R = v/(2B_{eff}) \qquad [m] \tag{12.1}$$

The range resolution (before any compression) effectively determines the short-range performance and the ability of the radar to locate objects or structures immediately below the surface. A short transmitted pulse means the system can be switched very quickly into receive mode for close-range observations. A long pulse (or other long-duration waveform) means that either the radar must be elevated above the surface or some form of eclipsing loss must be suffered—this is the signal loss that occurs when part of an echo has already returned before the receiver can be switched back to full sensitivity.

The requirement for good range resolution and near-surface performance means that short-pulse/high-B_{eff} systems are needed. This, in turn, implies a high carrier frequency, because conventional radio engineering requires the carrier frequency to be $\geq 100 \times B_{eff}$. Unfortunately, as Fig. 12.1 shows, high frequencies do not propagate well underground, and the dilemma of needing high frequencies for good resolution and low frequencies for good propagation has led to the development of carrier-free radar in order to enjoy the benefits of both.

12.3 CARRIER-FREE RADAR

A sudden electrical impulse, such as a flash of lightning, generates a wide range of radio frequencies and, indeed, thunderstorms cause interference up to the FM radio band and beyond. A radar based on the principle of connecting a discharge device to an antenna, rather than of modulating a carrier frequency, would therefore have an intrinsically very wide bandwidth, which could be exploited to give exceptional range resolution.

The ideal of using signals with a large relative bandwidth has been around for some time, and several books have been published on the subject; see for example Harmuth[2]. Relative bandwidth η has been defined by Harmuth as

$$\eta = \frac{f_H - f_L}{f_H + f_L} \qquad [\quad] \tag{12.2}$$

where f_H, f_L = highest and lowest frequency of interest. For a pure sinusoid

$\eta = 0$; for a typical radar (or communication system) $\eta \simeq 0.01$; but for wideband impulses η can be close to unity. In general, carrier-free radar is regarded as having $\eta \geqslant 0.8$. A well known example of a carrier-free system is a time-domain reflectometer, a device for transmitting short pulses down a cable to locate faults by detecting signals reflected from the discontinuities.

In conventional signals and systems analysis, any signal can be represented by a Fourier series of sinusoids, and for smoothly varying signals the series converges to the signal in a few terms. However, this not a good representation of signals that are rectangular, or contain straight edges, and it may take 40 or 50 terms in a series to get a reasonable approximation to the original signal shape. Signals can also be represented as the sum of a series of other orthogonal expressions, such as exponentials or Legendre polynomials, but the obvious representation of rectangular-type signals is to use a series of rectangular functions, the most common of which are known as *Walsh functions*. In fact, rather than thinking of signal analysis in terms of the Fourier series and frequency, it can be considered in terms of Walsh series and *sequency* (the analogue to frequency, which is a sinewave concept).

Various experiments have been carried out to transmit data using Walsh transmitters and receivers, and it has been suggested by Harmuth[3] that parts of the radio spectrum could be simultaneously shared by all users adopting orthogonal Walsh functions. In general, these ideas have not been widely adopted because of practical engineering difficulties, but the Walsh transform approach remains a useful and widely used concept for the analysis of carrier-free radar waveforms. It may be useful to read up about them if you are going to do any signal or data processing in this field.

A block diagram of a typical carrier-free radar system is shown in Fig. 12.2. There is no RF oscillator and the transmitter is usually a simple device such as an avalanche semiconductor, connected across a high-voltage ($\geqslant 250$ V DC) power supply and triggered into conduction by a pulse from the PRF generator. Because ranges are short, high PRFs are used, sometimes as high as 1 MHz. The transmitter waveform is a pulse with risetime of 1 ns or less, and a decay of perhaps 25 ns as shown in Fig. 12.3. Other types of transmitter can be constructed using pulse-sharpening diodes.

It is difficult and expensive to build T–R switches fast enough for carrier-free radar, and consequently most practical systems use separate antennas for transmitting and receiving. The antennas may be either widely separated to reduce mutual coupling, or co-located using some geometrical arrangement (such as crossed dipoles) to reduce the coupling. Fast T–R switches have been built, but special-purpose Schottky barrier diodes are needed, which must have a low forward impedance, a low reverse capacitance and the capability to withstand the high reverse voltages imposed by the transmitted impulse.

The receiver must have a wide bandwidth, a good dynamic range, and must also be non-dispersive in order to have a fast risetime. Generally

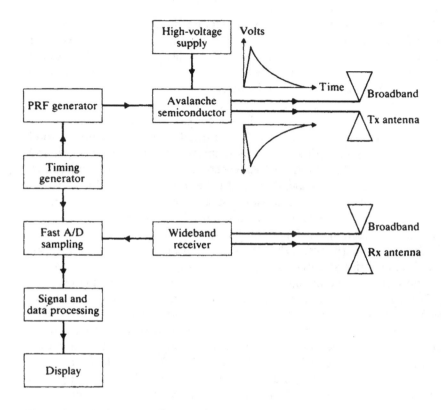

Figure 12.2 Typical carrier-free radar system.

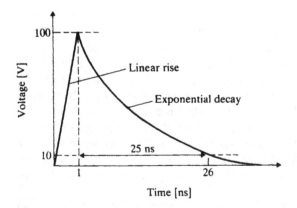

Figure 12.3 Carrier-free radar transmitter impulse waveform.

speaking, below 1 GHz, this does not present too many engineering difficulties and commercial units are available. Sampling the receiver output is much more of a problem. With spectral components up to 1 GHz, direct A/D conversion would be prohibitively expensive with today's technology, and some method of slowing down the data rate is needed. One method commonly used amounts to a 'stroboscopic' or aliasing technique; the echo pulse from a given range gate is successively sampled a few nanoseconds later on each PRF sweep so that the shape of the echo pulse is gradually revealed, as shown in Fig. 12.4. Some time-domain reflectometers work in this way. There is no inherent integration gain in the stroboscopic process, and the SNR is less than for direct sampling. However, since subsurface targets are stationary, there is plenty of time to integrate after stroboscopic sampling, provided the radar is not on a fast-moving platform.

An example of a recording made by a relatively simple carrier-free radar with a stroboscopic receiver is shown in Fig. 12.5a. Physical investigation of the ground confirms that the major features have been revealed correctly, as shown in Fig. 12.5b.

There has been some military interest in the use of carrier-free radar technology. Irregular wideband impulses are not easily detected and located by ESM (electronic support measures) receivers (see Chapter 14), and so these radars have some immunity to countermeasures. It is also difficult to

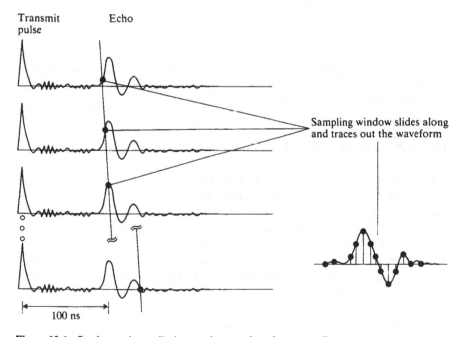

Figure 12.4 Stroboscopic or aliasing receiver to slow down sampling rate.

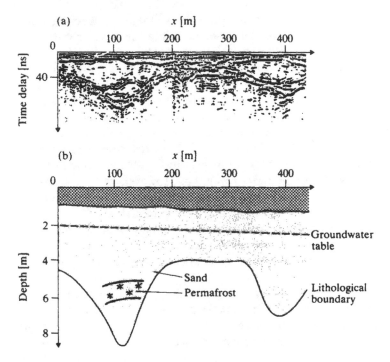

Figure 12.5 (a) Carrier-free radar profile of the ground made from a helicopter at a height of 50 m and travelling at 50 km h^{-1} (after Finkelstein at the Riga Institute of Civil Aviation, Latvia). (b) Geological investigation confirms the observations.

'Stealth' targets against a wideband signal because the common types of radar-absorbing materials (see Chapter 14), such as dielectric and magnetic absorbers, offer significant absorption only over a limited frequency range.

12.4 ANTENNA DESIGNS

The antenna is the weak point in a carrier-free radar system. If a simple dipole antenna were used, the shock excitation of the impulse would cause it to resonate at its fundamental frequency (plus harmonics) and the transmitted pulse would be reduced to a sinusoidal 'ringing' waveform.

A further problem concerns the impulse response of the antenna in a more general sense. The time-domain form of the echo received from the ground is a convolution of the transmitter pulseshape with all of the following:

- The impulse response of the antenna.
- The transmission properties of the ground.

- The shape of the target.
- The ground and antenna on the return path.
- The impulse response of the receiver.

This is shown diagrammatically, in both the time and frequency domains, in Fig. 12.6. If the shape of the target is to be recovered from the shape of the received waveform, there is a need to deconvolve these effects at the data processing stage. This requires the various system impulse responses to be very well characterized.

While the impulse response of the receiver (and T–R switch if fitted) can be measured in the laboratory, the impulse response of the antenna depends upon its reactive coupling with the ground, and will change with height about the ground or if the terrain changes.

The best solution to the ringing and the impulse response problems is to develop very wideband antennas, known as *frequency-independent antennas.* Although truly frequency-independent antennas cannot be made, there are several common design approaches that are used to approximate them (see for example Rumsey[4]).

Resistively loaded dipoles In a conventional dipole, the resonance occurs because the transmitted current travels down the wire, is reflected from the end with a phase reversal, and arrives back at the terminals in phase with the transmitted signal. This resonance can be avoided by loading the dipole resistively along its length, by resistive end-loading or by using radar-absorbing materials along its length. Basically, these techniques are designed so that the impulse travels down the antenna wire and disappears.

Resistively loaded dipoles have been found to be the most useful type of antenna for ground-probing radar, and the inevitable losses associated with the technique have been minimized by applying an increasing exponential taper to the resistive loading along the dipole arms.

Biconical antennas An infinitely long biconical antenna behaves as a perfect transmission line with a real impedance and so becomes an ideal radiator. Practical biconicals do suffer standing waves, but the bandwidth is better than a conventional dipole and improves as the cone angle increases. The two-dimensional version of a biconical, often called a bow-tie or butterfly antenna, is particularly useful in some applications, such as on the wings of an air-borne probing system.

Angle principle If the dimensions of a simple antenna are doubled and the wavelength is also doubled, the performance remains the same. This means that the impedance, polarization, antenna pattern, etc., are invariant to a change of scale in proportion to wavelength[4]. If the shape of the antenna is determined entirely by angles, the performance has to be independent of

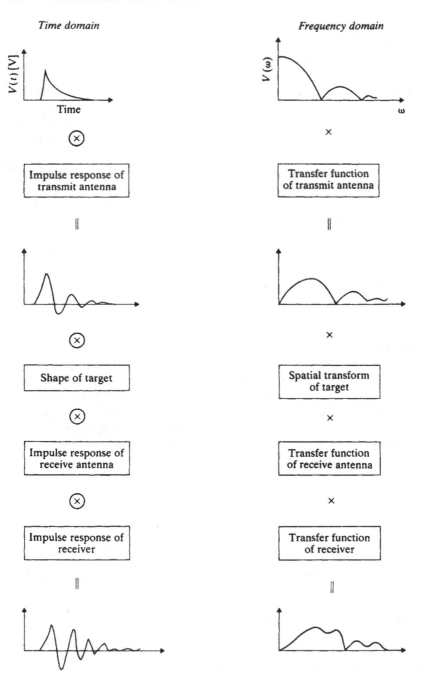

Figure 12.6 The signal engineering in a carrier-free impulse radar.

frequency. In practice, there is an upper and lower limit, but between them a very large bandwidth can be achieved. The most commonly used designs are planar and helical spirals, and log-periodic structures.

Horns At the higher frequencies, horn antennas can be used, and the bandwidth may be extended by careful design of the shape, including internal ridges, and by dielectric loading.

With some of these antenna designs there is an effect in which the active region of the antenna moves to different parts of the antenna as the frequency is changed, causing a frequency-dependent delay. Further details can be found in many antenna books, and for an interesting history of the development of frequency-independent antennas try Chapter 7 of Weeks[5].

12.5 DATA PROCESSING

Some types of modulation, such as frequency-modulated continuous wave (FMCW), Barker or pseudo-noise coding and chirp, may require some initial signal processing to compress the waveform and extract range information. After this, all ground-probing radar systems need to carry out a lot of data processing, often more than with conventional radar, in order to reduce clutter and remove defocusing effects that arise from the propagation. Also, many ground-probing radars are aimed at producing a three-dimensional map of the subsurface, and some form of imaging software is needed.

Clutter from discrete surface scatterers and discontinuities in the surface are a problem for ground-probing radar because propagation in free space is much better than in the ground. *Surface clutter* effects can be minimized by placing the antenna close to or on the ground, because reactive coupling tends to focus the beam into the ground and reduce the stray horizontal radiation. Any motion of the radar can be used to help identify clutter because the phase of front and rear surface echoes changes linearly with distance while subsurface echoes have non-linear phase variations. One disadvantage of the carrier-free impulse approach is that, while there is time delay information, there is no equivalent phase information available (in the time domain) because phase is usually considered to be a single-frequency concept. It may, however, be possible to make some progress with this problem by using the phase of the spectral components of the signal in the frequency domain.

Other receiving problems include *subsurface clutter* (i.e. underground targets not of interest), *transmitter breakthrough* into the receiver (usually due to ringing in the antenna), which causes close-range problems similar to clutter, and radio interference, often from the television bands.

A point scatterer located below the surface will be detected at several

(a)

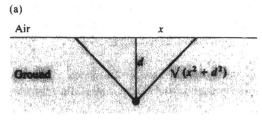

(b)

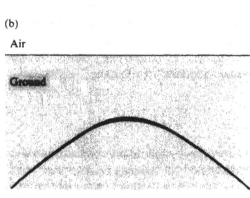

Figure 12.7 Response of a buried scatterer: (a) geometry showing apparent increase of depth with horizontal distance; (b) characteristic hyperbolic shape observed. In practice, the picture is more complicated, especially for elevated antennas, because of diffraction at the surface and interaction between space-wave and surface-wave effects.

different ranges by a moving radar, because of the finite width of the antenna pattern (see Fig. 12.7a). The actual range is given by

$$R = \frac{2}{v}\sqrt{(x^2 + d^2)} \qquad [\text{m}] \qquad (12.3)$$

where d = depth [m], v = propagation velocity and x = horizontal distance from scattering object [m]. Range is normally assumed to be depth, and this causes the plot of a scattering object to have a characteristic hyperbolic shape, as shown in Fig. 12.7b, when plotted on an 'echo-sounder' type of display without further processing. The exponential increase in propagation loss with range means that the resolution of the radar is somewhat better than might be expected from the beamwidth; moving to higher frequencies increases the attenuation further and improves the resolution.

The hyperbolic defocusing of an object can be corrected in the data processing software. The focusing depends on a knowledge of the propagation velocity v and the rate of attenuation, but methods of estimating these parameters from the echo have been evolved. Most systems use some form of linear focusing algorithm, usually performed in the frequency domain, and the software is sometimes extended to include full synthetic aperture processing and high-resolution mapping. It has been found that the length of a synthetic aperture need only be a quarter of the target depth or less. Radio holographic techniques have also been used to image the subsurface; more details and appropriate references can be found in reference 1. It may also be possible to use super-resolution techniques to increase the effective bandwidth of ground-probing radar and improve resolution.

12.6 SUMMARY

Ground-probing radar is one of the few ways of inspecting geological features and locating hidden objects and structural flaws. The technique is not new, but it is only since the advent of fast digital processing and microprocessors that sufficient low-cost signal and data processing has been available to solve many of the problems raised by this type of radar.

It is these very problems that make ground probing such an interesting field of radar. The technique is relatively inexpensive, has many and varied applications and raises abundant deconvolution and mapping problems for signal and data processing engineers to work on. There are further interesting issues to study such as novel antenna design, autonomous control of radar mapping vehicles and methods of three-dimensional data display. There is also plenty of scope for adapting the radar modulation to the type of target expected, and one form of modulation, carrier-free impulses, is becoming a specialized field in its own right, with potential military applications.

An interesting future development concerns a proposed unmanned mission to Mars, which includes a ground-probing radar to explore the geophysics and search for subsurface water–ice. The radar is to be built into the guide-rope of a balloon; during the Martian day the balloon will float through the Martian atmosphere, but during the night it will trail the guide-rope along the ground. The guide-rope itself will form the single antenna of the system. The development of this radar system has led to a lightweight, low-power, single-antenna system that has many applications for air-borne Earth resources mapping, including the search for water in arid regions.

12.7 REFERENCES

1. Special issue on 'subsurface radar', *IEE Proc.—F*, **135**(4), 1988. [A very useful collection of papers.]
2. *Nonsinusoidal Wave for Radar and Radio Communication*, H. F. Harmuth, Academic Press, New York, 1981.

3. *Transmission of Information by Orthogonal Functions*, H. F. Harmuth, Springer-Verlag, Berlin, 1972.
4. *Frequency Independent Antennas*, V. H. Rumsey, Academic Press, New York, 1966.
5. *Antenna Engineering*, W. L. Weeks, McGraw-Hill, New York, 1968.
6. Underground mapping of utility lines using impulse radar, R. Caldecott, M. Poirier, D. Scofea, D. E. Svoboda and A. J. Terzuoli, *IEE Proc.-F*, **135**(4), 343–353, 1988.

12.8 PROBLEMS

12.1 An article in the special issue of the *IEE Proceedings-F* on subsurface radar describes an impulse radar for mapping underground utility lines[6]. The 3 ns transmitter pulse has an amplitude of 200 V and the receiving system uses a preamplifier and a sampling oscilloscope. If the noise level at the sampling head is 10 mV, calculate the minimum gain of the preamplifier if objects are to be detected to a depth of 4.3 m. Assume a round-trip attenuation of 23 dB m^{-1} and 17 dB losses, mainly in the antenna system. Remember that voltage ratios are converted to decibels by $20 \log (V_1 / V_2)$.

12.2 What is the maximum unambiguous PRF that could be used if the radar in problem 12.1 were to be instrumented for a maximum range of 4.3 m? Assume the phase velocity of radio waves in the ground to be $c/3$.

12.3 If the system in problem 12.1 were to pass over a metal grate in the road, what do you think would happen?

12.4 If the radar in problem 12.1 were to be used for probing very dry rock, such as on Mars, how much greater might the penetration into the ground be? Use Fig. 12.1 to estimate the attenuation.

12.5 If a low-frequency version of the radar in problem 12.1 were designed specifically for geological applications and operated at a frequency near 10 MHz, what depths might be probed?

12.6 Calculate the maximum unambiguous PRF and the likely antenna size for the radar in problem 12.5.

THIRTEEN

MULTISTATIC RADAR

- The earliest form of radar
- Basic system configurations
- Ways of locating targets
- Applications

Multistatic radars have some special uses, particularly as a counter to jamming and anti-radar munitions.

13.1 INTRODUCTION

In a monostatic radar system the transmitter and receiver are located at the same place, sometimes sharing a single antenna. If the transmitter and receiver are widely separated, by a baseline typically one-third of the distance to the target, then the system is said to be *bistatic*. If there are several widely distributed receivers associated with a single transmitter, or (more rarely) several transmitters, then the system is *multistatic*. Bistatic radar is thus a subset of multistatic radar, but is by far the most common form used.

The earliest radars had to be bistatic since it was not practicable to build pulse waveform radars and T–R switching. An early example of a bistatic radar was built by Dr Albert Taylor and his assistant, Leo Young, at the US Naval Research Laboratory in 1922. Using a 500 Hz CW modulated transmitter, and a receiver placed in a car, Taylor and Young drove across the Potomac River and detected a variety of targets, including a wooden ship (for more details see Swords[1]).

It is not always necessary to have your own *cooperative transmitter* in a bistatic radar. During World War II the Germans made use of the British 'Chain Home' radar as a *transmitter of opportunity* for their 'Kleine-Heidelberg' receiving system, and were able to give warning of Allied bombing raids over the channel without giving away the position of their ground sites. Since then, bistatic and multistatic techniques have evolved into a sophisticated branch of radar engineering.

13.2 MULTISTATIC CONCEPTS

Figure 13.1 shows the basic arrangement of a multistatic radar system. The transmitter illuminates a sector either by a scanning *pencil beam* or by *floodlighting* the whole area. Usually the transmitter plays no further part in the radar, although it could be part of a monostatic radar in its own right (as was Chain Home during World War II).

The first signal detected by the receiver is the direct pulse, which has travelled along the baseline from the transmitter. This signal initiates the receiver timing, from which the target range will be determined. The receiving antenna system uses either a floodlight beam, a *staring array* of many fixed beams, or a scanning pencil beam carefully synchronized to a scanning transmitter beam.

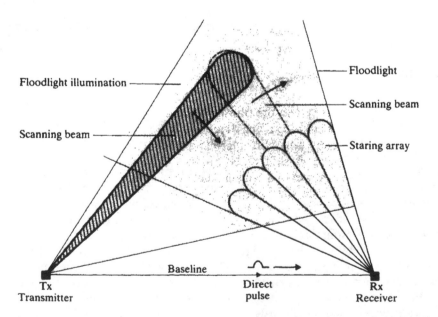

Figure 13.1 Basic antenna arrangements used by bistatic and multistatic radar systems.

Most technical aspects of multistatic systems are the same as for monostatic radar, but there are certain advantages in using bistatic systems, which make them suitable for particular tasks:

- The receiving site cannot be located by electronic warfare receivers (see Chapter 14) and is safe from attack by anti-radiation missiles and highly directional jamming.
- The receiver can be sited in a favourable radio-quiet location, perhaps somewhere where no transmitters are allowed, such as near flammable-liquids stores, gas terminals, etc.
- The transmitter can be placed in a radio-noisy location, for example in a city.
- The receiver requires little or no protection from the transmitter pulse and, because there are no large-amplitude, close-range echoes, the dynamic range requirement is less than for a monostatic radar.
- No T–R switch or duplexer is required; these devices are lossy, expensive and heavy (for air-borne radar applications).
- With certain configurations, less transmitter power is needed to detect targets bistatically than monostatically.
- High PRFs can be used because a bistatic system does not suffer the same range blindness as the equivalent monostatic system.
- Several receivers in a multistatic system increase the probability of detecting a given target since it is unlikely that RCS fading and jamming will affect all the receivers at the same time.
- If the target angle can be measured at both sites, as well as the bistatic range, data can be checked for self-consistency to remove false alarms.
- Bistatic radar measures a doppler component toward neither the transmitter nor the receiver, but along a line between them (actually, along a tangent to the confocal hyperbola); in Sec. 13.5 we discuss bistatic doppler further. If the bistatic doppler component is combined with a monostatic measurement, the full two-dimensional velocity vector can be derived, e.g. the direction and speed of a ship. If a third site is added into the system, true three-dimensional velocity vectors can be derived.

There are, of course, disadvantages to using multistatic radar, which prevent it from being found in more widespread use:

- The geometry of target location is more complicated.
- The synchronization between transmitter and receiver, which is trivial in a monostatic radar, may be quite complicated.
- Multibeam receivers are usually required; these are expensive.
- Two radar sites are required. In some ways this is an advantage because each site may be smaller than for a complete radar and better adapted to the functions of transmitting or receiving. On the other hand, two

radar sites (two aircraft, two pieces of coastline, etc.) generally cost more than a slightly larger single site.

- A further complication in using two sites is that both must be able to see the target. For low-flying targets this increases the problem of terrain obscuration.
- There is usually a 'dead zone' between the sites in which it is difficult to detect targets; see Sec. 13.5.

(a)

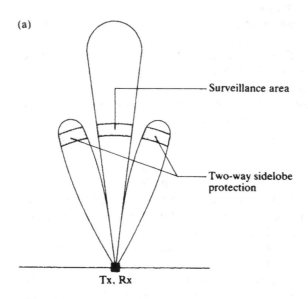

Surveillance area

Two-way sidelobe protection

Tx, Rx

(b)

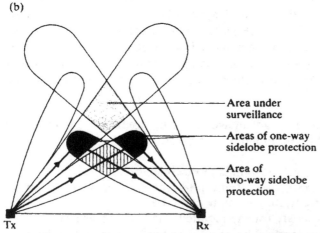

Area under surveillance

Areas of one-way sidelobe protection

Area of two-way sidelobe protection

Tx   Rx

Figure 13.2 (a) The clutter in the sidelobes of a monostatic radar antenna is attenuated on both transmit and receive. (b) With a bistatic system, there are clutter patches illuminated by the sidelobe of one antenna and the mainlobe of the other.

● As a rule, there is only one-way sidelobe protection against clutter. This is illustrated in Fig. 13.2; in Fig. 13.2a the clutter in the sidelobes of a monostatic radar antenna is attenuated on both transmit and receive; but in the case of a bistatic system Fig. 13b, there are clutter patches illuminated by the sidelobes of one antenna and the mainlobe of the other.

13.3 THE BISTATIC RADAR EQUATION

Using the same notation as in Chapter 1, the power flux density arriving at the target is

$$\text{Flux density} = \frac{P_t G_t L_t}{4\pi R_t^2} \quad [\text{W m}^{-2}] \qquad (13.1)$$

where L_t = losses in the transmitting system [ ] and R_t = distance from the transmitter to the target [m]. The power re-radiated towards the receiver by a target of bistatic RCS σ_b is

$$\text{Power re-radiated} = \frac{P_t G_t L_t \sigma_b}{4\pi R_t^2} \quad [\text{W}] \qquad (13.2)$$

The power flux density at the receiving site is therefore

$$\text{Flux density} = \frac{P_t G_t L_t \sigma_b}{4\pi R_t^2} \times \frac{1}{4\pi R_r^2} \quad [\text{W m}^{-2}] \qquad (13.3)$$

where R_r = distance from the target to the receiver [m]. The power received by an antenna of effective aperture $(G_r \lambda^2 / 4\pi)$ [m^2] at the receiving site is

$$P_r = \frac{P_t G_t G_r \sigma_b \lambda^2 L_t L_r}{(4\pi)^3 R_t^2 R_r^2} \quad [\text{W}] \qquad (13.4)$$

where L_r = receiver losses [ ].

When R_t and R_r are about the same as R for a monostatic radar, there is little difference between the two techniques, especially since experimentally the RCS of many targets is found to have similar bistatic and monostatic values (in fact, the bistatic RCS is often close to the monostatic value measured along a line bisecting the bistatic angle). However, there are large gains to be made if either the transmitter or the receiver can be moved closer to a target.

Worked example Compare the power received by a monostatic radar at a range of 100 km from a target, and by a covert receiver, tuned to the same transmitter, at a range of 10 km from the target.

SOLUTION Assuming all other system factors are equal, the R^4 losses are the important factor. For the monostatic case

$$R^4 = 10^{20} \, [\text{m}^4] \equiv 200 \, [\text{dB m}^4]$$

and for the bistatic case

$$R_t^2 R_r^2 = 10^{18} \, [\text{m}^4] \equiv 180 \, [\text{dB m}^4]$$

In the second case, a transmitter of only 1/100th the power of the monostatic radar is required for the same SNR to be received. The bistatic receiver is relatively safe from attack because it is electronically silent, and the transmitter is safer because the transmissions are weaker and consequently harder to detect.

The contour of equal signal strength for a monostatic system, which determines the detection range of the radar, is set by the R^4 factor and is just a sphere centred on the radar. For a bistatic system, the R^4 factor must be replaced by $(R_t \times R_r)^2$, and this leads to the dog's bone type of shape shown in Fig. 13.3a. The intersection of this shape with a plane gives the well known oval of Cassini (Fig. 13.3b). A target moving round this Cassini oval would give a constant SNR at the radar.

(a)

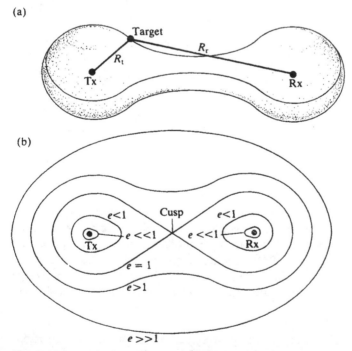

(b)

Figure 13.3 (a) Surface of constant SNR for a bistatic system. (b) The intersection of the surface with a plane gives the ovals of Cassini.

The shape of bistatic constant signal strength ellipsoids can be compared by normalizing them to the baseline $2a$ and defining an ellipticity factor e given by

$$e = \frac{\sqrt{(R_t \times R_r)}}{2a} \quad [\quad] \tag{13.5}$$

If $e \gg 1$, then a is small compared to the target distance and the oval of Cassini is produced. At even larger values of e, these ovals become nearly circular as the radar approaches a monostatic configuration (see Fig. 13.3b).

If $e = 1$, a figure-of-eight shape is created, known as the lemniscate of Bernoulli.

If $e \ll 1$, then small ovals are generated about the transmitter and receiver. As e gets smaller, the target is constrained to lie somewhere on a tiny near-circle close to one or other site. This would be the case in the worked example above, where e would be of the order of 0.16.

13.4 MULTISTATIC TARGET LOCATION

The first piece of information used for target location is the time delay between the transmission of a pulse and the reception of an echo (analogous to range in a monostatic radar). After finding out how to use this, we go on to examine the different possible geometries that can be used for target location.

The range measured by a monostatic radar defines a circle (an *iso-range contour*) centred on the radar, upon which the target must lie (see Fig. 13.4a). In the case of a bistatic system, the total path length $(R_t + R_r)$ defines a three-dimensional elliptical shape shown in Fig. 13.4b. On any given flat plane you can construct the appropriate ellipse by sticking two pins into a sheet of cardboard, tying a piece of cotton between them of a suitable length to represent $(R_t + R_r)$, and drawing out the shape with a pencil while keeping the cotton tight, as in Fig. 13.4c.

The importance of the elliptical path shown in Fig. 13.4b is that usually R_t and R_r cannot be measured independently because the transmitter–receiver timing information only gives the combined distance $2r = (R_t + R_r)$. Thus we know only the total length of the cotton, and not the two separate distances from the pins to the pencil. Note that these iso-range ellipses are not coincident with the constant SNR ovals of Cassini. A target flying around a constant-range ellipse would thus have a different SNR at each place on the contour; this is not true monostatically.

If only time-delay information were available, we could draw the appropriate ellipse, but we would not know where the target lay on that ellipse (in the same way, knowing only the range in a monostatic radar puts the target somewhere on a circle). Angular information is needed to make further progress, and there are several ways in which this may be introduced.

(a)   (b)

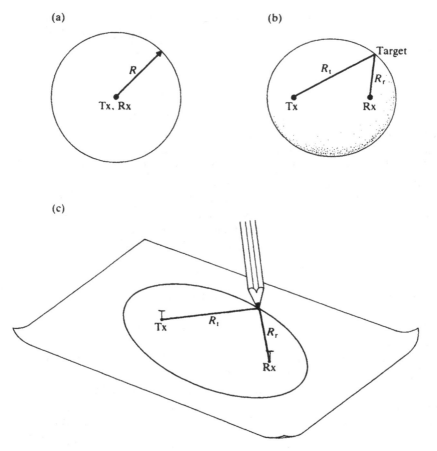

(c)

Figure 13.4 (a) Monostatic iso-range, or constant-range, contour. (b) Three-dimensional bistatic iso-range ellipsoid. (c) How to draw a two-dimensional bistatic range contour.

Elliptical Geometry

This is the obvious, and most common, method of target location (see Fig. 13.5a). It is necessary to know: $2a$, the baseline; $2r = (R_t + R_r)$ the round-trip distance; and θ_r, the target azimuth, measured at the receiver. The range R_r of the target from the receiver can be found by using the cosine formula to solve the triangle in Fig. 13.5a, giving

$$R_r = \frac{r^2 - a^2}{r - a \cos \theta_r} \quad [\text{m}] \tag{13.6}$$

Knowing the range and the azimuth from the receiver, the target position is found in the same way as for a monostatic radar. The target height can be determined by introducing the elevation angle in a similar way.

(a)

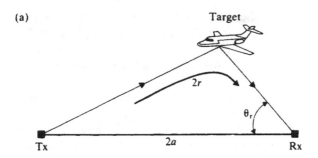

(b)

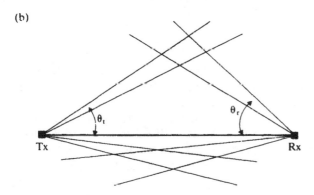

(c)

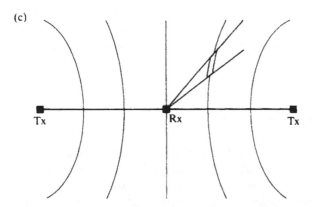

Figure 13.5 Methods of target location: (a) elliptical geometry; (b) theta–theta geometry; (c) hyperbolic geometry.

Theta–Theta Geometry

This is used when the target azimuth is also measured at the transmitter site. The target can be located by the intersection of two beams (see Fig. 13.5b) in much the same way as a radio 'fix' is obtained. The technique is useful when the transmitter is also a monostatic radar in its own right and is measuring the transmitter azimuth. This method can also be employed if the target jams the radar; in this case the radar transmitter is turned off and the target is tracked passively from the two angular measurements made on the jamming signal.

Theta–theta target location requires: $2a$, the baseline; and θ_t, θ_r, the target azimuth measured at the transmitter and receiver.

Note that, as Fig. 13.5b shows, there are places near the baseline where the accuracy is poor. It is generally true to say that the advantages of multistatic radar are maximized when the target is at intermediate range and near-broadside to the baseline.

Hyperbolic Geometry

This is used with two transmitters and one receiver. Here the target is located from the time *difference* between the two transmitter signals arriving at the receiver, which places the transmitter on a hyperbola. The receiver beam is then used to fix the position on the hyperbola, as shown in Fig. 13.5c. The information needed is: $2a$, the baseline; $2r$, the difference in the round-trip distances; and θ_r, the azimuth of the target, measured at the receiver.

From these measurements, the target can be located from the azimuth and the distance to the receiver given by

$$R_r = \frac{a^2 - r^2}{r + a \cos \theta_r} \qquad [\text{m}] \qquad (13.7)$$

A comprehensive table of target location equations, extended to three dimensions, is given by Skolnik[2].

13.5 BISTATIC DOPPLER

Understanding the meaning of bistatic doppler frequency shifts is more difficult than for the monostatic case. The change in the radio frequency arises from the rate of change of the total path length travelled by the signal. For monostatic radar this change in path length equals a change in range, but for a bistatic system the doppler shift f_d is given by

$$f_d = -\frac{1}{\lambda} \frac{d(R_t + R_r)}{dt} \qquad [\text{Hz}] \qquad (13.8)$$

The negative sign indicates a decrease in doppler frequency for an increasing path length.

An interesting case is that of the zero doppler contour, or *isodop*. Figure 13.6a shows an aircraft flying in a circle around a monostatic radar.

(a)   (b)

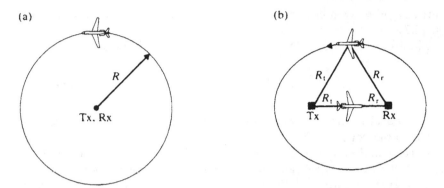

Figure 13.6 Zero isodop contours for (a) monostatic case and (b) bistatic case.

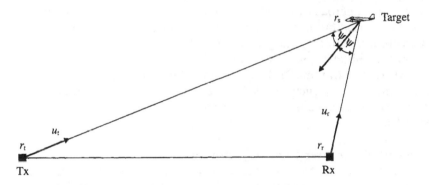

Figure 13.7 Geometry for evaluating the doppler shift.

Since the range does not change, there is no doppler shift and the aircraft would be hidden by ground clutter (although it would not pose much of a threat). In the case of a bistatic system, the iso-range ellipses represent the general zero isodops, but there is also the particular case of an aircraft flying along the baseline from the transmitter to the receiver; as R_t increases, R_r decreases by the same amount, and (as there is no net change in the total path) we see zero doppler again. In general, the doppler frequency represents a component of the target motion along a line roughly bisecting the bistatic angle 2ψ as shown in Fig. 13.7.

When the transmitter and receiver platforms are moving, as in the case of air-borne or ship-borne radar, the doppler frequency is modified by the component of platform velocity along R_t and R_r. The best way to express this mathematically is to define unit vectors u_t and u_r along R_t and R_r respectively, such that the doppler shift becomes

$$f_d = \frac{1}{\lambda}[\dot{r}_t \cdot u_t + \dot{r}_r \cdot u_r - \dot{r}_s \cdot (u_t + u_r)] \qquad [\text{Hz}] \qquad (13.9)$$

where r_s, r_t, r_r = position vectors of the target, transmitter and receiver (see Fig. 13.7).

Monostatically, Eq. (13.9) reduces to the normal formula

$$f_d = \frac{2}{\lambda}(\dot{r}_t - \dot{r}_s) \cdot u_t = -\frac{2v}{\lambda} \quad [\text{Hz}] \qquad (13.10)$$

There is no distinction between target and system platform speed contributions to the doppler shift.

In general, the doppler shift of a target can be evaluated from Eq. (13.9), although the mathematics can get quite complicated for air-borne bistatic radar when transmitter, receiver and target are all moving in different directions at different heights. Bistatic clutter is more difficult to determine, especially when wide beamwidths cause *clutter spreading* in the doppler spectrum. One approach to predicting the masking of targets by clutter is to plot clutter isodops and calculate the position on the locus of a target track where the clutter and target speeds will merge.

A further bistatic complication concerns the synchronization of the receiver with the transmitter in both time (in order to measure range information) and frequency (needed to determine the doppler shift). There are three ways of accomplishing the synchronization, although all have their drawbacks:

● Locking the receiver local oscillator to the direct transmitter signal received along the baseline. Platform motion at either end will cause errors because this reference signal will be doppler-shifted when it arrives.

● Locking both transmitter and receiver oscillators to an external time signal or beacon. This renders the radar dependent on an external signal, which might fail, or which might be attacked in wartime.

● Using a rubidium frequency standard at both ends. This is expensive, but for many applications the frequency stability is adequate for doppler processing requirements. Relativistic effects due to platform motion are rarely a problem.

One must not become too gloomy about the complications posed by bistatic doppler and the problems of doppler spreading of clutter. First, this is a subject amenable to computer simulation and prediction. Secondly, the transmitter and receiver platform speeds can be arranged for the benefit of the radar user. An example of this is *clutter tuning*, which concerns air-borne detection of surface vehicles such as tanks; normally this is a difficult task for monostatic radar because the high platform speed of the aircraft and the finite width of the antenna beam causes a significant clutter spreading problem (see problem 3.5 at the end of Chapter 3). Using a bistatic arrangement, the speeds of the transmitter and receiver aircraft can be adjusted to give zero platform doppler shift over the area of interest and the capability of being very sensitive to slowly moving land targets. This procedure is known as engaging *complementary motion*.

13.6 APPLICATIONS

Perhaps the classic application of multistatic radar is in air defence, because of the invulnerability of the receivers. A common design approach is to have a monostatic radar with a narrow scanning beam illuminating a given sector. A second (possibly concealed) receiver is used to increase the coverage area and to evade jamming and attack. This second receiver needs to be equipped with a staring array of antenna beams and to switch rapidly through them at the appropriate rate in order to follow the region of sky illuminated by the transmitter pulse—a process known as *pulse chasing* (see Fig. 13.8). More receivers may be used to increase further the coverage and the resistance to jamming.

The additional doppler information provided by a multistatic system can be valuable for sea sensing. The problems of inverting the doppler spectrum to find the sea state, discussed in Chapter 11, are eased somewhat by using two radars to probe the sea surface from different directions. If the two radars are also operated as a bistatic pair, then a third view is generated crossing the bistatic angle, which provides more information for very little extra cost.

Bistatic techniques are used extensively for targeting missiles; typically, a monostatic radar illuminates a target and the missile carries an additional

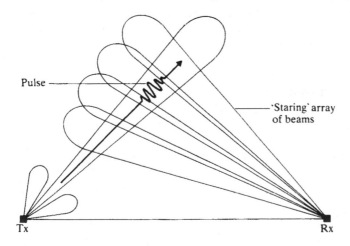

Figure 13.8   Pulse chasing.

receiver so that it can keep watch on the target while following an optimal trajectory that carries it outside the radar beam. Further details are given in Chapter 3.

All these applications require bistatic and multistatic tracking algorithms to be developed. Transferring the measurements onto cartesian coordinates means that non-linear filtering is necessary, which in turn means more computer processing power and complexity. If the target is detected by more than one receiver simultaneously, then tracking accuracy can be improved by integrating all the data into the filter.

13.7 SUMMARY

Bistatic and multistatic radars differ in a number of ways from equivalent monostatic systems. Some of the differences are unwanted and merely add to the system cost and engineering complexity, but others can be exploited to produce significant operational advantages.

Perhaps the biggest single advantage of multistatic operation is the reduced vulnerability of the receiver to jamming or physical attack by anti-radiation missiles. Another big advantage, in the case of mobile systems, is that clutter tuning may be employed to increase the sensitivity of the system to slowly moving targets. A higher PRF can be used with bistatic radar than with monostatic, which also aids sub-clutter visibility.

The original reason for the application of bistatic techniques in the 1920s, which was to isolate the receiver from the transmitter when CW waveforms were used, remains valid today and most large HF over-the-horizon radars have transmitters and receivers separated by over 100 km (see Chapter 10). These radars also follow the typical bistatic format of a single transmitter beam and several narrow receiver beams. The cost of multibeam receiving antennas, and other system complexities, means that multistatic radar will never replace monostatic radar in general usage, but in certain applications it remains a powerful technique.

Key equations

● The bistatic radar equation:

$$P_r = \frac{P_t G_t G_r \sigma_b \lambda^2 L_t L_r}{(4\pi)^3 R_t^2 R_r^2} \quad [\text{W}]$$

● Distance from target to receiver for elliptical geometry:

$$R_r = \frac{r^2 - a^2}{r - a \cos \theta_r} \quad [\text{m}]$$

● The bistatic doppler frequency shift:

$$f_d = -\frac{1}{\lambda} \frac{d(R_t + R_r)}{dt} \quad [\text{Hz}]$$

13.8 REFERENCES

1. *Technical History of the Beginnings of RADAR*, S. S. Swords, Peter Peregrinus for the IEE, Stevenage, Herts, 1986.
2. *Radar Handbook*, Ed. M. E. Skolnik, McGraw-Hill, New York, 1990.

13.9 FURTHER READING

Bistatic Radar, N. J. Willis, Artech House, Norwood, MA, 1991.
The geometry of bistatic radar systems, M. C. Jackson, *IEE Proc.-F*, **133**(7), 604–612, 1986.
[This is a particularly useful summary of bistatic geometry, which includes coverage, bistatic resolution cells and PRF.]

13.10 PROBLEMS

13.1 A bistatic short-range air surveillance radar operates at a frequency of 3 GHz. An aircraft flies directly towards the transmitter at 100 m s^{-1}.

(a) At the point where its path brings it at right angles to the receiving beam, what doppler shift is recorded?

(b) If the transmitter is a monostatic radar in its own right, what doppler shift would it detect?

(c) Explain the difference.

13.2 Assume that the radar in problem 13.1 has a baseline of 30 km. An echo from a target arrives at the receiver at an angle of 30° to the baseline and 200 μs after the direct pulse from the transmitter arrives.

(a) What is the round-trip distance $2r$?

(b) Calculate the range of the aircraft from the receiver.

(c) Calculate the range of the aircraft from the transmitter.

Neglect the curvature of the Earth and the aircraft height and treat this as a two-dimensional problem.

13.3 (a) In the problem above, if the target were travelling parallel to the baseline at 100 m s^{-1}, what doppler shift would be recorded?

(b) Calculate the bistatic angle.

13.4 A 1 GHz, 1 W transmitting buoy is dropped in a shipping channel such that ships must pass within 1 km of it. Overhead, an aircraft is listening for the echoes with an antenna of gain 20 dB. Assume that the buoy has no useful antenna gain (assume a gain of unity) and the aircraft receiver requires a signal strength of -130 dB W for detection.

(a) What is the maximum height at which the aircraft could detect a ship of bistatic RCS 30 dB m^2 (neglect losses).

(b) Would the air-borne receiver have a dynamic range problem with the direct signal arriving only a few microseconds before the echo?

(c) If the aircraft had its own monostatic radar on board, would this need more, or less, power than the transmitter on the buoy?

FOURTEEN

ELECTRONIC WARFARE

- The jargon of electronic warfare
- The effectiveness of electronic warfare
- Measures and countermeasures
- Stealth

The jamming of radar systems has been going on since radar was invented, and is now very sophisticated.

14.1 OBJECTIVES AND DEFINITIONS

A modern war is won by the side that best exploits the electromagnetic (EM) spectrum. During the Gulf War in 1991, Iraqi radars were blinded by electronic countermeasures and the opposing Allied forces were able to gain control of the air. Once air superiority has been achieved, victory is often just a matter of time. One of the key elements in a modern battle is how well the attacking and defending radars of each side stand up to the electromagnetic onslaught from the other.

The objectives of *electronic warfare* (EW) systems are simple:

- To *deny* the enemy the effective use of the EM spectrum.
- To *protect* friendly EM systems against EW attack.

There are three basic divisions of electronic surveillance:

1. Electronic support measures (ESM). This cryptic name describes the whole field of passive electronic eavesdropping using special-purpose *intercept receivers.* Within ESM, there are two subclasses, *COMINT* (gathering communications intelligence) and *ELINT* (gathering electronic intelligence), with which we are concerned. ESM has quite a long history, and during World War I, the interception of army field telephone and naval radio traffic was exploited successfully by both sides.

2. *Electronic countermeasures* (ECM). This is the active disruption of enemy EM communications and surveillance, and is often known as *jamming.* ECM also dates back to World War I, when wireless communications were deliberately interfered with by jammers. The difficulties with ECM are that it not only presents a possible interruption to your own EM signals, but it gives away your hostile intentions and can act as a beacon for enemy munitions to home in on.

3. *Electronic counter-countermeasures* (ECCM or anti-ECM). For every measure there is a countermeasure, and indefinite counter-counter-measures. We will discuss some ECCM techniques in Sec. 14.3, but the subject can be summarized as the art of making radars and other electronic systems *ECM-resistant* by trying to null out the effects of jammers or decoys.

A typical radar EW scenario is shown in Fig. 14.1. A coastal air defence radar illuminates a flight of incoming enemy bombers, which protect

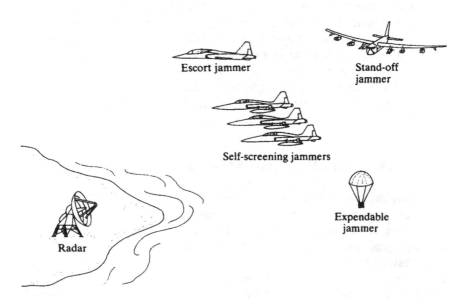

Figure 14.1 A typical radar electronic warfare scenario.

themselves by launching an *anti-radiation missile* (ARM) at the radar and by turning on their *self-screening jammers*. ARMs are sometimes designed to be fired at the sidelobes of a radar transmitting antenna to make the detection of the tiny missile itself even more difficult. The defence against ARMs includes separate radars to detect them, decoy transmitters, or even the use of bistatic systems using relatively inexpensive transmitters that are easily replaced.

The attacking flight is accompanied by an additional aircraft, which carries no bombs at all; this *escort jammer* (EJ) is packed solely with EW equipment to provide even more sophisticated confusion to the radar operator.

The radar uses its ECCM capabilities to evade these jammers and may fire *home-on jammer* (HOJ) missiles at the aircraft. HOJs use a similar technology to the ARMs. There is one weapon the defensive system cannot reach; at long range, within the safety of its own territory, a large enemy aircraft loiters at high altitude providing refined high-power jamming and other EW support; this is known as a *stand-off jammer* (SOJ).

One last type of ECM that may be used is an *expendable jammer*, which can be dropped by aircraft, carried by a remotely piloted vehicle (RPV), fired by artillery or even delivered by enemy agents, who may park an innocuous looking car, full of ECM electronics, near to a radar installation. Although expendables are short-term devices, they can provide enough time for a target to escape.

Unfortunately, the whole of this branch of electronic engineering is riddled with the type of abbreviations and acronyms we have seen above. You really do need to know the meaning of many of them if you wish to converse knowledgeably about military radar systems. A more complete classification of EW is given in Johnston in reference 1, and a catalogue of current EW systems may be found in *Jane's Radar*[2].

The effectiveness of all these EW techniques and weapons depends upon the ingenuity of their design and the performance of the jammer versus the echo power from the radar signal.

14.2 NOISE JAMMING AND THE RADAR EQUATION

Swamping radar or communication signals by transmitting high-power noise is one of the earliest and most widely used forms of jamming. In a typical application, an air-borne ESM receiver is used to detect hostile EM emissions and the signals are analysed to identify the radar. If the radar is thought to pose a threat, the pilot is warned by a *radar warning receiver* (RWR) display in the cockpit. The RWR display can give a simple indication of the target threat sector on the head-up display, a range-bearing display (as on the F15 Eagle) or an alphanumeric display (as used on the Tornado F2). The usual response to a threat is for the aircraft ECM system to begin automatically jamming the radar by transmitting noise on the same frequency.

When a target is detected clear of clutter problems, receiver noise becomes the factor limiting the radar performance. The effect of the jamming signal is to raise the noise floor and reduce the operating range. Since the radar signal suffers R^4 losses and the jamming signal only has to travel one way and experience R^2 losses, the advantage is usually held by the jammer.

Worked example If an L-band surveillance radar of the type designed in Chapter 2 used a peak transmitter power of 200 kW and an antenna gain of 36 dB, at what range could it detect an aircraft with an RCS of 10 dB m^2? What increase in noise power would reduce the radar detection range to one-third? If the jammer antenna on the aircraft had no useful gain, how powerful does the ECM transmitter need to be?

SOLUTION The maximum detection range is given by

$$R_{max} = \left(\frac{P_t G_t G_r \sigma \lambda^2 L_{sys}}{(4\pi)^3 N (SNR)} \right)^{1/4}$$

Putting in values:

$$
\begin{aligned}
P_t &= & 53 \ \text{dB W} & \\
G_t G_r &= & 72 \ \text{dB} & \\
\sigma &= & 10 \ \text{dB m}^2 & \\
\lambda^2 &= & -12.7 \ \text{dB m}^2 & \qquad \lambda = 0.231 \ \text{m} \\
L_{sys} &= & -7.0 \ \text{dB} & \qquad \text{Typical} \\
1/(4\pi)^3 &= & -33.0 \ \text{dB} & \\
1/N &= & +145.8 \ \text{dB W}^{-1} & \qquad 3 \ \mu\text{s pulse, 3 dB noise figure} \\
1/(SNR) &= & -13.0 \ \text{dB} & \\
\hline
R_{max}^4 &= & 215.1 \ \text{dB m}^4 & \\
R_{max} &= & 238.5 \ \text{km} &
\end{aligned}
$$

If R_{max} is to be reduced by a factor 3, the noise power must be increased by a factor of 81 because of the $(\text{noise})^{-1/4}$ dependence above. The noise floor must therefore be raised by 19.1 dB to -126.7 dB W^{-1}. Try putting this noise level into the radar equation to convince yourself that the detection range drops to just under 80 km.

The one-way jamming signal received by the radar is given by

$$P_r = \frac{P_t G_t L_t}{4\pi R^2} \times \frac{G_r \lambda^2 L_r}{4\pi}$$

and so

$$P_t = \frac{P_r(4\pi R)^2}{G_t G_r \lambda^2 L_{\text{total}}}$$

The jammer transmitter therefore needs a power of

$$
\begin{aligned}
P_r &= -126.7\,\text{dB W} \\
(4\pi)^2 &= 22.0\,\text{dB} \\
R^2 &= 107.5\,\text{dB m}^2 \qquad \text{Aircraft range of 238.5 km} \\
1/G_t &= 0\ \ \text{dB} \\
1/G_r &= -36\ \ \text{dB} \\
1/\lambda^2 &= 12.7\,\text{dB m}^{-2} \\
1/L_{\text{total}} &= 7.0\,\text{dB}
\end{aligned}
$$

$$P_t = -13.5\ dB\,W \qquad \equiv 45\,\text{mW}$$

COMMENT It is quite staggering to realize that a surveillance radar with such a large power × aperture product can be so effectively disabled by so little jamming power. As the aircraft moves nearer to the radar, it can still keep the detection range below its own range because the jammer R^2 term always dominates the radar R^4 term.

14.3 TYPES OF ELECTRONIC COUNTERMEASURES AND ELECTRONIC COUNTER-COUNTERMEASURES

The objectives of an ECM system are either to *deny* the radar an opportunity to detect a target, as in the worked example above, or to *deceive* the radar into following either the wrong target (such as a chaff cloud) or a non-existent target that has been generated electronically.

The simplest method of denial is *spot jamming*, in which the noise power is concentrated into the radar receiver bandwidth. The radar can counter this by hopping in frequency from pulse to pulse (which also increases the probability of detection because it decreases target fading effects).

The counter to *frequency hopping* is for the ECM system to use *multiple spot jamming* or to begin *barrage jamming* over an entire radar band. There are at least two ways the radar can counter barrage jamming: by using a look-ahead receiver to find the weakest point in the barrage (because jamming antennas are not perfectly frequency-independent) or by adopting *frequency diversity* and changing to an entirely different band (sometimes to a different radar system in fact). Frequency diversity is more effective with higher-frequency radars because the antenna problems are easier to solve. One possible ploy is to operate in one radar band and mode (e.g. pulse) during peacetime and to switch suddenly to an entirely different frequency band and mode (e.g. FMCW) in the event of war.

Dispensing *chaff* has been a successful ECM denial technique since World War II, although it is now only effective for slow-moving targets because of the development of doppler processing. If a radar has locked on to a slow-speed target, such as a ship or tank, it may still be possible to eject a cloud of chaff along the line of sight such that the radar initially tracks both target and chaff, but remains locked on to the latter as it drifts away in the wind. Possibly the best ECCM measures for dealing with chaff are improved doppler processing and the use of polarimetry to desensitize the radar system to the long, thin shape of chaff while maximizing the echo from the expected target shape. Polarimetry is not without problems, however, since different polarizations rarely propagate equally.

Sometimes it is not the radar target that undertakes electronic counter-measures, but a separate stand-off jammer, which attacks the sidelobes of the radar antenna. The ways of countering this threat are:

- To build antennas with very low sidelobes.
- To use a separate omnidirectional antenna to pick up the jamming signal for use in a coherent cancelling device.
- To use adaptive nulling, and effectively put 'holes' in the antenna pattern in the direction of the jammer.

Other general defences against denial jamming including firing HOJ missiles at the jammer, trying to *burn through* the jamming (by increasing the power × aperture × integration time product in the direction of the target) and turning the radar off and trying to follow the jamming signal using passive tracking techniques.

Noise jamming is relatively easy to detect and, as we have seen, there are a variety of countermeasures, some of which can be dangerous for the jammer. For this reason, *deception jamming* (to fool the radar or its operator with 'spoof' transmissions) has become more popular.

An active method of deception jamming is a *repeater jammer*, which generates false echoes by receiving a radar pulse, delaying it in a digital memory and retransmitting it to appear as targets at other ranges. A *transponder jammer* uses the incoming pulse to trigger a suitably modulated reply to the radar, which may be delayed until after the radar beam has swept past so that it appears at the wrong azimuth.

One defence that a target can use, when it has been detected, is a *range gate stealer* and deliberately strengthen its own echo with a larger radiated pulse. The timing of this artificially enhanced echo is then slowly shifted to change the apparent range; hopefully the radar tracker continues to follow the larger 'target'. The range gate stealer is often countered by jittering the PRF in a quasi-random manner. Similarly, a *velocity gate stealer* can be engineered to falsify the target speed by tuning the reply frequency.

There are many other ECM tricks designed to break the tracking performance of radars. Provided one knows exactly how the enemy radar works, it is relatively easy to defeat it. A good guide on the use of ELINT to analyse enemy radar is given by Wiley[3]. The key to ECCM is for a radar not to be detected in the first place and to have a *low probability of intercept* (*LPI*). There are many ways of achieving LPI, but they are all based on spectral spreading of the radar signal by such techniques as FMCW, long pseudo-random codes, etc. Radar systems can accurately compress these signals on reception to a narrow final bandwidth. The intercepting ECM receiver, not knowing the compression algorithm, must receive these signals over the full swept bandwidth and so suffers a large noise penalty. Another defence against ELINT and ECM is to use separate peacetime and wartime operational modes, and to transmit deliberately misleading surveillance-type signals during peacetime to try to get the enemy to develop the wrong type of jammer.

One final deception system that should be mentioned here is the use of decoy RPV aircraft. These are relatively inexpensive and can be flown in large numbers at the enemy until their air defence weapons are exhausted. The possibility that the enemy radar can recognize the decoys and ignore them can be countered by placing a nuclear or biological weapon on the occasional one to ensure that they must all be treated seriously.

The ECM–ECCM game can be played indefinitely. There are many sophisticated modifications that can be made to radar transmitter waveforms, antenna patterns and receiver techniques to defeat denial and deception jamming. However, it is probably also true to say that the countermeasures to a radar are always less expensive to build than the radar itself. The final arbiters in electronic surveillance are probably, therefore, espionage (to know how the countermeasure will work in wartime) and money (to fund the development of a counter to it). All of this expensive game can be invalidated if targets are not visible to radar in the first place; this has been the driving force behind the *Stealth* programme.

14.4 STEALTH APPLICATIONS

Stealth is the art of concealing targets from detectors. For submarines, it is important to be acoustically Stealthy; ships and aircraft probably need to balance acoustic, visual, infrared and radio Stealth in proportion to the threat posed by the sensors in each of these spectra. The threat of radar detection is usually a major concern to an attacking force and a lot of effort is put into trying to reduce radar signatures.

The subject of Stealth arose from the analysis of Allied bomber losses during World War II in which the tiny, but very fast, wooden Mosquitoes (carrying a bombload almost as large as a B17) fared very much better than contemporary larger and slower metal bombers. After the war, the Northrop

'Flying Wing' and the Avro Vulcan bomber showed that the aircraft shape, as well as the construction material, was important in the reduction of the RCS.

A change in tactics also occurred after the war; the large defensive bomber formations, and escorting fighters, were replaced by solo nuclear missions in which single bombers were required to navigate through enemy defences alone. Escaping enemy ground fire by flying at very high altitudes was eventually abandoned after the U2 of Francis Gary Powers was shot down in 1960. Flying at low altitudes was dangerous because of the accuracy of radar-controlled anti-aircraft artillery, unless the aircraft flew *very* low to get beneath the radar defences. The solution lay in adopting an idea tried in World War I, to make aircraft 'invisible' by covering the airframes in translucent materials, and then extending the concept to include the radar spectrum (see Sweetman[4], for example).

The maximum radar detection range (Eq. (1.15)) can be summarized as

$$R_{max} \propto \sigma^{1/4} \quad [m] \tag{14.1}$$

The RCS σ must be changed dramatically if the radar performance is to be seriously affected. Computer modelling of airframes and ship structures shows that most of the scattering comes from flat surfaces and corner reflectors, formed where different parts of the structure join together. Table 2.1 gave the RCS of flat plates and corner reflectors; try putting a few numbers in to get a feel for the values that can occur. The RCS of most structures can be dramatically reduced by choosing an inherently low RCS design and then using smooth, continuously curved shapes and avoiding any right angles. In the case of an aircraft, this avoidance of right angles may mean that a single fin cannot be used and the aircraft must be designed without one (the flying wing approach used in the B2 bomber) or with two inclined fins (as used on the F117 fighter).

After shaping, any remaining 'hot spots', such as the reflections from engine air intakes on aircraft, can be covered with *radar-absorbing material* (RAM). RAM is generally some form of lossy dielectric material built into a wafer, which is tuned for maximum absorption in a particular radar band. Care also needs to be taken to ensure that the radar antennas on ships and aircraft do not act as RCS hot spots.

In a battle scenario, Stealth is a very effective counter to radar, especially when other ECM activities are going on to reduce the radar efficiency. There are, however, a few antidotes to Stealth. Wideband carrier-free impulse radar may be used to try to exploit RCS weaknesses in the radio spectrum (see Chapter 12) and, in general, radar needs to move to lower frequencies to combat Stealth. Over-the-horizon radar (Chapter 10) may be effective because the scattering mechanism is more related to radar-induced currents flowing throughout the structure of the target than from the individual scattering facets on the surface that dominate the RCS at microwave frequencies. Skywave OTH radar is particularly useful because it looks down on air targets, the aspect from which they present the largest scattering cross-section. Bistatic radar has also been cited as a possible countermeasure[4] because it is hard to design Stealth shapes against an unknown radar geometry.

14.5 SUMMARY

The very success of radar at detecting and tracking military targets means that it has become the target of electronic attack itself. As soon as a military radar system is turned on, its EM emissions will be detected unless it uses spread spectrum LPI techniques to evade detection. Any radar signals detected will be analysed by ESM receivers and an ECM strategy formulated; this may include the use of Stealth to reduce the RCS of the target. While electronic countermeasures are effective, and expensive to defend against using ECCM techniques, the balance at present probably resides on the side of radar, provided the waveforms are controlled intelligently.

14.6 REFERENCES

1. *Modern Radar Techniques*, Ed. M. J. B. Scanlan, Collins, Glasgow, 1987. [Chapter 4 is an organized guide to ECM and ECCM.]
2. *Jane's Radar and Electronic Warfare Systems 1989–90*, Jane's Information Group, London, 1989.
3. *Electronic Intelligence: The Analysis of Radar Signals*, R. G. Wiley, Artech House, Dedham, MA, 1982.
4. *Stealth Bomber*, B. Sweetman, Airlife Publishing, Shrewsbury, 1989.

14.7 PROBLEMS

14.1 Consider a radar speed trap based on a 10 GHz 'unity' radar system (1 W, unity antenna gain, zero losses, 1 μs pulse). What is the SNR for a single-pulse detection of a car at a range of 100 m if the car has a radar cross-section of 10 m²?

14.2 If the motor car in problem 14.1 contained a spot jammer, how much would this need to increase the noise level in the radar receiver to reduce the range to 10 m for the same received SNR?

14.3 What noise power would the jammer have to radiate to achieve the result in problem 14.2, assuming the noise was all concentrated in the same bandwidth as the radar signal?

14.4 If the radar trap above were to use random frequency hopping over the range 10–30 GHz, and the jammer were to counter by barrage jamming the entire band, by how much would the total jammer output noise power have to be raised?

14.5 If, as an alternative, the radar in problem 14.1 were to attempt to burn through the spot jamming using (perfect) coherent integration, how many pulses would need to be integrated?

14.6 Suggest a suitable PRF for the radar and estimate whether the integration time determined by problem 14.5 would give adequate doppler resolution to book a motorist for speeding.

14.7 What would seem to be the most economical strategy to avoid a fine for speeding:
  (a) to build an in-car radar warning receiver;
  (b) to develop a radar jammer;
  (c) to cover the car in Stealth materials?

FIFTEEN

RECENT DEVELOPMENTS

- Electronic beam steering
- Transmitting arrays
- Combining radar ouptuts
- Improving resolution

Electronic methods of steering beams, communication between radars and advanced processing give modern radar systems considerable flexibility.

15.1 INTRODUCTION

In this chapter we look at current radar technology and find out how it is being used. We will try to predict the future of radar in the next chapter.

The antenna has been the subject of some of the most interesting radar developments in recent years. Domestic satellite dishes are representative of the old level of antenna technology because they must be mechanically pointed at their target. Radar versions of these parabolic reflectors are rotated by heavy and expensive (but reliable) turning gear, and a target can only be observed once every few seconds when the dish faces that direction.

At the centre of a reflecting antenna, such as a satellite dish, is the device that illuminates it, called the feed. Feeds can be dipoles, microwave horns or other devices for radiating or receiving electromagnetic waves. The purpose of the reflector is merely to increase the collecting area, and hence the gain, of the antenna. However, a large collecting area can also be achieved in an entirely different way. It is possible to arrange many dipoles on a flat plate and connect them together in a phase-coherent manner to form a *phased array*. In the UK, one domestic system for receiving satellite broadcasts was constructed in this way.

Phased arrays are often considerably more expensive than reflector antennas but they have the advantage that the beam can be steered electronically rather than mechanically. If the phasing of a receiving antenna is carried out digitally by computer calculation, rather than by analogue devices (which may be controlled by a computer), the process is known as *digital beamforming*, and there is considerable flexibility for beamshaping and the formation of multiple beams.

Phased arrays are now in widespread use for receiving, but transmitting versions are only just coming into service because of the difficulties of building, and controlling, large arrays of small transmitting modules. These *active arrays* represent the current level of technology for single radar systems.

An entirely separate development area has been concerned with netting independent radars (and other sensors) together to improve target detection and tracking accuracy. We briefly examine this field, which is sometimes known as *multihead radar*.

Finally, we take a brief overview of the recent advances made in signal and data processing, which have enabled radars to go beyond their original function of radio detection and ranging.

15.2 PHASED ARRAYS

An array of radiating elements can be one-dimensional (*linear array*) or two-dimensional (*planar array*). Arrays work the same way on both transmit and receive, but it is often easier to follow the theory for the transmitting case because of the analogy in optics with Young's double slit experiment and diffraction gratings.

A simple linear array, with element spacing d, is shown in Fig. 15.1. If all the elements are excited in phase, then a wavefront is radiated broadside to the array. If a successive time delay is introduced across the array, causing a phase change ψ between elements, then the wavefront will be radiated at an angle θ given by

$$\frac{\psi}{2\pi} = \frac{d \sin \theta}{\lambda} \quad [\ ] \qquad (15.1)$$

The field $E_a(\theta)$ from N elements can be found by summing across the array. Using a little trigonometry the summation comes out as

$$E_a(\theta) = \frac{E_1(\theta)}{\sqrt{N}} \left(\frac{\sin[N\pi(d/\lambda)\sin\theta]}{\sin[\pi(d/\lambda)\sin\theta]} \right) \quad [\text{V m}^{-1}] \qquad (15.2)$$

where $E_1(\theta)$ = field due to a single element [V m^{-1}].

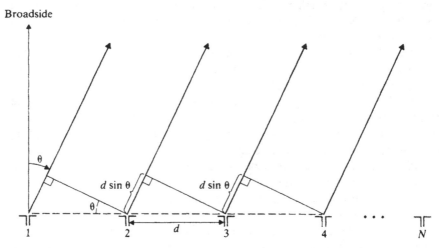

Figure 15.1 A linear array of dipole elements.

In optics, the field arising from a diffraction grating has two components; the diffraction pattern from a single slit and the interference pattern created by the interaction of coherent signals from different slits. It is exactly the same with antenna arrays. Each element of the array acts as a source with its own diffraction pattern $E_1(\theta)$, and the radiation from the sources interacts to form an interference pattern, which is usually called the *array factor* (AF) given by

$$AF = \frac{E_a(\theta)}{E_1(\theta)} = \sqrt{N}\left(\frac{\sin[N\pi(d/\lambda)\sin\theta]}{N\sin[\pi(d/\lambda)\sin\theta]}\right) \quad [\ ] \quad (15.3)$$

The radiating elements are usually simple devices such as dipoles, slots, patches or sometimes small horns. All these elements have fairly wide diffraction patterns and the antenna pattern is determined by the much narrower array factor. Even so, care must be taken not to steer the array beam outside the pattern of the individual radiating elements. A normalized array pattern is shown in Fig. 15.2 in comparison with the diffraction pattern of a single half-wave dipole.

Part of the purpose in rearranging Eq. (15.2) to get Eq. (15.3) is because the term in large parentheses in Eq. (15.3) can be shown to have a maximum value of unity. This means that the field strength has been increased by $\sqrt{N}$ times the field from a single element, and therefore the *power gain of the array* has been increased by N. Some other properties of array patterns are now outlined.

The *half-power beamwidth* of the pattern steered off-broadside through an angle θ can be found from Eq. (15.3) by setting the expression in large

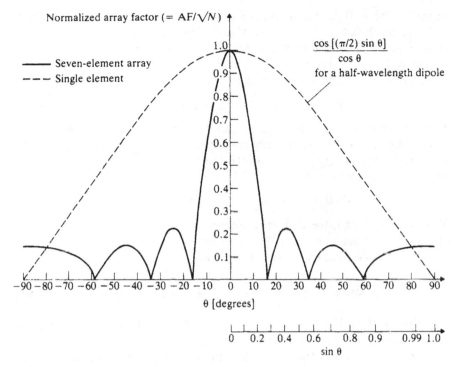

Normalized array factor ($= AF/\sqrt{N}$)

——— Seven-element array
– – – Single element

$$\frac{\cos\left[(\pi/2)\sin\theta\right]}{\cos\theta}$$

for a half-wavelength dipole

θ [degrees]

$\sin\theta$

Figure 15.2 The normalized array factor for a seven-element array, of spacing $d = \lambda/2$, compared with the diffraction pattern of a single element.

parentheses to $\sqrt{0.5}$. For large N, this leads to

$$\Delta\theta = \frac{0.886\lambda}{Nd\cos\theta} \quad \text{[radians]} \tag{15.4}$$

For the broadside pattern this reduces to

$$\Delta\theta = 0.886\lambda/D \quad \text{[radians]} \tag{15.5}$$

where D = overall length of the array = Nd. This, of course, corresponds to the λ/d approximation for antenna beamwidth that we have seen earlier in the book. The width increases by $1/\cos\theta$ as the beam is steered away from broadside because the effective length of the array is reduced by $\cos\theta$ when viewed from an angle θ off-bore-sight.

A convenient approximation of Eq. (15.5) in degrees, to remember for rough calculations, is

$$\Delta\theta \sim 51\lambda/D \quad \text{[degrees]} \tag{15.6}$$

The beamwidth factor (the aperture in wavelengths needed for a one-degree beam) is about 51 for the uniform illumination assumed in Eq. (15.6), but is between 60 and 70 for tapered distributions, according to the sidelobe level required. The beamwidth factor is also higher for apertures that are round, rather than rectangular.

Nulls occur in the pattern at angles θ_{null} given by

$$\sin \theta_{null} = \pm k\lambda/Nd \qquad k = 1, 2, \ldots \qquad [\quad] \qquad (15.7)$$

Often, element spacings near $\lambda/2$ are used, giving only one mainlobe (plus one to the rear, if the array is not mounted above a reflecting plane). For wider element spacings, *grating lobes* (other principal maxima, or 'secondary beams') occur when

$$\sin \theta_{gl} = \pm k\lambda/d \qquad k = 1, 2, \ldots \qquad [\quad] \qquad (15.8)$$

With careful design, the individual element pattern can sometimes be used to null out grating lobes near 90°, as shown in Fig. 15.2.

Sidelobes (secondary maxima) occur at

$$\sin \theta_{sl} = \pm \frac{(2k + 1)\lambda}{2Nd} \qquad k = 1, 2, \ldots \qquad [\quad] \qquad (15.9)$$

The first sidelobe has a power gain of -13.4 dB with respect to the peak gain of the array. This is a well known result arising from a rectangular illumination function. Those of you familiar with Fourier transform (FT) theory will realize that the FT of a rectangular function (in this case the aperture) gives rise to a $(\sin x)/x$ function of a similar form to the array factor. Equation (15.3) can be approximated, for small values of θ, by an expression of the form

$$E_a(\theta) \simeq \sqrt{N} \, \frac{\sin[\pi(D/\lambda)\sin\theta]}{[\pi(D/\lambda)\sin\theta]}$$

$$\equiv \sqrt{N} \, \frac{\sin x}{x} \qquad [V\,m^{-1}] \qquad (15.10)$$

Thus the far-field pattern of the array can be represented by the FT of the spatial illumination function, provided that the element spacing is close enough to avoid grating lobes. For this reason, FTs are used extensively in antenna theory.

Some of you may be confused at this point because you have only used the FT to transform between the time domain and the frequency domain. Table 15.1 shows you how to relate this new way of using the FT to the more usual time/frequency form. Bracewell[1] is particularly good at covering this use of FTs. These days, there are some very readable books on antenna theory, and on arrays in general. If you wish to explore further, try references 2–4, especially 4.

Table 15.1 Use of Fourier transforms in antenna theory

Waveforms	Antennas
Time, t	Aperture distance in wavelengths, $N\pi d/\lambda$
Waveform, $x(t)$	Aperture field distribution, $E(N\pi d/\lambda)$
$\Updownarrow$	$\Updownarrow$
Angular frequency, ω	Direction, $\sin\theta$
Spectrum, $X(\omega\Omega)$	Field radiation pattern, $E(\sin\theta)$

$$X(\omega) = \tau\,\frac{\sin(\omega\tau/2)}{(\omega\tau/2)}$$

$$E(\sin\theta) = E_1(\sin\theta)\,\frac{\sin[N\pi(d/\lambda)\sin\theta]}{N\pi(d/\lambda)\sin\theta}$$

If all elements are fed equally, an array has a rectangular illumination creating sidelobes only 13.4 dB smaller than the mainlobe, as discussed above. Sidelobes this large are generally fatal for a radar system, owing to the difficulty of separating small targets in the mainlobe from larger targets in a sidelobe.

Low sidelobes are achieved through accurate element and array design and the application of a *weighting* or *window* function, such as triangular, raised cosine, Dolph–Chebyshev polynomials, etc., to the array. The effect of weighting is such that the elements at the centre of the array are fully utilized but those towards the edge are increasingly attenuated, thereby causing a wider beam and a loss of gain and SNR; unfortunately, these are necessary afflictions.

Worked example Consider the air-borne surveillance radar, shown in Fig. 15.3, operating in look-down mode to detect cruise missiles flying across the sea surface below. If the doppler shift of main-beam echoes, caused by the velocity V of the aircraft, are compensated in the signal processing, what would be the apparent speed of a large oil-rig in the forward sidelobe? How can the problem be corrected?

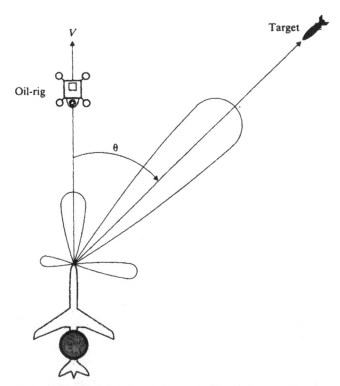

Figure 15.3 An air-borne surveillance radar and antenna pattern showing sidelobe structure.

SOLUTION Stationary targets in the main beam will have an apparent motion of $V \cos \theta$ towards the aircraft, which must be removed in the processing. The echo from an oil-rig appearing in the forward lobe will have a relative velocity of V. After processing, this velocity will appear as $V - V \cos \theta$. The oil-rig will therefore appear to be moving towards the radar and will be detected as a possible hostile target.

There are two possible solutions to the problem. One is to use the tracker to show that the apparent doppler shift of a target does not correspond to its rate of change of position when plotted over many scans. The problem with this approach is that, when flying near land, an antenna with large sidelobes may detect hundreds of 'moving' targets, which completely swamp the tracker, thereby allowing the smaller, real target to escape detection. The other solution is to use antenna arrays with very low sidelobe levels so that sidelobe echoes occur infrequently enough for the tracker to cope.

COMMENT Perhaps the best known air-borne surveillance radar is the Westinghouse E-3 air-borne warning and control system (AWACS). This uses a large phased array mounted inside a saucer shaped radome mounted above the fuselage of the aircraft and rotating once every 10 s. The S-band array employs over 4000 elements and has exceptionally low sidelobes. AWACS has been so successful that an improvement programme is under way that will take until 1998 to complete (more details can be found in *Jane's Radar*[5]).

The processes of *beamforming* and *beamsteering* are identical. A progressive time delay is applied to the elements, creating the tilted wavefront shown in Fig. 15.4. A beam steered through an angle θ_s has an array factor of

$$AF = \sqrt{N} \left\{ \frac{\sin[N\pi(d/\lambda)(\sin \theta - \sin \theta_s)]}{N \sin[\pi(d/\lambda)(\sin \theta - \sin \theta_s)]} \right\} \qquad [\quad] \qquad (15.11)$$

As an alternative to using a time delay, at a single frequency the steering can be applied to incrementing the phase between elements by

$$\psi = 2\pi(d/\lambda) \sin \theta_s \qquad [\text{radians}] \qquad (15.12)$$

Worked example Calculate the beam pattern for a five-element array with an element spacing of one wavelength. Is this a useful design? Hint: try calculating the pattern for a beam steered $50°$, before deciding.

SOLUTION The normalized array factor is shown in Fig. 15.5a. Grating lobes appear near $\pm 90°$. It might be thought that the individual element diffraction pattern would largely eliminate these grating lobes, allowing the wide element spacing to be used to give an extra-narrow mainlobe.

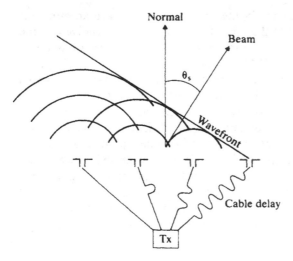

Figure 15.4 Beam steering by the introduction of a progressive time delay between array elements.

Unfortunately, if Eq. (15.11) is used to calculate the steered array factor the resulting grating lobe, shown in Fig. 15.5b, is fatally large.

COMMENT The difficulties of steering array patterns to wide angles effectively limits most systems to $\pm 60°$. The angle scanned can be increased by using two arrays, for example on adjacent sides of a building. This arrangement is used by the huge US Pave Paws early-warning radar for detecting ballistic missiles. Each array uses 2000 elements arranged on a circular plane 30 m in diameter, and the two systems cover a 240° sector[5]. The UK version of Pave Paws, on Fylingdales Moor, uses three arrays giving a 360° capability.

Other solutions to the problem of limited scan include circular arrays and conformal arrays. A circular array of elements can be used to give good 360° azimuth coverage, but it has the difficulty that amplitude as well as phase has to be controlled as the beam scans if a reasonable sidelobe level is to be obtained. Also, it becomes defocused at high elevations unless the phasing is changed. Curved arrays, conformal to the fuselage of an aircraft, are also an attractive idea in principle, but difficult to engineer in practice.

Grating lobes can be avoided, even when steering, if the element spacing is restricted. Rearranging Eq. (15.8) gives the spacing limit for a steering angle θ_s:

$$d < \lambda \left(\frac{1}{1 + \sin \theta_s} - \frac{1}{N} \right) \quad [\text{m}] \qquad (15.13)$$

(a)

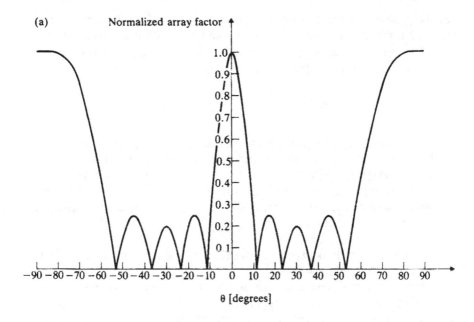

(b)

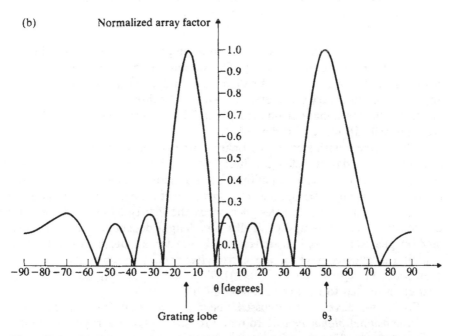

Figure 15.5 The beamshape for a five-element array with $d = \lambda$: (a) the broadside pattern; (b) steered 50° off-boresight.

For the problem above, this gives a spacing of near λ for a broadside beam, near $\lambda/2$ if the beam is steered to $\pm 30°$ and near $\lambda/4$ if the beam is to be steered to $\pm 60°$. When weighting is used, a slightly smaller element spacing is needed and a suitable approximation is given by Radford[6]:

$$d < \lambda \left(\frac{1}{1 + \sin \theta_s} - \frac{1.5}{N} \right) \quad [\text{m}] \qquad (15.14)$$

Both time-delay and phase-delay beamforming and steering are used. Analogue time-delay beamformers are somewhat slow and clumsy, involving the use of relays to switch lengths of cable in and out, but there are no frequency restrictions and simultaneous operation over a wide band is possible. Phase delays can be introduced by a variety of means, illustrated in Fig. 15.6, which include:

- Diodes to switch between alternative path lengths or to change the reactance of loading stubs.
- Ferrite phaseshifters; these use a cylindrical ferrite rod, inserted in a waveguide, which can be magnetized.
- An array of hybrids, known as a Butler matrix, arranged in a similar way to an FFT algorithm.
- A preset resistor matrix combining signals of different phases to give the appropriate value.

Surprisingly, it is not necessary to maintain precise phase and amplitude values at each element on the array. The summation of the field from many elements tends to cancel out small errors, and consequently phases are approximated in binary steps, often with as few as three bits giving $45°$ steps in phase. The loss in main-beam gain from a 3-bit phaseshifter is only 0.23 dB, falling to 0.01 dB for a 5-bit device[6].

The problem with steering by phase, rather than by time delay, is that changing the radar wavelength alters the steering angle. This frequency dependence has been exploited in the past as a method of electronic scanning (a given frequency corresponds to a given direction), but with modern electronic warfare the radar designer needs the liberty to use frequency in a much freer way. The upshot of this wavelength dependence is that most modern phased array radars are restricted in instantaneous bandwidth, with a limit (as a percentage of the carrier frequency) roughly equal to the beamwidth in degrees (see Eq. (10.2)). This limit equates to a pulse length restriction related to the aperture size, as in Eq. (10.3).

So far we have only discussed linear arrays, while most radars use two-dimensional *planar arrays*. In many respects, planar arrays are similar to linear arrays. The simplest way to develop a planar array is to form a stack of linear array planks and vary the phase between the planks to

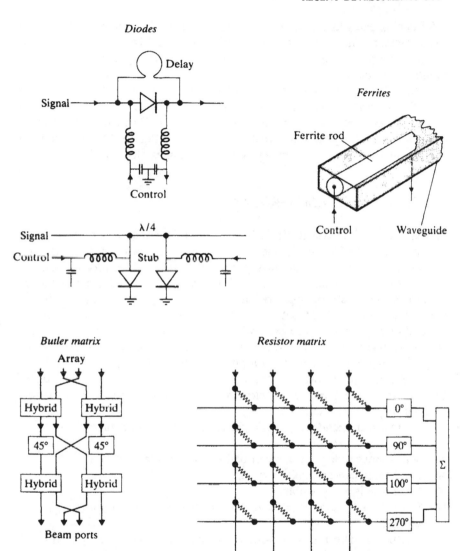

Figure 15.6 Methods of phaseshifting between array elements.

beamform in the second dimension. Other designs use triangular or rectangular lattice layouts of elements on the plane. The increased complexity of these arrangements requires larger computational resources to control the phaseshifters and creates something of a wiring distribution problem.

The wiring problem has been tackled in recent years by using fibre optics to control the phaseshifters and, in a further development, optical beamformers have been constructed. Optical fibres avoid most of the problems of RF cables, being lightweight, lost-cost, low-loss and immune to interference and crosstalk. These attractive properties of optical transmission in manipulating microwave signals have given rise to an emerging technology known as *microwave optics*.

One of the additional problems of planar arrays is that moving away from the normal causes a form of conical distortion because azimuth and elevation measurements are no longer independent. Corrections have to be made when tracking off-axis targets, which again adds to the complexity of the system and restricts performance at large angles off-bore-sight.

15.3 DIGITAL BEAMFORMING

In general, the increase in flexibility of radar antennas is matched by an increase in cost. The early mechanically rotating designs are the least expensive, followed by mechanically rotated arrays that beamform electronically in elevation. A full three-dimensional radar (a two-dimensional array plus range) using an analogue beamformer is very expensive and the beams are relatively difficult to set up. But there is still one further degree of adaptability (and cost) to be gained from using arrays. The ultimate flexibility is to have one receiver and A/D converter per element and to form beams by performing calculations in a computer on the A/D samples. This is digital beamforming. Sometimes, the amount of hardware needed can be reduced by combining local groups of elements to form sub-arrays, and using a receiver for each of these.

There are tremendous advantages to digital beamforming:

- Many independent beams can be formed simultaneously.
- The beams can be made very agile.
- Long dwells are possible.
- Sharp nulls can be created in the beam pattern to suppress jamming signals.
- The beam pattern can be adapted to enhance angular resolution.
- If the A/D samples from each array element are stored, beams can be formed off-line, at a later time.

The disadvantages of digital beamforming are high cost and performance limitations. The latter are determined mainly by the technology of the A/D converters. Slow converters with few bits make an array narrowband with a limited dynamic range. As technology improves, so does the performance of this type of array.

A digital beamforming system usually has a relatively simple receiver for each element, which down-converts the frequency into I and Q (in-phase and quadrature) channels for the A/D converter (see Fig. 15.7). Real-time

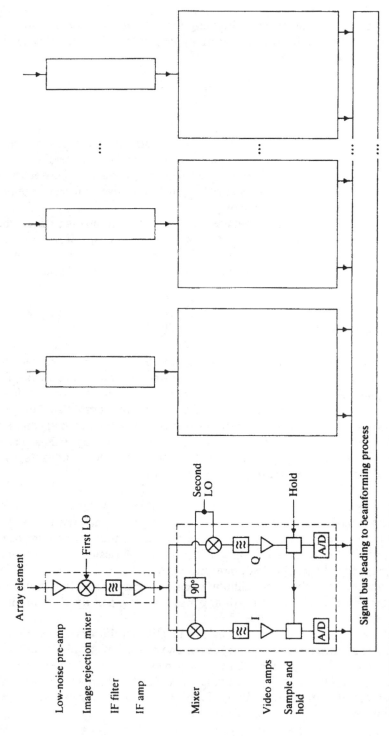

Figure 15.7 Block diagram of a typical digital beamforming system.

309

beamforming takes place by multiplying these complex pairs of samples by appropriate weights in multiply/accumulate integrated circuits (ICs). The array output is formed from

$$\text{Array output} = \sum_{n=0}^{N-1} V_n W_n \, e^{-j2\pi n(d/\lambda)\sin\theta} \, C_n \qquad [\text{V}] \qquad (15.15)$$

where V_n = complex signal from nth channel, W_n = weighting coefficient, $e^{-j2\pi n(d/\lambda)\sin\theta}$ = steering phaseshift and C_n = correction factor. Corrections are necessary for several reasons. These include errors in the position of the element, temperature effects and the difference in behaviour between those elements embedded in the array and those near the edge, caused by coupling effects with neighbouring elements.

Digital beamforming is now used extensively for receiving arrays, where sufficient funding is available, and it can be found in use from small HF sea-sensing systems up to large-scale surveillance radars. An example of a system in service is the Patriot radar system, described in Chapter 3.

15.4 ACTIVE ARRAYS

When an array is used for transmitting, as well as receiving, it is said to be an *active array*, and represents the most up-to-date (and highest-cost) form of antenna. An active array has a number of separate modules, each with its own amplifier and phase control. There are considerable advantages to the use of an array of small transmitting elements (rather than one large transmitter), including the flexibility common to receiving arrays plus the possibility of very large power × aperture products. The disadvantages are those of yet further increase in cost and complexity.

One of the main difficulties of active arrays is that not all phaseshifters (ferrites, for example) are reciprocal, and a different phase is received from that transmitted. Furthermore, the transmitting beam may need to be wider than the receiving beams to encompass a block of four monopulse beams. The end result is that transmitting and receiving phaseshifters need to be set up independently, leading to designs similar to that shown in Fig. 15.8. These modules require solid-state gallium arsenide transmitter ICs as well as the receiving chip set. The power output per module is generally of the order of a few watts.

An example of an active array is the UK multifunction electronically scanned adaptive radar (MESAR) developed by the UK Admiralty DRA(M) and Siemens–Plessey Radar. This is an S-band array of 918 elements using 4-bit phaseshifters. Each element uses GaAs ICs for the microwave circuitry and silicon ICs for control. Array thinning is used to improve the transmit sidelobe level.

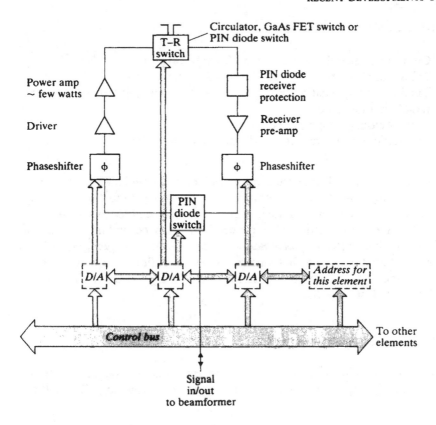

Figure 15.8   Block diagram of an active array element.

The MESAR design demonstrates the current level of technology. Generally speaking, very-high-speed silicon ICs (VHSICs) are needed to carry out all the calculations and gallium arsenide microwave monolithic ICs (GaAs MMIC) are needed for the RF components and fast A/D converters. The development of these technologies, and of fibre optic control, will be crucial to the future of both digital beamforming and active arrays.

15.5 MULTIFUNCTION RADAR

Given the flexibility of phased array radar, the traditionally separate roles of surveillance, tracking and even radio communications can be taken over by a single multifunction radar system. The system's resources (beams, receivers, computers) must be allocated efficiently to carry out all the required functions, which might include:

- Continually searching the horizon for new targets.
- Taking confirmatory looks at potentially new targets.
- Tracking many existing targets, with a scan rate adjusted to meet the target characteristics.
- Tracking out-going missiles.
- Providing high-power, narrow-beam communications.

Of course, the problem is the same as with every surveillance radar—there is never enough time to do everything, and a priority list must be drawn up for the beam scheduling. Making such complex, weighted decisions in a computer is quite difficult and is generally held to be within the area of artificial intelligence (AI). Fortunately, modern software languages, such as ADA, are structured such that antenna array operations can be written as software modules, which can be called and run by the scheduler with a predetermined priority list.

15.6 MULTIHEAD RADAR

When several monostatic radars can see the same target, as shown in Fig. 15.9, their outputs can be combined to improve the tracking. This method

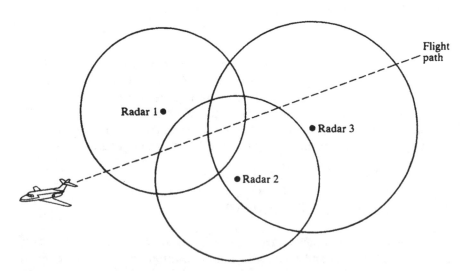

Figure 15.9 When the coverage of surveillance radars overlaps, their outputs may be combined to improve the overall tracking performance.

of netting independent radars together is sometimes known as multihead radar and has become popular since the development of digital communications enabled reliable low-cost data links to be formed.

Multihead radar is especially applicable for air defence and ATC radars with overlapping cover, but other situations where it may be applied include naval systems where several radars, and other sensors, may be combined within the area of a fleet.

The outputs from several sensors must be sent to a data processing centre for combining. There are then three ways in which the data may be combined:

1. Individual plots from each radar can be transmitted to the data processing centre and the tracker can make use of all the data simultaneously. This gives the best results, but requires high-bandwidth data links and a lot of computer processing power.
2. Individual plots can be sent to the centre and combined together before trackforming. This requires a wideband data link but it reduces the computational load and still gives quite good results.
3. Tracks can be formed locally at each radar and sent to the centre for combining; this requires only a low-capacity data link but the improvement in tracking accuracy is modest.

There are other advantages to multihead radar in addition to improved tracking accuracy. Low-flying aircraft in mountainous country will only be detected by radars in short glimpses, but netting several radars together, to view the target across different terrain, increases the probability of detection and of forming some sort of assessment of the threat. A multihead radar system is also more jammer-resistant because each radar views the target from a different angle.

In some ways, multihead radar networks are more attractive than multistatic arrangements. Each radar operates independently and causes only a relatively small degradation to the system if it should fail or be jammed. Multihead is also much simpler than multistatic because no synchronization is required and different radars, from different manufacturers, can readily be netted together.

Merging data in multihead systems is not without difficulties. For example, data are usually combined by giving each measurement a weight that is inversely proportional to its variance. In this way the tracker takes more account of good data than bad. A problem occurs when one radar correctly detects a target manoeuvre and uses a filter with a larger variance in order to keep track. The data from this radar will receive less weight than those radars that have not yet spotted the manoeuvre. This is typical of the type of quandary that can be so interesting for a radar engineer to work on.

15.7 HIGH RESOLUTION RADAR TECHNIQUES

Modern radar has gone beyond 'radio detection and ranging' to incorporate target classification and recognition algorithms. These facilities require improvements in resolution to identify distinguishing features on targets. We have seen throughout this book how the resolution depends on the system parameters, but we have not yet mentioned the contributions that can be made by developing the hardware, the signal processing and the data processing software.

There are many methods of increasing the resolution of radar systems. Often, techniques were developed in other fields of research and adapted for use with radar. Some of the current approaches involving changes in hardware, or data collection strategies, are discussed below.

Wideband radar systems High range resolution can sometimes be used to identify features on a target, if a sufficiently large bandwidth can be obtained. Wideband techniques include carrier-free radar (Chapter 12), pulse to pulse frequency modulation, noise modulation, wideband chirp, etc. The main disadvantage is that the increased noise bandwidth usually restricts this technique to short range applications.

Multifrequency illumination Rather than having a single radar channel of wide bandwidth, it is possible to select a number of narrower bandwidth channels and correlate the outputs to look for certain spatial properties of the target. This subject has been extensively investigated by Gjessing (reference 7) and can be quite powerful when *a priori* information on the target shape is available, although it is somewhat sensitive to aspect angle.

Multipolarization techniques It has been realized that more information on many types of target can be improved if multipolarization techniques are used. This arises because the polarization signature of targets has, in principle, five degrees of freedom (three amplitudes and two relative phases). In contrast, frequency diversity yields only a single extra degree of freedom. Evaluation of the polarization characteristics of targets is an active area of research, and radars to exploit these polarization signatures are under development.

Quadpolarized systems Use of several frequencies and full polarization information is a characteristic of the latest generation of airborne remote sensing radars. These quadpolarized systems are capable of recognizing the different scattering mechanisms contributing to the backscatter. This can be utilized in target recognition and classification. Some of the most significant progress in understanding the scattering properties of extended targets and in the application of SAR are likely to stem from this advance.

Interferometry A further advance in the use of SAR is in an interferometric mode. In this mode, the phase difference between SAR images gathered on closely spaced orbits can be used to construct a map of the topography of the imaged area. Tests on Seasat and ERS-1 data have indicated that it should be a practicable technique. Very curious results were also obtained over flat agricultural land, and there is clearly more investigation needed. These results have led to a flurry of development of multi-antenna interferometric systems for aircraft operation, and SAR processors giving high phase accuracy.

Below we describe some of the software and mathematical techniques that can be used to increase radar resolution.

Inverse synthetic aperture radar ISAR is used to increase angular resolution in the same way as SAR (see Chapter 11) but with the difference that the radar remains stationary and advantage is taken of the target motion.

Improved doppler spectrum analysis There are many modern spectral estimators, such as the maximum likelihood method, maximum entropy method, minimum eigenvector method, etc. These can be used to improve the doppler resolution and to identify targets by resolving such features as turbine blade rotation, or the pitch, roll and yaw of an aircraft or ship. Improved spectral estimators are also applied to phased arrays to improve the angular information.

Radar cross-section fluctuation analysis When a target is composed of a limited number of individual scatterers that characterize it, the target may be identified by statistical or spectral analysis of the RCS amplitude fluctuations.

Super-resolution Even when the shape of a target is smaller than the range resolution, some information may be recovered from the echo by carrying out a deconvolution process known as super-resolution, provided that the 'point spread function' (= impulse response) of the system is known. Essentially, the radar bandwidth is artificially increased by restricting the possible choice of target shapes.

Tomography The structure of a given target maps into a particular shape in the ambiguity diagram (the convolution of the target structure with the pulseshape in range and doppler). Different chirp rates correspond to different angles on the ambiguity plane and so may be used to gain information on the target shape, in a similar way to tomographic imaging used in medicine.

Many of these techniques provide good results in computer simulations, but in practice the effects of noise and phase errors due to unresolved motions may lead to degradation of performance. The best strategies often involve the combination of several of the techniques above. If the objective is to identify a small target, such as an aircraft or ship, it can be useful to include a 'library' or 'catalogue' of possible target shapes. Some means of relating high resolution observations to the library is then required and these days there is a lot of interest in neural nets which can be trained to recognize features in the data that relate to certain types of target.

15.8 SUMMARY

This brief review of recent developments in radar is intended to help you identify the key areas to study if you wish to pursue a career in modern radar. Phased array theory and high-resolution techniques are particularly important because of their use in other fields such as sonar. Some of the useful phased array formulae to remember are given below.

Key equations

- Phase change between elements in a phased array:

$$\psi = 2\pi(d/\lambda)\sin\theta \qquad [\text{radians}]$$

- Array factor:

$$\text{AF} = \sqrt{N}\left(\frac{\sin[N\pi)(d/\lambda)\sin\theta}{N\sin[\pi(d/\lambda)\sin\theta]}\right) \qquad [\quad]$$

- Half-power beamwidth (for large N):

$$\Delta\theta = \frac{0.886\lambda}{Nd\cos\theta} \qquad [\text{radians}]$$

- Approximate half-power beamwidth for broadside pattern:

$$\Delta\theta \sim 51\lambda/D \qquad [\text{degrees}]$$

- Positions of nulls in pattern:

$$\sin\theta_{\text{null}} = \pm k\lambda/Nd \qquad [\quad]$$

where $k = 1, 2, \ldots$.

- Positions of grating lobes:

$$\sin\theta_{\text{gl}} = \pm k\lambda/d \qquad [\quad]$$

- Positions of sidelobes:

$$\sin \theta_{sl} = \pm \frac{(2k + 1)\lambda}{2Nd} \quad [\quad]$$

- Spacing limit to avoid grating lobes for steering angle θ_s:

$$d < \lambda \left(\frac{1}{1 + \sin \theta_s} - \frac{1.5}{N} \right) \quad [\text{m}]$$

- Array factor for beam steered through angle θ_s:

$$\text{AF} = \sqrt{N} \left(\frac{\sin [N\pi(d/\lambda)(\sin \theta - \sin \theta_s)]}{N \sin [\pi(d/\lambda)(\sin \theta - \sin \theta_s)]} \right) \quad [\quad]$$

- Array output for digital beamforming:

$$\text{Array output} = \sum_{n=0}^{N-1} V_n W_n e^{-j2\pi n(d/\lambda) \sin \theta} C_n \quad [\text{V}]$$

15.9 REFERENCES

1. *The Fourier Transform and its Applications*, R. N. Bracewell, McGraw-Hill, New York, 1986.
2. *Antennas*, F. R. Connor, Edward Arnold, London, 1972 [Short and very readable.]
3. *The Handbook of Antenna Design*, vol. 2, Eds A. W. Rudge, K. Milne, A. D. Olver and P. Knight, IEE Electromagnetic Wave series, Peter Peregrinus, Hitchin, Herts, 1983. [Long but still readable.]
4. *Radio Wave Propagation and Antennas*, J. Griffiths, Prentice-Hall, Englewood Cliffs, NJ, 1987.
5. *Jane's Radar and Electronic Warfare Systems 1989–90*, Jane's Information Group, London, 1989.
6. Electronically scanned antenna systems, M. F. Radford, *Proc. IEE*, **125**(11R), 1100–1112, 1978. [A particularly lucid review.]
7. *Target Adaptive Matched Illumination Radar*, D. T. Gjessing, IEE Electromagnetic Waves series 22, Peter Peregrinus, Hitchin, Herts, 1986.
8. The MU radar with an active phased array system: 1. Antenna and power amplifiers, S. Fukao, T. Sato, T. Tsuda, S. Kato, K. Wakasugi and T. Makihira, *Radio Sci.*, **20**(6), 1155–1168, 1985.

15.10 FURTHER READING

Practical Phased-Array Antenna Systems, Ed. E. Brookner, Artech House, 1991. [A promising new book, not yet released at the time of writing.]

15.11 PROBLEMS

15.1 The Japanese MU (middle and upper atmospheres) MST radar uses an active antenna array composed of 475 crossed Yagi antennas (each having a gain of 7.24 dB) arranged on a circular plane of diameter 103 m. Each antenna has its own solid-state transmit–receive module, with a peak output power of 2.4 kW (see Fukao *et al.*[8]). The operating frequency is 46.5 MHz. What is the antenna gain and beamwidth?

15.2 The beam in the above can be steered electronically up to 30° off-vertical. If the elements were laid out on a rectangular grid, what would be a suitable array spacing? (The authors[9] tell us that 'neither spatial nor electrical tapering is incorporated'.)

15.3 The triangular lattice arrangement of antenna elements (see solution to 15.2) has a maximum separation of 1.2λ in some directions. Where might the first grating lobes be expected to appear?

15.4 We are told that it is possible to steer the beam in each inter-pulse period, which may be as short as 400 μs. Would you expect the necessary phaseshifts to be calculated each time?

SIXTEEN

THE FUTURE OF RADAR

- Where is radar going?
- Increasing the bandwidth
- More complex systems
- More complex processing

The increasing sophistication of computers and software will give rise to a new generation of intelligent radar systems.

16.1 INTRODUCTION

Has radar been developed as far as it is going to go, or does it have a bright future with many further applications and capabilities to be discovered? As with any branch of science or engineering, the way to answer this type of question is to consider whether the system is limited by physical laws, by cost or by technology.

The laws of physics applying to radar are well understood, but do not yet limit our systems. We cannot escape the range resolution being limited by the effective bandwidth of the system, for example, but we can apply processing power to such techniques as super-resolution to extend the effective bandwidth.

Cost is a serious limitation to what can be achieved; there are many cases where phased array radars are desirable but mechanical turning gear is used instead because it is less expensive. However, the influence of large military research programmes has continued radar development, and the falling cost of components, especially computing power, continues to help progress.

Technology is therefore the limiting factor. Faster A/D converters, gallium arsenide transmit and receive modules, better simulation software, more advanced mathematics for data processing algorithms—these are the tools needed to build better radars in the future. As we show below, the overriding need is to increase the bandwidth of radar systems. The other benefits will then follow as a matter of course.

16.2 DEVELOPING THE CONCEPT OF BANDWIDTH

There are three principal performance requirements for a radar sensor: sensitivity, resolution and data rate. A military radar has a fourth requirement of survivability (see Radford[1]). The fate of radar will be determined by how well these requirements can be fulfilled.

In future, radars will be required to be more sensitive so that smaller targets can be detected in more hostile environments. Microlight aircraft may need tracking near civil airlanes and military radars will be required to detect increasingly Stealthy targets. Meanwhile, the radio environment deteriorates as radio interference and the sophistication of jammers continue to increase.

The sensitivity of a radar can be thought of in the following way. If a solid angle of sky Ω is required to be searched in time T, then

$$\frac{T}{t} = \frac{\text{time to search the whole sector}}{\text{dwell time in each beam position}}$$

$$= \frac{\Omega}{\Delta\theta \, \Delta\phi} \quad \left(= \frac{\text{whole sector}}{\text{beamwidth}} \right) \quad [\quad] \quad (16.1)$$

where t = the dwell time in each position [s]. Rearranging Eq. (16.1) gives

$$t = (T \, \Delta\theta \, \Delta\phi)/\Omega \quad [\text{s}] \quad (16.2)$$

The final radar bandwidth after processing is $1/t$ and the noise power in the receiver is therefore

$$N = kT_0 F\left(\frac{1}{t}\right) = \frac{FkT_0\Omega}{T \, \Delta\theta \, \Delta\phi} \quad [\text{W}] \quad (16.3)$$

Putting Eq. (16.3) into the radar equation, and replacing G_t by $4\pi/(\Delta\theta \, \Delta\phi)$ and G_r by $4\pi A_e/\lambda^2$ gives

$$\text{SNR} = \frac{P_t A_e \sigma TL}{4\pi R^4 \Omega F k T_0} \quad [\quad] \quad (16.4)$$

which is the formula developed by Radford[1]. Note that the expression is frequency-independent, although some terms, such as the RCS, will vary with frequency. The importance of expressing the radar equation as in Eq. (16.4) is that only the power × aperture product $P_t A_e$ is under the control of the radar designer, the others being set by the requirement or the environment.

Survivability and cost mitigate against larger power × aperture products, so where is the improved performance to be found? The answer lies in improved resolution and data rate, and both of these require an increase in the fundamental limiting factor of radar performance, the bandwidth.

Bandwidth can be thought of in terms of the rate of searching radar resolution cells. The total number of beam positions n_1 to be searched is

$$n_1 = \frac{\Omega}{\Delta\theta\,\Delta\phi} \quad [\ ] \tag{16.5}$$

The number of range bins n_2 is given by

$$n_2 = \frac{\text{interval between pulses}}{\text{pulse duration}} = \frac{1/\text{PRF}}{\tau}$$

$$= \frac{1}{(\text{PRF}) \times \tau} \quad [\ ] \tag{16.6}$$

and the maximum number of doppler filter channels n_3 is set by the number of pulses in the dwell time as

$$n_3 = (\text{PRF}) \times t \quad [\ ] \tag{16.7}$$

Combining Eqs (16.5) and (16.7) with Eq. (16.2) gives the total number of resolution cells n to be searched as

$$n = n_1 n_2 n_3$$

$$= \frac{\Omega}{\Delta\theta\,\Delta\phi} \times \frac{1}{(\text{PRF}) \times \tau} \times (\text{PRF}) \times t$$

$$= T/\tau \quad [\ ] \tag{16.8}$$

Remembering that $1/\tau$ is the bandwidth B of the receiver, we can rearrange Eq. (16.8) as

$$n/T = B \quad [\text{Hz}] \tag{16.9}$$

Thus the maximum number of resolution cells that can be searched every second is set by the bandwidth of the system.

The simplest way to increase the bandwidth of a radar system is to use multiple antenna beams and receivers; the effective bandwidth B_{eff} is then

$$B_{\text{eff}} = mB \quad [\text{Hz}] \tag{16.10}$$

where m = number of simultaneous beams and receivers.

Worked example The Marconi Radar System Martello S723 is a long-range L-band surveillance radar using a single transmit beam, but eight receive beams stacked in elevation. If the pulse length (after compression) is 0.25 μs, and the antenna rotation rate is 10 RPM, how many resolution cells could be searched?

SOLUTION If the bandwidth of a single channel is 4 MHz, the effective bandwidth of the system is 32 MHz; if this represents the number of cells inspected every second, and one antenna revolution takes 6 s, the total number of cells to be searched is about 2×10^8.

COMMENT Although 2×10^8 seems a very large number of cells to be inspected, it looks more realistic if you work out the values of n_1, n_2 and n_3 separately. This large number of resolution cells reinforces the point made in Chapter 2 that front-end radar data rates are high.

Increasing the bandwidth of a radar brings several benefits:

● Increasing the receiver bandwidth decreases the range resolution cell size and cuts down the clutter.
● The use of more simultaneous beams implies that longer integration times can be used in each beam position. This increases the radar sensitivity.
● Longer integration times mean that the doppler resolution and clutter rejection can be improved.
● The data processing rate for each target can be increased.

In this way, all our requirements can be met without increasing the power × aperture product and exposing the radar to an increased risk of ECM attack. Generally speaking, the analogue parts of a radar system are not the factor limiting the bandwidth, but the computational power available. This situation is changing with the arrival of parallel processing, very large-scale integration (VLSI) signal processing integrated circuits and optical methods of solving integrated circuit interconnection problems. Soon, large amounts of low-cost computing power will be available, and we must decide how to make best use of it in the design of the next generation of radars. One of our main aims must be to make radars more adaptable to the tasks they have to carry out.

16.3 ADAPTIVITY

In the past, radars have been rather unintelligent devices, perhaps more so than was necessary. Early radars often did not know if they were being

jammed and yet the noise level in the receiver, measured continuously as
the antenna rotated, could have been used to draw a shape on a PPI display
representing the maximum detection range for targets of various RCS. For a
radar that is internally noise-limited, these contours would be perfect circles
on a PPI display, but if external noise was received from any particular
direction, this would cause a dint in the circle and warn the operator that
the radar was not working well in that direction.

Imagine, now, a phased array radar in which all aspects of the system
are controlled by the data processing computer, as shown in Fig. 16.1. The
beam scheduling is controlled adaptively and well behaved targets are
inspected less frequently than those which have been found to be manoeuvring.
If the transmitted power is variable, this may also be controlled adaptively
to illuminate small long-range targets more strongly. It may be more
important to control the transmitter waveform adaptively and reserve long
pulse and wide swept bandwidths for difficult situations. When allocating
air defence resources, for example, it may be important to find out whether
a single echo represents one attacking aircraft or a very tight formation of
several planes; this may be possible by allocating extra integration time,
bandwidth and computing resources to the problem until it is resolved, while
other continuing searches are maintained using shorter pulses.

Adaptive doppler filtering and tracking algorithms may be one method
by which other information can be integrated into the system, in a process
known as *data fusion*. A modern warship, for example, has many sensors
physically close to each other and it is relatively easy (electrically) to
connect them together. If a target is identified by other means (e.g. visual
identification), the radar data processing can be allocated on the basis of
the known characteristics of the target. A modern fly-by-wire aircraft would

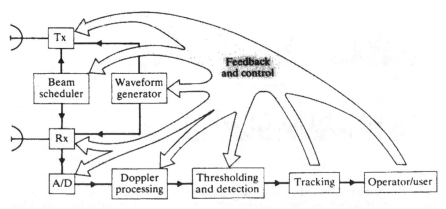

Figure 16.1 Adaptive phased array radar, in which all aspects of the system are controlled by
the data processing.

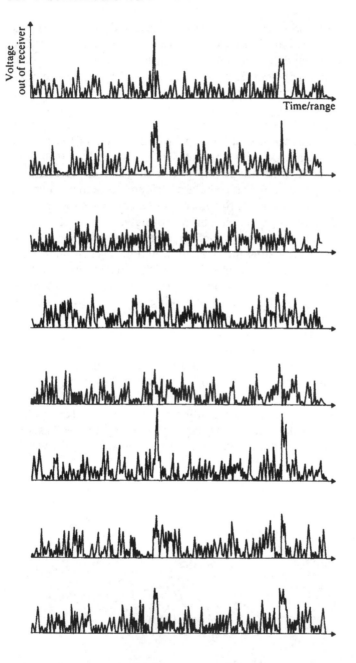

Figure 16.2 Searching for targets as a pattern among a series of scans can be more productive than scanning each threshold independently. These scans were generated artificially using white noise and a double pulse generator, set to give a low SNR. To find the targets, try looking at the diagram with your eyes nearly level with the bottom of the page.

need more frequent inspection than a B52 bomber, because of the fighter's greater manoeuvrability.

Data fusion is one of the areas where *artificial intelligence* (AI) techniques can be used to make radars more intelligent. There is no magic in AI, but there are some very useful programming ideas. The process of track initiation could be improved by moving away from simple 'hit or miss' thresholding to the utilization of 'soft thresholding' and fuzzy logic processing. In addition, pattern recognition algorithms could be used to determine whether a series of poor-quality hits represents the presence of a real track or not. This process is illustrated in Fig. 16.2; try looking at these range sweeps with your eye almost level with the page and the two targets (both at very low SNR) should become obvious. In fact, *historical* displays, which progressively reveal the last few hits in a sector, have been used in the past, but it was the human operator who made the decisions; AI will allow us either to give extra aid to radar operators or to replace them by software, which is more reliable and can detect targets at lower SNR for the same probability of detection and false alarm as at present.

16.4 SUMMARY

The future of radar does not lie in larger and more powerful systems, but rather in slightly smaller systems that are more agile, intelligent and difficult to detect because of the larger bandwidths that will be used. The resolution of radars, and the number of targets that can be tracked, can be expected to increase as large amounts of low-cost computer power become available.

The factor limiting radar performance is likely to remain technology (rather than any law of physics) for some time to come. Until recently, it was the ability to process the large amounts of data that created an upper limit to performance, but with the advent of parallel processing the weakest link in the chain is almost certain to become the A/D converter because of the problem of increasing the sampling rate while maintaining a high dynamic range. The recent trend of moving the point at which the A/D conversion takes place further and further up the receiving chain towards the antenna only exacerbates this problem.

Radar seems certain to provide challenging engineering, mathematical and computational problems to be solved for years to come. We hope that this book has conveyed the main ideas and helped you to understand the underlying principles of radar. We hope also that you have gained an appreciation of the importance of radar in many diverse areas, and sensed some of the excitement of working in this field.

16.5 REFERENCE

1. Whither radar, M. F. Radford, *GEC J. Res.*, 3(2), 137–143, 1985.

I

SYMBOLS, THEIR MEANING AND SI UNITS

These are the basic symbols used. Some have several subscripts, e.g. L, L_s, L_{sys}, etc.

A	signal scaling factor	[ ]
A	antenna area	$[m^2]$
A_c	clutter area	$[m^2]$
$2a$	bistatic baseline	$[m]$
B	bandwidth	$[Hz$ or rad $s^{-1}]$
C	array element correction coefficient	[ ]
C_n^2	structure constant	$[m^{2/3}]$
C_r^2	layer scattering factor	[ ]
C_s^2	volume scattering factor	[ ]
c	speed of light	$[m\ s^{-1}]$
$c(t)$	clutter voltage	$[V]$
D	antenna directivity	[ ]
D	diameter of a water droplet	$[m]$
D	ambipolar diffusion coefficient	$[m^2\ s^{-1}]$
D	length of an antenna array	$[m]$
d	linear antenna dimension	$[m]$
d	array element spacing	$[m]$
d	distance between synthetic aperture radar and scatterer	$[m]$
E	energy	$[J]$
$E[\ \ ]$	expectation value	$[units]$
E	field from N elements of an array	$[V\ m^{-1}]$

e	partial pressure of water vapour	[millibars]
e	bistatic ellipticity factor	[ ]
F	receiver noise figure	[ ]
F_a	effective antenna noise factor	[ ]
f	frequency	[Hz]
f	over-the-horizon propagation factor	[a]
f_d	doppler frequency shift	[Hz or rad s^{-1}]
Δf_d	doppler resolution	[Hz or rad s^{-1}]
δf_d	doppler uncertainty	[Hz or rad s^{-1}]
Δf	frequency separations or swept frequency	[Hz or rad s^{-1}]
G	antenna gain	[ ]
g	acceleration due to gravity	[m s^{-1}]
H	atmospheric scale height	[m]
$H(\omega)$	system transfer function	[ ]
h	height	[m]
h	over-the-horizon height factor	[ ]
$h(t)$	system impulse response function	[ ]
k	Boltzmann's constant	[J K^{-1}]
k	threshold multiplicative factor	[ ]
k	effective Earth radius factor	[ ]
L	loss factor	[ ]
L	length of a sea wave	[m]
L_0	turbulence scale size	[m]
l	characteristic dimension of a target	[m]
M	number of observations to be averaged	[ ]
M	fast Fourier transform length	[ ]
m	over-the-horizon radar cross-section modification factor	[ ]
$m(t)$	system response to noise input	[V]
N	noise power	[W per unit bandwidth]
N	radio refractivity	[ ]
N_e	electron density	[electrons m^{-3}]
n	refractive index	[ ]
n	number of observations	[ ]
$n(t)$	noise voltage	[V]
P	power	[W]
P_d	probability of detection	[ ]
P_{fa}	probability of false alarm	[ ]
P_r	transmitted power	[w]
P_t	received power	[w]
$p(\sigma)$	radar cross-section probability density function	[ ]

p	air pressure	[millibars]
q	electron line density	[electrons m^{-1}]
R	range	[m]
ΔR	range resolution	[m]
δR	range uncertainty	[m]
R_e	radius of the Earth	[m]
$R_m(\tau)$	autocorrelation function	[ ]
r	precipitation rate	[mm h^{-1}]
$2r$	bistatic transmitter–target–receiver distance	[m]
r_a	synthetic aperture radar along-track resolution	[m]
r_e	effective radius of the electron	[m]
r_i	meteor trail initial radius	[m]
$S(f$ or $\omega)$	power spectral density	[W per unit bandwidth]
$S_m(\omega)$	power spectrum of $m(t)$	[W rad^{-1}]
s	convolution integration variable	[s]
T	inter-pulse period (reciprocal of pulse repetition frequency)	[s]
T	interval between observations	[s]
T	temperature	[K]
T_0	system noise temperature	[K]
T_u	underdense echo delay time	[s]
t	time	[s]
t	integration time	[s]
$U(\omega)$	Fourier transform of $u(t)$	[V s rad^{-1}]
$u(t)$	radar signal voltage	[V]
V	voltage	[V]
V_c	clutter volume	[m^2]
V_{res}	volume of a resolution cell	[m^3]
v	velocity	[m s^{-1}]
δv	velocity uncertainty	[m s^{-1}]
$v(t)$	system response to signal input	[V]
W	array element weighting coefficient	[ ]
x	target position	[m]
$y(t)$	receiver output voltage	[V]
Z	radar reflectivity factor	[mm^6 m^{-3}]
α	track position smoothing factor	[ ]
α	angle of refraction	[radians or degrees]
β	track speed smoothing factor	[ ]
β	synthetic aperture radar antenna real beamwidth	[radians]
β_s	synthetic aperture radar synthetic beamwidth	[radians]

$\delta(t)$	delta function	[ ]
ε	antenna efficiency factor	[ ]
ε_r	relative permittivity	[ ]
η	average radar cross-section per unit volume	$[\mathrm{m}^{-1}]$
η	relative bandwidth	[ ]
θ	azimuth angle	[radians or degrees]
$\Delta\theta$	azimuth beamwidth	[radians or degrees]
$\delta\theta$	azimuth uncertainty	[radians or degrees]
λ	wavelength	[m]
ρ	radius of curvature	[m]
$\rho_u(t)$	autocorrelation function of $u(t)$	[V]
σ	radar cross section	$[\mathrm{m}^2]$
σ	standard deviation	[units]
σ^2	variance	$[\mathrm{units}^2]$
τ	pulse duration	[s]
τ_d	two-way propagation delay to target	[s]
ϕ	elevation angle	[radians or degrees]
$\Delta\phi$	elevation beamwidth	[radians or degrees]
$\delta\phi$	elevation uncertainty	[radians or degrees]
$\chi(t, \omega_d)$	ambiguity function	[ ]
ψ	signal phase	[radians]
$\delta\psi$	phase difference between signals	[radians]
Ω	solid angle	[steradians]
ω_d	doppler frequency	$[\mathrm{rad\ s}^{-1}]$

II

ACRONYMS AND ABBREVIATIONS

AC	alternating current
ACF	autocorrelation function
A/D	analogue-to-digital (converter)
AI	artificial intelligence
ARM	anti-radiation missile
ATC	air traffic control
CAT	clear-air turbulence
CCIR	International Radio Consultative Committee
CFAR	constant false-alarm rate
chirp	linear frequency sweep during a pulse
COMINT	communications intelligence
CW	continuous wave
DC	direct current
ECCM	electronic counter-countermeasures
ECM	electronic countermeasures
EJ	escort jammer
ELINT	electronic intelligence
EM	electromagnetic
ERP	effective radiated power
ERS-1	European remote sensing satellite
ESM	electronic support measures
EW	electronic warfare
FFT	fast Fourier transform

FIR	finite impulse response
FM	frequency modulation
FMCW	frequency-modulated continuous wave
FMICW	frequency-modulated interrupted continuous wave
FOM	figure of merit
fruit	false replies unsynchronized in time
FT	Fourier transform
GHz	gigahertz
gw	groundwave
HF	high frequency
HOJ	home-on jammer
IC	integrated circuit
ICAO	International Civil Aviation Organization
IF	intermediate frequency
IFF	identify, friend or foe
ISAR	inverse synthetic aperture radar
kHz	kilohertz
lidar	light detection and ranging
LPI	low probability of intercept
LVA	large vertical aperture
MHz	megahertz
MST	mesosphere–stratosphere–troposphere
MTI	moving-target indicator
OTH	over-the-horizon
PDA	probabilistic data association
PDF	probability density function
PPI	plan position indicator
PRF	pulse repetition frequency
radar	radio detection and ranging
RAM	radar-absorbing material
RCS	radar cross-section
RF	radio frequency
RMS	root mean square
RPV	remotely piloted vehicle
RWR	radar warning receiver
Rx	receiver
SAR	synthetic aperture radar
SCV	sub-clutter visibility
SNR	signal-to-noise ratio

SOJ	stand-off jammer
SSR	secondary surveillance radar
ST	stratosphere–troposphere
sw	skywave
T–R	transmit–receive
Tx	transmitter
UHF	ultra high frequency
VHF	very high frequency
VLSI	very large-scale integration

USEFUL CONVERSION FACTORS

Nautical mile (n. mile) = 1/60 of a degree of latitude

$$= 6080 \text{ ft}$$

$$= 1852 \text{ m}$$
$$= 1.852 \text{ km}$$

This is the International nautical mile; the UK nautical mile is 1.853 18 km. We have adopted the symbol n. mile to avoid confusion with nanometres (nm).

Knot (kt) = 1 n. mile per hour

$$\simeq 0.514\,44 \text{ m s}^{-1}$$

$$= 1.852 \text{ km/h}^{-1}$$

Foot (ft) = 0.3048 m exactly

A foot is the unit of length in the old British system of units.

1000 ft = 0.3048 km

Speed of light (c) = 299 792 458 m s^{-1} exactly

Boltzmann's constant (k) = 1.380 658 $\times$ 10^{-23} J K^{-1}

This has 8.5 ppm uncertainty. When a standard temperature of 290 K is used:

$$kT_0 = 4.003\,908 \times 10^{-21} \text{ dB J}$$

$$= -204 \text{ dB J}$$

$$= -204 \text{ dB W Hz}^{-1}$$

Radius of the Earth R_e = 6378.160 km

The Earth's shape is an oblate spheroid with an equatorial bulge caused by its rotation. The figure above is the equatorial radius or semi-major axis. The polar radius, or semi-minor axis is 6356.775 km.

Some SI prefixes commonly used in radar are:

Prefix	Symbol	Power	Example
pico	p	10^{-12}	picosecond
micro	μ	10^{-6}	microsecond
milli	m	10^{-3}	millimetre
centi	c	10^{-2}	centimetre
kilo	k	10^{3}	kilowatt
mega	M	10^{6}	megahertz
giga	G	10^{9}	gigahertz

Reference source: *The McGraw-Hill Dictionary of Science and Technical Terms*, 4th edn, Ed. S. P. Parker, McGraw-Hill, New York, 1989.

USING DECIBELS

Decibels can be one of those little things that you can get a mental block about. It is not that there is anything difficult about the maths, but rather the concept of handling voltage and power ratios in this way. However, it is worth learning to use decibels, because they make life so much easier for the radar engineer.

The idea behind decibels is that the numbers used in radar calculations are often very large (e.g. R^4) or very small (e.g. P_r), and they must usually be multiplied and divided. This makes life difficult and can easily lead to mistakes. Converting these numbers to logarithms means that, first, they become a sensible size and, secondly, they only have to be added and subtracted. This greatly facilitates repeating calculations, when one of the parameters is varied.

The definition of decibels is that, if P_1 and P_2 are two amounts of power then the first is said to be n decibels greater than the second, where

$$n = 10 \log_{10}(P_1/P_2)$$

The ratio of two voltages V_1 and V_2 can be converted to decibels by

$$n = 20 \log_{10}(V_1/V_2)$$

In radar, it is more common to find the $10 \log_{10}$ power version used in calculations, rather than voltage ratios, but this is not necessarily so in other subjects. To escape from decibels and return to linear numbers, divide n by 10 (or 20 for voltage ratios) and then use the 10^x function on your calculator.

It can sometimes be difficult to understand negative decibels. These represent ratios less than unity. A figure of -10 dB means a ratio of $1/10$,

-20 dB $= 1/100$, -30 dB $= 10^{-3}$, etc. Inverting a ratio just changes the sign in decibels. For example, $(4\pi)^3 = 33$ dB and $1/(4\pi)^3 = -33$ dB. Note that if you have a noise power of -144 dB W and you *increase* the noise level by 10 dB, the new noise power is -134 dB W.

Although decibels are often used to denote straightforward ratios, they can also be used to express how large a number is compared to some reference level. Thus if we say a transmitter power is 30 dB W, we mean that it is 30 dB above a watt, i.e. it is a kilowatt. One reference level commonly used is the milliwatt, and powers are referred to this level using the dB mW. Thus a received power of -100 dB mW is the same as -130 dB W, or 10^{-13} W.

Some commonly used values are listed below.

Ratio	dB	Ratio	dB
$(4\pi)^{-3}$	-33.0	1	0
10^{-3}	-30.0	2	3.0
$(4\pi)^{-2}$	-22.0	4	6.0
10^{-2}	-20.0	5	7.0
10^{-1}	-10.0	10	10.0
$1/4$	-6.0	10^2	20.0
$1/2$	-3.0	10^3	30.0
		1 MHz	60 dB Hz
		1 GHz	90 dB Hz

$$\text{Boltzmann's constant } k = 1.38 \times 10^{-23} \text{ W Hz}^{-1} \text{ K}^{-1}$$

$$= 228.6 \text{ dB W Hz}^{-1} \text{ K}^{-1}$$

$$\text{Speed of light } c = 3 \times 10^8 \text{ m s}^{-1} = 84.8 \text{ dB m s}^{-1}$$

Historical note: The decibel was invented by American telephone engineers in the 1920s. The unit was named after Sir Alexander Graham Bell, who is credited with inventing the telephone on 2 June 1875 (although this was after a long and competitive research programme). The original unit, the bel, was too large to be useful.

Reference source: *The McGraw-Hill Dictionary of Science and Technical Terms*, 4th edn, Ed. S. P. Parker, McGraw-Hill, New York, 1989.

SOLUTIONS TO PROBLEMS

Chapter 1

1.1 (a) $\lambda/d = 0.03/(2.7 \times 0.8) = 0.014$ [rad] $= 0.8°$

(b) $\lambda/d = 0.10/(3.6 \times 0.8) = 0.035$ [rad] $= 2°$

(Racal–Decca quote 0.8° and 2° respectively in their specifications.)

1.2 $G = 32\,000/(\Delta\theta° \, \Delta\phi°) = 32\,000/(0.8° \times 23°) = 32.4$ dB

(Racal–Decca quote 32 dB.)

1.3 $\Delta R \sim c/2B \sim 37.5$ m

(Racal–Decca quote 35 m.)

1.4 $\Delta\theta = 0.032/3.4 = 0.54°$

$\Delta\phi = 0.032/0.75 = 2.4°$

(Siemens–Plessey quote 0.55° and 2.5° respectively in their data summary.)

1.5 $G = 4\pi A/\lambda^2 = 4\pi(3.4 \times 0.75)/(0.032)^2 = 45$ dB

(Siemens–Plessey quote a figure of 'not less than 41.5 dB' for this radar.)

1.6 The transverse resolution of the system $R \, \Delta\theta = 7400 \times 0.032/3.4 = 70$ m. This could be improved on considerably because of good SNR at such short ranges. For an airfield controller to use these measurements to position an aircraft with a mean error of 25 m appears to be very good. An error of 25 m at 4 n. mile ($= 7.4$ km) reduces to 6.3 m at 1 n. mile and in the evaluation the error did not in fact exceed 6 m. Assuming a runway length of 1 n. mile, this should be sufficiently accurate for a bad-weather landing.

1.7

Parameter	2690 BT marine radar	ACR 430 airfield control radar
P_t	44 dB W	47.4 dB W
$G_t G_r$	64 dB	83.0 dB
σ	10 dB m^2	10 dB m^2
λ^2	-30.5 dB m^2	-30.5 dB m^2
L_s	-5 dB	-5 dB
$1/(4\pi)^3$	-33 dB	-33 dB
$1/N$	$+131$ dB W	$+131$ dB W
$1/(\text{SNR})$	-13 dB	-13 dB
$R^4_{\max}$	167.5 dB m^4	189.9 dB m^4
$R_{\max}$	15.4 km	55.9 km
	8.3 n. mile	30.2 n. mile

A calculation worksheet included in the Siemens–Plessey report[5] gives the prediction of radar range as 30 n. mile. In practice, these detection ranges would be increased by the integration of many pulses; such integration would be necessary for the doppler processing, especially in the case of the marine radar.

Chapter 2
2.1 Duty cycle $= \tau/T = \tau \times (\text{PRF})$

Mode 1: Duty cycle $= 1.5 \times 10^{-6} \times 480 = 1/1389$

Mode 2: Duty cycle $= 1.0 \times 10^{-6} \times 691 = 1/1447$

The mean power $=$ peak power $\times$ duty cycle.

Mode 1: Mean power $= 864$ W

Mode 2: Mean power $= 829$ W

(Alenia quote 0.87 and 0.83 kW, respectively.)

2.2 We have

$$n = \frac{\Delta\theta}{360°} \times (\text{no. pulses/s}) \times (\text{time [s] of one revolution})$$

$$= \frac{\Delta\theta}{360} \times \frac{(\text{PRF})}{(\text{RPM})/60} = \frac{\Delta\theta \times (\text{PRF})}{6 \times (\text{RPM})}$$

$$n = \frac{1.2 \times 480}{6 \times 6} = 16 \text{ hits/scan}$$

2.3 Equating

$$R_{max} = \frac{c}{2 \times (\text{PRF})} \quad \text{with} \quad \frac{10^6}{12.4 \times (\text{PRF})}$$

and remembering that 1 nautical mile = 1853 m, tells us that the number 12.4 is a nautical mile expressed in microseconds. This is another way of expressing the '150 m per microsecond' rule as '12.4 μs for every nautical mile'.

2.4

$P_t =$	60.8 dB W	1.2 MW
$G_t G_r =$	73.0 dB	$G = 25\,000/1.2 \times 4.7$
$\sigma =$	3.0 dB m²	2 m² target
$\lambda^2 =$	-12.7 dB m²	
$L =$	-7.0 dB	
$1/(4\pi)^3 =$	-33.0 dB	
SNR $=$	-10.0 dB	
$(\text{Noise})^{-1} =$	$+145.8$ dB W	Assuming bandwidth $= \tau^{-1}$
		$= (1.5 \times 10^{-6})^{-1}$

$R_{max}^4 =$	219.9 dB m⁴	
$R_{max} =$	55.0 dB m	$\equiv 314 \text{ km} \sim 170 \text{ n. mile}$

The Alenia engineers' radar calculation worksheets use a more detailed procedure, but arrive at a similar conclusion; 164 n. mile for this case.

2.5 $$R_{unambig} = \frac{c}{2 \times (\text{PRF})} = \frac{1.5 \times 10^8}{480} = 312.5 \text{ km}$$

This is similar to the maximum detection range and so the system is nominally unambiguous in range. However, Alenia provides a 5 per cent PRF stagger to remove long-range ground clutter effects caused by tropospheric ducting.

2.6 Clutter cell size $= R \, \Delta\theta \, \Delta R$

$$= 300\,000 \, [\text{m}] \times 2.1 \times 10^{-2} \, [\text{rad}] \times 225 \text{ m} \, (\equiv 1.5 \, \mu s)$$

Using a land reflectivity of 1/100, this implies an RCS for the clutter in each resolution cell of 14 000 m² $\equiv$ 41.5 dB m². If the target RCS is 3 dB m² and the system noise is 13 dB lower, the minimium dynamic range requirement is 51.5 dB m². An A/D converter nominally gives 6 dB of dynamic range per bit, suggesting that a minimum of 9 bits is required. In practice 6 dB/bit is not achieved because the sensitivity must be set such that the quantization noise is small compared to the system noise.

(The ATCR-44K radar uses a 12 bit A/D converter sampling every 1.5 μs.)

2.7

$$
\begin{aligned}
P_t &= && 43.0 \, \text{dB W} && 20 \, \text{kW} \\
G_t G_r &= && 32.0 \, \text{dB} \\
\sigma &= && 3.0 \, \text{dB m}^2 \\
\lambda^2 &= && 7.6 \, \text{dB m}^2 \\
L &= && -5.0 \, \text{dB} \\
1/(4\pi)^3 &= && -33.0 \, \text{dB} \\
(\text{SNR})^{-1} &= && -13.0 \, \text{dB} \\
(\text{Noise})^{-1} &= && +147.0 \, \text{dB W}
\end{aligned}
$$

$$
\begin{aligned}
R_{\text{max}}^4 &= && 181.6 \, \text{dB m}^4 \\
R_{\text{max}} &= && 45.4 \, \text{dB m} && \equiv 34.5 \, \text{km} \equiv 18.6 \, \text{n. mile}
\end{aligned}
$$

This is much lower than the modern Alenia radar.

To be fair to the Freya system, there would probably have been considerable incoherent gain involved in the display process, which would have improved the detection range over this single-pulse calculation. Nonetheless, the calculation serves to illustrate that the improvement in the λ^2 factor over a modern L-band radar is not sufficient to compensate for the much smaller power $\times$ aperture product.

Chapter 3

3.1 $(\text{PRF}) \times 60/(\text{RPM}) = 25$ pulses/scan

3.2 The radar equation indicates a SNR of 0.3 dB for a 120 m^2 target at 100 km. In one minute, 1500 angular measurements are made, which would increase the SNR of the measurement by up to 31.8 dB. The best that might be achieved is $2°/\sqrt{(2 \times \text{SNR})} = 0.04°$, although the spinning of the reflector as it dangled below the balloon would cause RCS variations that would degrade this figure. (Siemens–Plessey quote better than 0.15° for this system.)

3.3 The beamwidth is roughly $2.5° \times 4°$, making 40 positions to be searched. Each requires 9.6 ms and the search could nominally be completed within 0.4 s. In practice, additional time would be required to change beam position, but it would still be sufficient to maintain target contact during a dog-fight. Some of the parameters used in this problem are similar to those used by the Westinghouse APG-66 radar fitted to the F-16 aircraft—the 'fighting falcon'.

3.4 We would expect an accuracy ~ 3 minutes of arc. At 1 km range this corresponds to a tangential error ~ 1 m (accuracies of 3 minutes RMS in angle and 5 m in range are claimed for this radar). With these accuracies it would be dangerous for an air target to approach nearer than 1 km, and the ST802 can in fact be used for controlling guns of calibres from 20 to 76 mm.

3.5 (a) V_r (beam centre) $= 300 \cos 10° \cos 45° = 208.9$ m s^{-1}
$\qquad V_r$ (beam edge) $= 300 \cos 10° \cos 47° = 201.5$ m s^{-1}
The difference of 7.4 m s^{-1} corresponds to a doppler shift of ± 493 Hz across the beam.
$\qquad$ (b) 10 Hz $= 0.15$ m s^{-1} $\sim 0.04°$. These parameters are similar to those used on the GEC–Ferranti Blue Vixen radar recently fitted to the UK Sea Harrier aircraft. The Blue Vixen is a multimode fire control radar, but with the capability of doppler beam sharpening. (GEC–Ferranti suggest that a resolution of $1/20°$ or better is achievable with this technique.)
$\qquad$ (c) V_r (ahead) $= 300 \cos 10° \cos 0° = 295.4$ m s^{-1}
$\qquad V_r$ (beam edge) $= 300 \cos 10° \cos 2° = 295.3$ m s^{-1}
The difference in velocity is insufficient to give good resolution and doppler beam sharpening is not normally used within $\pm 15°$ of the aircraft track.

Chapter 4
4.1 If mean RCS $= s$, PDF is

$$p(x) = (1/s)\,e^{-x/s} \qquad x \geqslant 0$$

Thresholding 10 dB above the mean gives

$$P_{fa} = \int_{10s}^{\infty} (1/s)\,e^{-x/s}\,ds = e^{-10} \sim 4.54 \times 10^{-5}$$

So expected number of false alarms is $512^2 \times P_{fa} \sim 12$. For gaussian PDF, we must solve

$$\tfrac{1}{2} - \Phi(k) = 4.54 \times 10^{-5}$$

which gives $k = 3.91$.
$\qquad$ In the exponential case we are nine (additive) standard deviations above the mean; in the gaussian case only 3.91.

4.2 (a) $\qquad\qquad$ Output $= \rho(t) = \displaystyle\int_{-\infty}^{\infty} u(s)u(s - t)\,ds$

This is the product integrated in the overlap (see diagram):

$\qquad\qquad |t| > 2T$; no overlap $\qquad\qquad \rho(t) = 0$
$\qquad\qquad 0 \leqslant t \leqslant 2T$ (as shown) $\qquad \rho(t) = V^2(2T - t)$

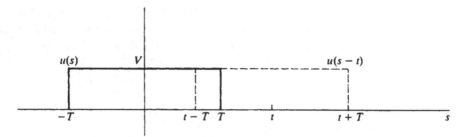

By symmetry

$$p(t) = \begin{cases} 0 & |t| > 2T \\ V^2(2T - |t|) & |t| \leqslant 2T \end{cases}$$

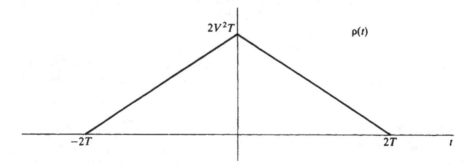

(b) Energies are the same, therefore there is the same detection performance after matched filtering. Output in case (i) is as above with $T = 10^{-3}$. Output in case (ii), using diagrams as above, is:

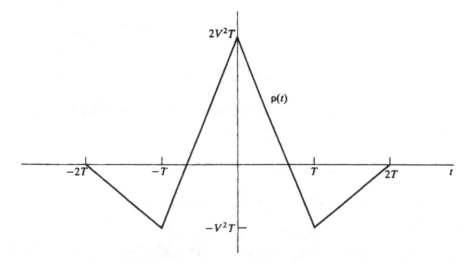

Both outputs have the same maximum value at 0; this is the energy $2V^2T$. The second waveform gives a sharper main lobe; this has implications for system resolution, as discussed in Chapter 6.

4.3 By Eq. (4.33), noise power $= kT_0E_h/2$, and

$$E_h = \int_{-T/2}^{T/2} \cos^4\left(\frac{\pi t}{T}\right) dt = \frac{3T}{8}$$

(use $\cos^2 x = \frac{1}{2}[1 + \cos(2x)]$ twice). Therefore,

$$\text{Noise power} = 7.5 \times 10^{-25} \text{ W} \sim -241 \text{ dB W}$$

Energy in signal $= E = 10^{-16} \times 3T/8$. Hence

$$\text{SNR} = E/N$$

$$= 10^{-16} \times \tfrac{3}{8} \times 10^{-3} \times \frac{1}{4 \times 10^{-21}}$$

$$= 9.375$$

$$\sim 9.7 \text{ dB}$$

4.4

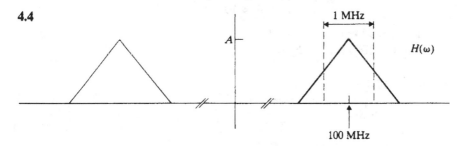

Use

$$E_h = \frac{1}{2\pi} \int_{-\infty}^{\infty} |H(\omega)|^2 \, d\omega$$

Since the integral of each triangular function squared will not change if it is moved to the origin, we can use the diagram below:

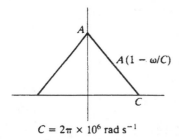

$C = 2\pi \times 10^6 \text{ rad s}^{-1}$

The function is even, so energy in each triangular function is

$$2 \times \frac{1}{2\pi} \int_0^C A^2(1 - \omega/C)^2 \, d\omega = \frac{A^2C}{3\pi}$$

Using both sidebands gives

$$E_h = \tfrac{4}{3} \times 10^6 \times A^2$$

So

$$\text{Output noise power} = \tfrac{2}{3} \times 10^6 \times A^2 N$$

If we use the second definition of bandwidth, we must use $C = \pi \times 10^6$, giving half the output noise power above.

4.5 Energy in returned pulse $= \tau P_r$. So

$$E = \frac{\tau P_t A_e^2 \sigma L_s}{4\pi\lambda^2 R^4}$$

Therefore,

$$\text{SNR} = \frac{E}{N} = \frac{E}{4 \times 10^{-21}} = \frac{5 \times 10^{21}}{\pi R^4}$$

(a) $R = 50$ km: $P_d = 1$
(b) $R = 100$ km: $P_d = 0.81$

4.6 In Chapter 1,

$$\text{SNR} = \frac{\text{signal power}}{\text{noise power}}$$

$$= \frac{\text{signal power}}{B} \times \frac{1}{\text{noise power/unit bandwidth}}$$

For simple pulses, $1/B \sim \tau$ where $\tau = $ pulse duration. Therefore, first term is signal energy.

4.7 (a) $k = 4.75$, and so,

$$0.9 = 0.5 - \Phi(4.75 - \sqrt{(2 \times \text{SNR})})$$

$$\Phi(-4.75 + \sqrt{(2 \times \text{SNR})}) = 0.4$$

$$\text{SNR} = 18.2 \qquad (12.6 \, \text{dB})$$

(b)

$$A = \ln(0.62 \times 10^6)$$

$$B = \ln 9$$

$$\text{SNR} = 20.6 \qquad (13.1 \, \text{dB})$$

4.8 SNR for integrated pulses = 15.6 (Sec. 4.6). Therefore,

$$SNR_1 = 15.6/16 = 0.975$$

and

$$P_d = \tfrac{1}{2} - \Phi(4.75 - \sqrt{1.95}) = 0.0004$$

4.9 Matched filter $= u^*(-t) = \exp(jat^2) \qquad |t| \leqslant T$

$$\rho_u(t) = \int_{-T}^{T} u(s+t)u^*(s)\,ds$$

$$= \int_{-T}^{T-t} \exp[-ja(s+t)^2]\exp(jas^2)\,ds \qquad 2T \geqslant t \geqslant 0;\ \text{see diagram}$$

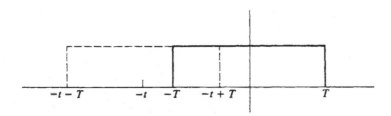

$$= \exp(-jat^2) \int_{-T}^{T-t} \exp(-2jast)\,ds$$

$$= -\frac{\exp(-jat^2)}{2jta}\{\exp[-2ja(T-t)t] - \exp(2jaTt)\}$$

$$= \frac{1}{at}\sin[(2T-t)at]$$

When $t = 0$, this gives the value $2T$.

A similar calculation for $-2T \leqslant t \leqslant 0$ gives

$$\rho_u(t) = \begin{cases} \dfrac{1}{at}\sin[(2T-|t|)at] & |t| \leqslant 2T \\[2mm] 0 & |t| \geqslant 2T \end{cases}$$

This is plotted in the diagram.

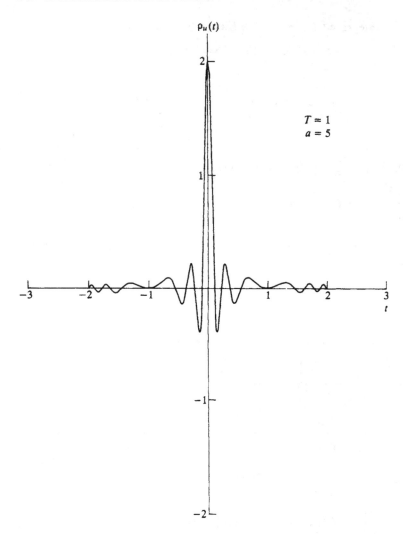

$$\rho_u(t)$$

$$T = 1$$
$$a = 5$$

4.10 After averaging, an SNR of 15.6 is needed. Therefore,

$$\text{Single-pulse SNR} = \frac{15.6}{32}$$

$$\text{SNR} = \frac{\tau P_t A_e^2 \sigma L_s}{4\pi \lambda^2 R^4 k T_0} \sim 0.0155 P_t$$

Hence $P_t \geqslant 31.4\,\text{W}$.

Chapter 5

5.1 FFT gain is 15 dB plus a maximum incoherent averaging gain of 3 dB. Manoeuvring not only alters the RCS, causing a loss of integration gain, but also spreads the echo across several doppler cells, thereby making detection difficult.

5.2 30.1 dB gain; position uncertainty reduced by a factor of $32(=\sqrt{1023})$. When an aircraft turns, its wings have a much larger vertical component, which increases the RCS. Despite doppler spreading, it is often possible for a surface-wave HF radar to follow an aircraft as it turns.

5.3 Equation (4.78) gives $k = 4.8$, i.e. a margin of 6.8 dB must be added to the mean. Equation (4.81) gives a SNR of 19.9 or 13.0 dB.

5.4

Observation/time	1	2	3	4	5	6
α	1.0	1.0	0.83	0.7	0.6	
β	1.0	1.0	0.5	0.3	0.2	
c_n	0	35	88	118	158	
x_p	0	35	70	128.9	162	**199.5**
x_s		35	84.9	121.3	159.6	
v_n		35	44	40.7	39.9	

5.5 Track error is 0.6 times measurement error after five observations. The standard deviation is about 4 m, and Eq. (5.29) suggests an association gate size of 6 m.

Chapter 6

6.1 (a)

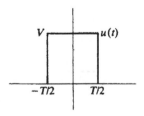
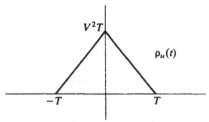

Output from matched filter

$$\Delta\tau_d = \frac{2\int_0^T V^4 T^2 (1 - t/T)^2 \, dt}{V^4 T^2} = \frac{2T}{3}$$

(b) $u(t) = V \exp(-t^2/T^2)$

$$\rho_u(t) = \int_{-\infty}^{\infty} V^2 \exp[-(s+t)^2/T^2] \exp(-s^2/T^2)\, ds$$

$$= V^2 \exp(-t^2/2T^2) \int_{-\infty}^{\infty} \exp[-2(s+t/2)^2/T^2]\, ds$$

$$= V^2 T \sqrt{(\pi/2)} \exp(-t^2/2T^2)$$

Using

$$\frac{1}{\sigma\sqrt{(2\pi)}} \int_{-\infty}^{\infty} \exp(-s^2/2\sigma^2)\, ds = 1$$

$$\Delta\tau_d = \int_{-\infty}^{\infty} \exp(-t^2/T^2)\, dt = T\sqrt{\pi}$$

$$\Delta\tau_\omega = \frac{2\pi \int V^4 \exp(-4t^2/T^2)\, dt}{[\int V^2 \exp(-2t^2/T^2)\, dt]^2} = \frac{2\pi\sqrt{(\pi T^2/4)}}{\pi T^2/2} = \frac{2\sqrt{\pi}}{T}$$

6.2 The outputs are as shown:

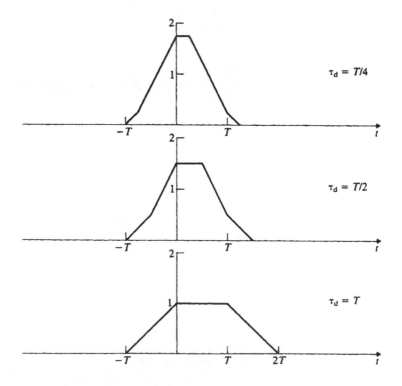

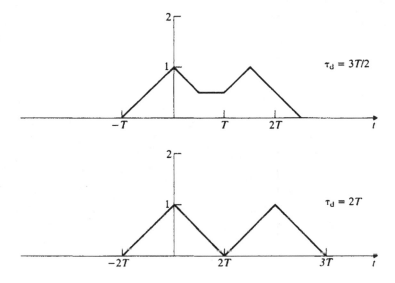

6.3 Use Eq. (6.19). Zero doppler:

$$\text{Output} = \chi(t,0) = \begin{cases} 1 - |t|/T & |t| \leqslant T \\ 0 & |t| > T \end{cases}$$

Doppler frequency $= 2\pi/T$:

$$\text{Output} = \chi(t, 2\pi/T) = \begin{cases} e^{j\pi t/T}\,(1/\pi)\,\sin(\pi|t|/T) & |t| \leqslant T \\ 0 & |t| > T \end{cases}$$

$$\text{Total output} = 1 - \frac{|t|}{T} + \frac{1}{\pi}\, e^{j\pi t/T} \sin\left(\frac{\pi|t|}{T}\right) \qquad |t| \leqslant T$$

Square law detecting this output yields

$$\left(1 - \frac{|t|}{T}\right)^2 + \frac{1}{\pi^2} \sin^2\left(\frac{\pi|t|}{T}\right) + \frac{2}{\pi}\left(1 - \frac{|t|}{T}\right)\cos\left(\frac{\pi t}{T}\right)\sin\left(\frac{\pi|t|}{T}\right)$$

$$= \left(1 - \frac{|t|}{T}\right)^2 + \frac{1}{\pi^2} \sin^2\left(\frac{\pi t}{T}\right) + \frac{1}{\pi}\left(1 - \frac{|t|}{T}\right)\sin\left(\frac{2\pi|t|}{T}\right) \qquad |t| \leqslant T$$

In practice, the values in this question are not too realistic for aircraft. What is the doppler velocity corresponding to $T = 1$ ms?

6.4 Using the notation of problem 4.2, and the answer to problem 6.1, gives $\Delta\tau_d = 4T/3$ (replace T by $2T$) for first pulse. For the second (see diagram)

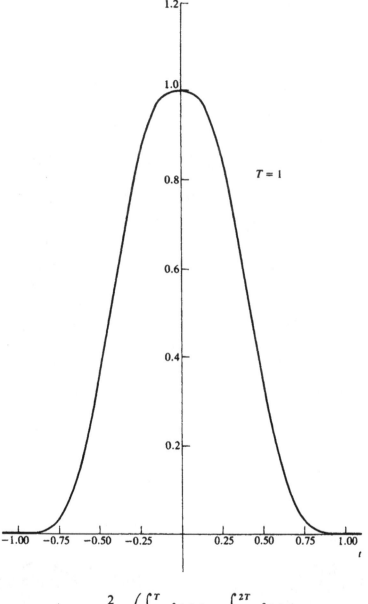

$$\Delta\tau_d = \frac{2}{4V^4T^2}\left(\int_0^T \rho^2(t)\,dt + \int_T^{2T} \rho^2(t)\,dt\right)$$

$$\Delta\tau_d = \frac{1}{2T^2}\left(\int_0^T (2T-3t)^2\,dt + \int_T^{2T}(t-2T)^2\,dt\right) = \frac{2T}{3}$$

This confirms our remark in the solution to problem 4.2 about the resolution of the second pulse.

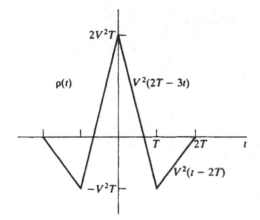

6.5 If there are M pulses, the energy both in $u(t)$ and in $u'(t)$ are M times greater than for a single pulse. Hence using Eq. (6.3), the effective bandwidth does not change.

$$u(t) = p(t) + p(t - T) + p(t - 2T) + p(t - 3T)$$

Therefore,

$$U(\omega) = (1 + e^{-j\omega T} + e^{-2j\omega T} + e^{-3j\omega T})P(\omega)$$

$$= \frac{1 - e^{-4j\omega T}}{1 - e^{-j\omega T}} P(\omega)$$

$$= \frac{e^{-2j\omega T}}{e^{-j\omega T/2}} \frac{(e^{2j\omega T} - e^{-2j\omega T})}{(e^{j\omega T/2} - e^{-j\omega T/2})} P(\omega)$$

$$= e^{-3j\omega T/2} \frac{\sin(2T\omega)}{\sin(T\omega/2)} P(\omega)$$

Plots (shown below) of $|P(\omega)|$ for $a = 1$ and $a = \frac{1}{2}$, of $|\sin(2T\omega)/\sin(T\omega/2)|$ and of their product show that the statement does not hold for any of the other commonly used definitions of bandwidth, e.g. 3 dB width or distance to first zero in the transform. To make the statement reasonable, the pulses

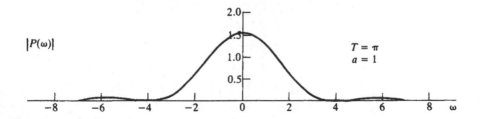

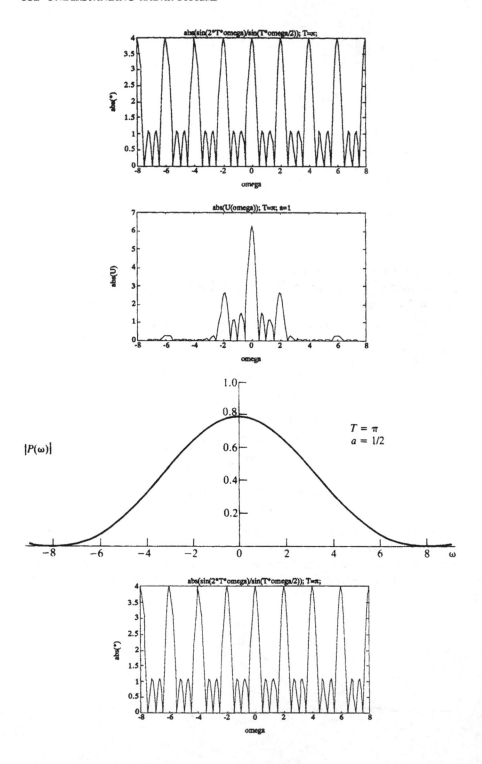

$|P(\omega)|$

$T = \pi$
$a = 1/2$

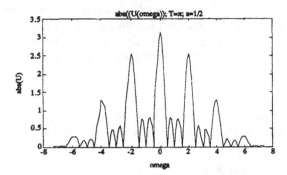

need to be rectangular, and of width equal to the inter-pulse period. Here

$$|P(\omega)| = \frac{aT}{2} \mathrm{Sa}^2\left(\frac{aT\omega}{4}\right)$$

where $\mathrm{Sa}\, x = (\sin x)/x$.

6.6 Use Eq. (6.37).

$$\frac{R\beta}{4} = \frac{R\lambda}{2V} f_{\mathrm{du}} = \frac{R\lambda}{2V} \frac{2V_r}{\lambda}$$

where V_r = radial velocity of ship. Therefore,

$$V_r = V\beta/4$$

The sign needs interpretation. For SAR, k in Eq. (6.33) is negative. Hence, if the ship's motion is towards the platform (ω_d is positive), the corresponding time is positive, i.e. the ship is imaged at a later time, so is moved in the along-track direction in the image (not the opposite way). So we know the ship is moving towards the radar; the component in the direction is 4.5 km h^{-1} (unless aliased). From the wake direction we can then calculate the full velocity vector.

6.7 (a) $-1, 0, 1, 4, 1, 0, -1$
   (b) $1, 0, -1, 4, -1, 0, 1$
   (c) $1, 0, 1, 0, 3, 0, -1, 0, 0, 0, 0, 8, 0, 0, 0, 0, -1, 0, 3, 0, 1, 0, 1$

Chapter 7
7.1 The power received by a unity-gain antenna at 256 n. mile ($\simeq 474$ km) would be ~ -99 dB W. This is nearly 2 dB greater than the -101 dB required to trigger the transponder, and the signal is therefore adequate when the aircraft is in the centre of the beam. The excess power (1.7 dB)

would ensure that, as the interrogation beam swept over the aircraft, the transponder would be triggered for about $3/4°$ either side of maximum illumination. This must be near the limit of satisfactory performance and is probably the reason why Cossor quote 256 n. mile as the maximum range of the system.

7.2 $P_r = -95.6$ dB W at the centre of the beam. Over the azimuth beamwidth of $2.45°$ the signal would equal, or exceed, -98.6 dB W.

7.3 $R_{max} = 616.6$ km $\simeq 333$ n. mile. Losses amount to 12 dB.

7.4 The power × aperture product of the uplink $(52.5$ dB W $+ 0$ dB$)$ is similar to the downlink $(24$ dB W $+ 27$ dB$)$. The main difference lies in the greater sophistication of the ground-based receiver being able to operate at signal levels $(-110$ dB W$)$ that are smaller than the transponder receiver on the aircraft $(-101$ dB W$)$.

7.5 From problem 7.2, $P_r = 95.6$ dB W and SNR $= 34.4$ dB. We could therefore expect the intrinsic resolution of $2.45°$ to be improved by $\sqrt{(2 \times \text{SNR})}$ to give an uncertainty of $0.03°$. In practice, monopulse can improve upon this figure when the aircraft is near the centre of the beam and the uncertainty might be nearer half this value. A good guide to the dependence of the angular error on the SNR, angle from bore-sight, etc., is to be found in Chapter 10 of Stevens[1].

Chapter 8

8.1 (a) 112.9 km, (b) 130.4 km, (c) 120.9 km, (d) 233.2 km.

8.2 35.1 m and 56.0 m respectively.

8.3 (a) The detection would be halved to 2250 m by the 6 dB diffraction loss.

(b) The detection range of a point target would be reduced by a factor of 0.71.

8.4 500 km. This principle is used in altimetry (see Chapter 11) to remove ionospheric effects and allow precise measurements of ocean surface height from space.

Chapter 9

9.1 $Z = 200(1)^{1.6} = 200 = 23 [\text{dB mm}^6/\text{m}^{-3}]$, so

Constant =	$-106.4 \,[\text{dB m s}^{-1}]$		$2.3 \times 10^{11} \text{ m s}^{-1}$
$P_t =$	$58.1 \,[\text{dB W}]$		650 kW
$G_t G_r =$	$74 \,\,[\text{dB}]$		37 dB each way
$Z =$	$23 \,\,[\text{dB mm}^6 \text{ m}^{-3}]$		
$\Delta\theta \, \Delta\phi =$	$-29.1 \,[\text{dB rad}^2]$		$2°$ beam
$\tau =$	$-57.0 \,[\text{dB s}]$		$2 \,\mu\text{s}$ pulse
$1/\lambda^2 =$	$19.7 \,[\text{dB m}^{-2}]$		2.9 GHz
$1/R^2 =$	$-86.0 \,[\text{dB m}^{-2}]$		20 km range

$$P_r = -103.7 \,[\text{dB W}]$$

The SNR would therefore be 33.3 dB. More detailed performance calculations by Met. Office engineers arrive at a similar figure of 32.6 dB.

9.2 $Z = 200(100)^{1.6} = 317 \times 10^3 = 55 \,[\text{dB mm}^6 \text{ m}^{-3}]$. This would increase the SNR by 32 dB (the Met. Office performance calculations likewise show a 32 dB enhancement).

9.3 Equation (9.7) can be reduced to

$$P_r \propto \frac{200(r)^{1.6}}{R^2}$$

The largest ratio is when $r = 64$ and $R = 1$ km, giving -8.1 dB. The smallest ratio is for $r = 1/8$ and $R = 200$ km, giving -97.5 dB. The dynamic range requirement is therefore nearly 90 dB. This can be relaxed by using range-dependent gain, and Siemens–Plessey do in fact use swept gain following a $1/R^2$ law to 200 km.

9.4 The R^2 term increases the loss by only 0.1 dB and this is almost compensated by an increase in the area of the layer illuminated by the off-vertical beam. The cause of the 10 dB drop in power when viewing off-vertical (an effect known as *aspect sensitivity*) may be an anisotropy of the turbulence at the edges of the layer, although scattering from steps in the refractive index may cause the same effect. Not all layers exhibit aspect sensitivity.

9.5 We have

$$\eta = 0.38 C_n^2 \lambda^{-1/3}$$

$$= 0.38 \times 10^{-17} \times 0.55 = 2.1 \times 10^{-18}$$

and so

$$C_s^2 = \Delta RG\eta$$

$$= 300 \times 1259 \times 2.1 \times 10^{-18} = 7.9 \times 10^{-13} = -121.0\,dB$$

Therefore,

$$
\begin{aligned}
P_t &= 57.8\,dB\,W \\
\lambda^2 &= 15.0\,dB\,m^2 \\
L_{sys} &= -7.0\,dB \\
C_s^2 &= -121.0\,dB \\
1/(16\pi^2) &= -22.0\,dB \\
1/R^2 &= -80.8\,dB\,m^{-2}
\end{aligned}
$$

$$P_r = -158.0\,dB\,W$$

9.6 The MU transmitter power is 2.2 dB larger than for the SOUSY system. The antenna gain, at 34 dB, is 3 dB larger, and so the SNR should be about 5.2 dB larger. This illustrates the efficiency of performing calculations in dB, rather than multiplying linear terms.

9.7 The PRF must be 6250 Hz and the total number of pulses transmitted in 10.5 s is 65 536 (see also Eq. (10.1) in the next chapter). Averaging these echo pulses in blocks of 512 leaves 128 samples for the FFT processing. Each FFT cell will be 1/10.5 Hz wide and, as there are 128 points, the doppler spectrum will be ± 6 Hz wide.

9.8 The component of the aircraft's horizontal velocity in the off-vertical beam is $300 \sin 12° = 62$ m s$^{-1} = 19$ Hz doppler shift. This is outside the doppler spectrum and the aircraft would not be observed. In the vertical beam the aircraft would be observed as a very large low velocity target, which would contaminate wind observations at that height.

9.9 Yes. The calculations in Section 9.4 hold for MST radar and show meteors to be very bright targets, although somewhat short-lived compared to the integration times usually used by MST radars. For a meteor to be observed in a narrow vertical beam of an MST radar, it must be travelling horizontally with a considerable path length through the atmosphere. Under these conditions, many small meteors burn up before they reach the radar beam and some larger meteors disintegrate through thermal shock. For these reasons, although MST radars are used to observe meteors, few meteor echoes are observed in the vertical beam.

Chapter 10
10.1 SRI quote the following figures: (a) 0.5°, (b) gain exceeds 30 dB between 8 and 28 MHz, (c) 93 dB W, (d) FOM of 104 dB J.

10.2 (a) $\Delta R = 3$ km, $R\,\Delta\theta = 13$ km, so area $= 39 \times 10^6$ m$^2 \equiv 75.9$ dB m^2.
(b) 52.9 dB m$^2 \equiv 195 \times 10^3$ m^2.

10.3 $N_{ext} = 60 - (2 \times 15) - 204 + 47 = -127$ dB W

10.4

$$
\begin{array}{rll}
4\pi = & 11.0 \text{ dB} \\
P_tG_tG_rt_i = & 104 \text{ dB J} \\
\sigma = & 52.9 \text{ dB m}^2 \\
m = & 6 \text{ dB} \\
l_s = & -15 \text{ dB} \\
1/F_a = & -30 \text{ dB} \\
1/kT_0 = & +204 \text{ dB W}^{-1} \text{ Hz} \\
1/(fh\lambda)^2 = & -275 \text{ dB m}^{-2} & \quad [\lambda/(4\pi R)^2]^2 = -265 \text{ dB m}^{-2} \\
\hline
\text{SNR} = & 57.9 \text{ dB}
\end{array}
$$

10.5 The doppler shift corresponds to a velocity of 150 m s^{-1}. The SNR is about 35 dB lower than the sea echo, giving an apparent RCS of 18 dB m^2; because of the height factor h, the true RCS is 24 dB m^2. This is therefore a large aircraft travelling slowly.

Even when the ionospheric losses are undetermined, the sea clutter can be used in the way above as an approximate method of calibrating the RCS of an unknown target.

Chapter 11
11.1 The dominant Bragg line should appear at ± 0.4 Hz in the doppler power spectrum. The extra shift of 0.3 Hz implies a tidal current of 3 m s^{-1} or nearly 6 knots. Ship speeds are generally restricted to 5 knots approaching a harbour, and hence the knowledge of a 6 knot cross-current could aid the pilot.

11.2 The travel time difference between two ranges separated by r_s is

$$\frac{2}{c}[(R + r_s) - R]$$

Hence

$$\frac{2r_s}{c} = \frac{1}{B}$$

The range difference between near and far edges of the beam (in slant range) is

$$R_M - R_m = \frac{h}{\cos(\theta + \beta)} - \frac{h}{\cos \theta}$$

$$\sim h\left(\frac{1}{\cos \theta - \beta \sin \theta} - \frac{1}{\cos \theta}\right)$$

$$\sim \frac{h\beta \sin \theta}{\cos^2 \theta}$$

Therefore

$$N \sim \frac{2B}{c} h\beta \tan \theta \sec \theta$$

Using values $B = 5 \times 10^8$, $\beta = 0.06/0.6 = 0.1$ rad, then

$$N \sim 8 \sec \theta \tan \theta$$

The variation of N with θ is shown in the plot below.

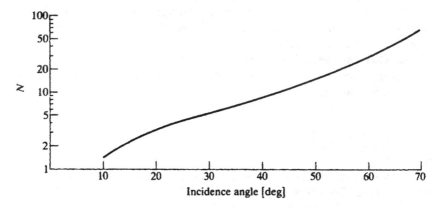

11.3 We have (see diagram)

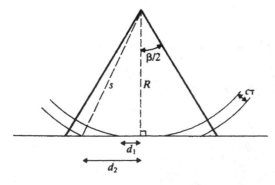

$$\text{Area} = \pi(d_2^2 - d_1^2)$$
$$= \pi[s^2 - (s - c\tau)^2]$$
$$= \pi c\tau(2s - c\tau)$$
$$\sim 2\pi c\tau s$$

Minimum value of s is R. Maximum value of s is

$$(R^2 + R^2\beta^2/4)^{1/2} \sim R(1 + \beta^2/8)$$

For a 2° beamwidth, s varies by less than 0.02 per cent. So

$$\text{Area} \sim 2\pi R c\tau$$

Time for leading edge to reach beam edge is

$$\frac{1}{c}\frac{R}{8}\beta^2$$

Hence condition given.

For ERS-1: $\beta = \lambda/d = 0.018$ rad or $1.04°$ and $R = 7.85 \times 10^5$ m. Thus, for pulse-limited operation,

$$\tau \leqslant 1.07 \times 10^{-7}\,\text{s}$$

The radar equation is

$$P_r = \frac{P_t G^2 \sigma A \lambda^2 L_s}{(4\pi)^3 R^4}$$

where $\sigma = $ RCS/unit area, $A = $ illuminated area, $G = 4\pi A_e^2/\lambda^2$ and $r = $ antenna radius. So

$$P_r = \frac{P_t \pi^2 r^4 \sigma c\tau \lambda^2 L_s}{2R^3}$$

For a simple pulse and matched receiver, noise power $= kT_0 B \sim kT_0/\tau$, giving

$$\text{SNR} = \frac{P_t \pi^2 r^4 \sigma c\tau^2 \lambda^2 L_s}{2R^3 kT_0}$$

$$= 0.0375 P_t \sigma$$

$$= 59.3 \qquad (17.7\,\text{dB})$$

for $P_t = 500$ W and $\sigma = 5$ dB m². In practice, the ERS-1 altimeter uses a 20 μs chirp with 50 W Tx power.

11.4 Number of complex multiplications/second for real-time operation is

$$M = \frac{4R\lambda V}{d^3}$$

Therefore, using subscripts a and s for air-borne and space-borne,

$$\frac{M_a}{M_s} = \frac{R_a}{R_s} \frac{V_a}{V_s} \left(\frac{d_s}{d_a}\right)^3$$

$$= \frac{1}{20} \times \frac{1}{135} \times 1000$$

$$\sim 0.37$$

At C-band, $\qquad M_a = \frac{2}{3} \times 10^6$

At L-band, $\qquad M_a = \frac{8}{3} \times 10^6$

11.5 The return from the slant swath will occupy a time duration

$$\frac{2}{c}(R_M - R_m) = \frac{2R_s}{c}$$

Successive pulses must be at least this far apart, even with no margin, i.e.

$$PRF \leqslant c/2R_s$$

But $PRF \geqslant 2V/d$, and so

$$R_s \leqslant cd/4V$$

For the aircraft:

$$R_s \leqslant \frac{3 \times 10^8 \times 2 \times 3600}{4 \times 300 \times 1000} = 180 \text{ km}$$

In practice, aircraft SARs normally use much narrower swaths. Air-borne SARs therefore do not usually have to trade swath width for resolution.
By contrast, for the spacecraft:

$$d \geqslant \frac{4 \times 7.5 \times 10^3 \times 10^5}{3 \times 10^8} = 10 \text{ m}$$

Hence best possible resolution is 5 m in the along-track direction. The trade-off between coverage and resolution is a serious design consideration.

11.6 Lay-over occurs when the top of a slope is nearer to the radar than the bottom (see diagram). The limiting case occurs when the top and bottom (and the whole of the slope) are imaged in the same place, i.e., are equidistant from the radar. This is the line OX in the diagram. At the long ranges used by SAR, this means that OS is nearly perpendicular to OX, or $\phi = \theta$. Hence any slope facing the satellite with a gradient exceeding 23° will suffer lay-over. Slopes of this magnitude are common in hilly areas.

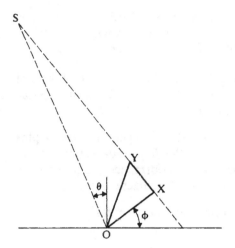

For shadowing, the radiation to X must be blocked by Y. This corresponds to a slope of 67° on the slope away from the satellite (steep, but not uncommon in mountainous terrain).

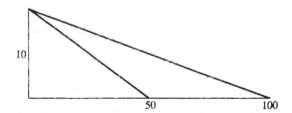

For the aircraft, incidence angles range from 78.7° to 84.3°. Hence lay-over will occur only in very steep terrain, but shadowing will be common. SAR operating at these angles can respond to very slight variations in topography, such as you see near sunset as shadows lengthen. Note that for the resolutions of the order of 1 m or so available to air-borne radar, lay-over is always encountered in images of urban areas.

Chapter 12
12.1 200 V/10 mV implies a dynamic range of 86 dB, reduced to 69 dB by the antenna losses. The attenuation rate and the depth requirement suggest a 99 dB dynamic range. The signal must therefore be boosted by at least 30 dB if targets are to be detected to depths of 4.3 m. These figures are, in fact, those given in reference 6, in which it is reported that objects were detected to 3.5 m (the limiting factor being mainly television interference).

12.2 11.6 MHz (PRFs this high are not used in practice).

12.3 The signal appearing in the receiver may be only 17 dB below 200 V, or as high as 28 V (reference 6). This would be sufficient to damage the receiver if protection methods (attenuators and diode limiters) were not used.

12.4 For two-way propagation losses of 200 dB/100 m (right-hand scale of Fig. 12.1), the 99 dB budget in the answer to problem 12.1 would enable the penetration to be extended to about 50 m.

12.5 At 10 MHz the losses are about 12 dB/100 m, suggesting that depths up to 825 m might be achieved. However, at this frequency, external noise would dominate internal noise and the system performance could be worse. On Mars, the noise level would be determined by galactic noise.

12.6 PRF = 60 kHz. A half-wave antenna would be 15 m long.

Chapter 13
13.1 (a) 1 kHz.
    (b) 2 kHz.
    (c) Monostatically the target motion decreases the path of the radio signal on both the outward and the return leg. In this bistatic case, only the outward path is shortening.

13.2 (a) 60 km (remember to allow for the time taken by the direct signal travelling along the baseline).
    (b) 39.7 km.
    (c) 20.3 km.

13.3 (a) 1082 Hz (plus or minus, depending on whether the aircraft is travelling towards, or away from, the receiver).
    (b) 47.5° (from the cosine formula).

13.4 (a) 6.7 km (about 22 000 ft). In practice, some pulse integration would be needed for the doppler processing and the extra gain from this would enable the aircraft to loiter at much higher altitudes.
    (b) The direct signal would be about 41 dB higher than the echo. This should cause no dynamic range problems for the receiver.
    (c) A radar on board the aircraft would be able to use the 20 dB antenna gain both ways. This would more than compensate for the additional path losses, allowing a smaller transmitter to be used (assuming the same RCS). However, it would give away the aircraft position.

Chapter 14
14.1 10.5 dB.

14.2 40 dB to −104 dB W.

14.3 70 mW.

14.4 The noise power spectral density of 70 mW in a 1 MHz bandwidth would have to be extended across a 20 000 MHz bandwidth. The total noise power would therefore have to be raised 43 dB to 1.4 kW.

14.5 10 000 pulses.

14.6 A suitable PRF might be 100 kHz (unambiguous range 1.5 km). The answer to problem 14.5 implies an integration time of 0.1 s at this PRF. This gives a speed resolution of 0.15 m s^{-1}, or 0.3 mph, sufficient to determine whether a car is speeding.

14.7 (a) Radar warning receivers (RWRs) covering several bands used by speed traps can be purchased, but they are expensive.

(b) A jamming device would be illegal and very expensive, unless the exact radar frequency was known.

(c) Stealth materials are also expensive and only really effective against a radar of known frequency. In any case, it would be difficult to conceal some parts of the car such as the headlights, driver, radio antenna, etc. Perhaps the cheapest solution is to keep to the speed limit. This also saves fuel.

Chapter 15

15.1 The antenna gain can be derived from either 475 (26.8 dB) plus 7.24 dB for each element, or from $G = 4\pi A/\lambda^2$, using the array area of 8330 m². Both methods give a gain of 34 dB. The beamwidth $\lambda/D = 6.45/103 = 3.6°$ is the value quoted in the article.

15.2 Using Eq. (15.13) (appropriate for no weighting) we would expect a spacing of a little over 0.6λ. In fact, a triangular lattice of spacing 0.7λ is used since triangular grids generally allow a wider range of antenna beam steering than rectangular grids with the same element density[9]. A further sophistication of the MU system is that groups of 19 antenna elements are collected into hexagonal sub-arrays.

15.3 Roughly 56° off-zenith for the vertical beam. The authors quote 'no grating lobe is formed at beam positions within 40° of the zenith'.

15.4 Calculating 475 phase angles and setting up 475 phaseshifters 2500 times per second would require over one million such operations per second. Instead, the MU system stores all phaseshift data required for beam steering in a ROM (read-only memory) installed in each transmit module. The appropriate values are then read out according to instructions from the module controller.

BIBLIOGRAPHY

Throughout this book we have tried to restrict references to books which are relatively easy to obtain. Occasionally we have referred to original publications in journals, but these can be found in most university and industrial libraries. To help with finding references, we list them for you (first author only) alphabetically below.

Abramowitz, M. *Handbook of Mathematical Functions*, Dover, New York, 1964.

Atlas, D. (ed.) *Radar in Meteorology*, Battan Memorial and 40th Anniversary Radar Meteorology Conference, American Meteorology Society, 1990.

Barker, R. H. Group synchronizing of binary digital systems, in *Communication Theory*, (ed.) W. Jackson, Academic Press, New York, pp. 273–287, 1953.

Barrick, D. E. Remote sensing of sea state by radar, chapter 12 in *Remote Sensing of the Troposphere*, (ed.) V. E. Derr, NOAA Environmental Research Laboratories, Boulder, Colorado, 1972.

Bar-shalom, Y. Tracking in a cluttered environment with probabilistic data association, *Automatica*, 11, 451–461, 1975.

Battan, L. J. *Radar Meteorology*, University of Chicago Press, Chicago, 1959.

—— *Radar Observation of the Atmosphere*, University of Chicago Press, Chicago, 1973.

Bean, B. R. *Radio Meteorology*, Dover Publications, New York, 1968.

Berkowitz, R. S. *Modern Radar: Analysis, Evaluation and System Design*, Wiley, New York, 1966.

Bracewell, R. N. *The Fourier Transform and its Applications*, McGraw-Hill, New York, 1986.

Brigham, E. O. *The Fast Fourier Transform*, Prentice-Hall, Englewood Cliffs, NJ, 1974.

Brookner, E. (ed.) *Practical Phased-Array Antenna Systems*, Artech House, 1991.

Caldecott, R. Underground mapping of utility lines using impulse radar, *IEE Proc.-F*, 135(4), 343–353, 1988.

CCIR, *World Distribution and Characteristics of Atmospheric Radio Noise*, CCIR Report 322, 1963.

—— *Reference Atmosphere for Refraction*, Recommendations and Reports of the CCIR, CCIR Rec. 369-3, V, ITU, Geneva, 1986.

Cole, H. W. *Understanding Radar*, BSP Professional Books, Oxford, 1985.

Colgrave, S. B. Track initiation and nearest neighbour incorporated into probabilistic data association, *IE Aust. IREE Aust.*, 6(1), 191–198, 1986.

Connor, F. R. *Antennas*, Edward Arnold, London, 1972.

Costas, J. P. A study of a class of detection waveforms having nearly ideal range–doppler ambiguity properties, *Proc. IEEE*, **72**, 996–1009, 1984.

Doviak, R. J. *Doppler Radar and Weather Observations*, Academic Press, New York, 1984.

Eastwood, Sir E. *Radar Ornithology*, Methuen, London, 1967.

Elachi, C. *Spaceborne Radar Remote Sensing: Applications and Techniques*, IEEE Press, New York, 1987.

Frank, R. L. Polyphase codes with good nonperiodic correlation properties, *IEEE Trans. Information Theory*, **IT-9**, 43–45, 1963.

Fukao, S. The MU radar with an active phased array system: 1. Antenna and power amplifiers, *Radio Sci.*, **20**(6), 1155–1168, 1985.

Georges, T. M. Real-time sea-state surveillance with skywave radar, *IEEE J. Ocean Eng.*, **OE-8**(2), 97–103, 1983.

Gjessing, D. T. *Target Adaptive Matched Illumimation Radar*, IEE Electromagnetic Waves series 22, Peter Peregrinus, Hitchin, Herts, 1986.

Gonzales, C. A. Pulse compression techniques with application to HF probing of the mesosphere, *Radio Sci.*, **19**, 871–877, 1984.

Gossard, E. E. *Radar Observations of Clear Air and Clouds*, Elsevier, Amsterdam, 1983.

Greenhow, J. S. Turbulence at altitudes of (80–100) km and its effects on long duration meteor echoes, *J. Atmos. Terr. Phys.*, 384–392, 1959.

Griffiths, J. *Radio Wave Propagation and Antennas*, Prentice-Hall, Englewood Cliffs, NJ, 1987.

Hall, M. P. M. (ed.) *Radiowave Propagation*, Peter Peregrinus for the IEE, Stevenage, Herts, 1989.

Harmuth, H. F. *Transmission of Information by Orthogonal Functions*, Springer-Verlag, Berlin, 1972.

—— *Nonsinusoidal Waves for Radar and Radio Communication*, Academic Press, New York, 1981.

Hinkley, E. D. (ed.) *Laser Monitoring of the Atmosphere*, Springer-Verlag, Berlin, 1976.

Hirsch, H. L. *Practical Simulation of Radar Antennas and Radomes*, Artech House, Norwood, MA, 1988.

Jackson, M. C. The geometry of bistatic radar systems, *IEE Proc.-F*, **133**(7), 604–612, 1986.

Jane's, *Jane's Radar and Electronic Warfare Systems 1989–90*, London, 1989.

Kolosov, A. A. *Over-the-Horizon Radar*, Artech House, MA, 1987.

Kovaly, J. J. *Synthetic Aperture Radar*, Artech House, MA, 1976.

Leonov, A. I. *Monopulse Radar*, Artech House, Norwood, MA, 1986.

Levanon, N. *Radar Principles*, Wiley, New York, 1988.

Lipa, B. J. Extraction of sea state from HF radar: mathematical theory and modelling, *Radio Sci.*, **21**(1), 81–100, 1986.

Long, M. W. *Radar Reflectivity of Land and Sea*, Lexington Books, Lexington, MA., 1975.

Lynn, P. A. *Radar Systems*, Macmillan, London, 1987.

Maffett, A. L. *Topics for a Statistical Description of Radar Cross Section*, Wiley, New York, 1989.

Manasse, R. Range and velocity accuracy from radar measurements, *MIT Lincoln Lab. Report*, 312–326, 1955.

Marcum, J. A statistical theory of target detection by pulsed radar, *IRE Trans.*, IT-6, 145–267, 1960.

McKinley, D. W. R. *Meteor Science and Engineering*, McGraw-Hill, New York, 1961.

Meeks, M. L. *Radar Propagation at Low Altitudes*, Artech House, MA, 1982.

Mischenko, Y. A. *Over-the-Horizon Radar*, Zagorizontnaya Radiolokatsiya, Military Publishing House, Ministry of Defence, Moscow, 1972.

Nathanson, F. E. *Radar Design Principles*, McGraw-Hill, New York, 1969.

Neal, B. T. CH—the first operational radar, *GEC J. Res.*, 3(2), 73–83, 1986.

Papoulis, A. *Probability, Random Variables and Stochastic Processes*, McGraw-Hill, New York, 1985.

Pedrotti, F. L. *Introduction to Optics*, Prentice-Hall, Englewood Cliffs, NJ, 1987.

Picquenard, A. *Radio Wave Propagation*, Macmillan, London, 1974.

Radford, M. F. Electronically scanned antenna systems, *Proc. IEE*, **125**(11R), 1100-1112, 1978.

———— Whither radar, *GEC J. Res.*, 3(2), 137–143, 1985.

Ramsay, D. A. The evolution of radar guidance, *GEC J. Res.*, 3(3), 92–103, 1985.

Rhodes, D. R. *Introduction to Monopulse*, McGraw-Hill, New York, 1959.

Rohan, P. *Surveillance Radar Performance Prediction*, Peter Peregrinus for the IEE, Stevenage, Herts, 1983.

Rice, S. O. Mathematical analysis of random noise, *Bell Syst. Tech. J.*, **23** and **24**, 1944; reprinted in *Selected Papers on Noise and Stochastic Processes*, (ed.) N. Wax, Dover, New York, 1954.

Rotheram, S. Theory of SAR ocean wave imaging, in *Satellite Microwave Remote Sensing*, Ellis Horwood, Chichester, 1983.

Rudge, A. W. (ed.) *The Handbook of Antenna Design*, vol. 2, IEE Electromagnetic Wave series, Peter Peregrinus, Hitchin, Herts, 1983.

Rumsey, V. H. *Frequency Independent Antennas*, Academic Press, New York, 1966.

Scanlan, M. J. B. (ed.) *Modern Radar Techniques*, Collins, Glasgow, 1987.

Shafer, G. *A Mathematical Theory of Evidence*, Princeton University Press, Princeton, NJ, 1976.

Shearman, E. D. R. Radio science and oceanography, *Radio Sci.*, **18**(3), 299–320, 1983.

Sherman, S. M. *Monopulse Principles and Techniques*, Artech House, Norwood, MA, 1984.

Schwartz, M. *Information Transmission, Modulation and Noise*, McGraw-Hill, New York, 1980.

Skolnik, M. I. *Radar Handbook*, McGraw-Hill, New York, 1970.

———— *Introduction to Radar Systems*, McGraw-Hill, New York, 1985.

———— *Radar Handbook*, McGraw-Hill, New York, 1990.

Stevens, M. C. *Secondary Surveillance Radar*, Artech House, Norwood, MA, 1988.

Stremler, F. G. *Introduction to Communication Systems*, Addison-Wesley, Reading, MA, 1982.

Sugar, G. R. Radio propagation by reflection from meteor trails, *Proc. IEEE*, **52**, 116–136, 1964.

Sweetman, B. *Stealth Bomber*, Airlife Publishing, Shrewsbury, 1989.

Swerling, P. Detection of fluctuating pulsed signals in the presence of noise, *IRE Trans.*, **IT-3**, 175–178, 1957.

———— Probability of detection for fluctuating targets, *IRE Trans.*, **IT-6**, 269–308, 1960.

Swords, S. S. *Technical History of the Beginnings of RADAR*, Peter Peregrinus for the IEE, Stevenage, Herts, 1986.

Toomay, J. C. *Radar Principles for the Non-specialist*, Van Nostrand Reinhold, New York, 1989.

Ulaby, F. T. *Microwave Remote Sensing: Active and Passive*, vols 1–3, Addison-Wesley, Reading, MA., 1981.

Wax, N. (ed.) *Selected papers on Noise and Stochastic Processes*, Dover, New York, 1954.

Weeks, W. L. *Antenna Engineering*, McGraw-Hill, New York, 1968.
Wiley, R. G. *Electronic Intelligence: The Analysis of Radar Signals*, Artech House, Dedham, MA, 1982.
Willis, N. J. *Bistatic Radar*, Artech House, Norwood, MA, 1991.
Woodward, P. M. A theory of radar information, *Phil. Mag.*, **41**, 1001, 1950.
——— *Probability and Information Theory, with Applications to Radar*, Pergamon Press, Oxford, 1953.

INDEX

Printed in the USA
CPSIA information can be obtained
at www.ICGtesting.com
JSHW011508221024
72173JS00005B/1243